Multilevel Converters

Scrivener Publishing
100 Cummings Center, Suite 541J
Beverly, MA 01915-6106

Publishers at Scrivener
Martin Scrivener (martin@scrivenerpublishing.com)
Phillip Carmical (pcarmical@scrivenerpublishing.com)

Multilevel Converters

Edited by
Salman Ahmad
Farhad Ilahi Bakhsh
and
P. Sanjeevikumar

Scrivener Publishing

WILEY

This edition first published 2024 by John Wiley & Sons, Inc., 111 River Street, Hoboken, NJ 07030, USA and Scrivener Publishing LLC, 100 Cummings Center, Suite 541J, Beverly, MA 01915, USA

For more information about Scrivener publications please visit www.scrivenerpublishing.com.

Wiley Global Headquarters
111 River Street, Hoboken, NJ 07030, USA

For details of our global editorial offices, customer services, and more information about Wiley products visit us at www.wiley.com.

Library of Congress Cataloging-in-Publication Data

ISBN 978-1-394-16632-9

Front cover images supplied by Pixabay.com
Cover design by Russell Richardson

Set in size of 11pt and Minion Pro by Manila Typesetting Company, Makati, Philippines

Printed in the USA

10 9 8 7 6 5 4 3 2 1

Contents

List of Contributors

Suresh Lakhimsetty, Department of Electrical Engineering, Sardar Vallabhbhai National Institute of Technology Surat, Surat, India. Email: suresh@eed.svnit.ac.in

Hareesh Myneni, Department of Electrical Engineering, National Institute of Technology Srinagar, Srinagar, India. Email: hari247.sol@gmail.com

Obbu Chandra Sekhar, Department of Electrical Engineering, National Institute of Technology Delhi, New Delhi, India. Email: obbuchandra@nitsri.net

Hasan Iqbal, Department of Electrical Engineering, Aligarh Muslim University, Aligarh, UP, India and Department of Electrical and Computer Engineering, Florida International University, Miami, FL, USA. Email:hiqba003@fiu.edu

Mohammad Tayyab, Department of Electrical Engineering, Aligarh Muslim University, Aligarh, UP, India and Electrical and Computer Engineering Department, Texas A & M University at Qatar, Doha, Qatar. Email: mtayyab1@myamu.ac.in

Arif Iqbal, Department of Electrical Engineering, Rajkiya Engineering College, Ambedkar Nagar (U.P.), India. E-mail: arif.iqbal.in@gmail.com

S. P. Singh, Indian Institute of Technology Roorkee, India. Email: spseefee@iitr.ac.in

Mohammad Ali, K.A. CARE Energy Research & Innovation Center, King Fahd University of Petroleum and Minerals, Dhahran, Saudi Arabia, Email: mohammad.ali@kfupm.edu.sa

Muhammad Khalid, Interdisciplinary Research Center for Renewable Energy & Power Systems, King Fahd University of Petroleum and Minerals, Dhahran, 31261, Saudi Arabia and Electrical Engineering Department, King Fahd University of Petroleum and Minerals, Dhahran, Saudi Arabia. Email: mkhalid@kfupm.edu.sa

Dhananjay Kumar, Dept. of Electrical Engineering, Government Engineering College, Siwan, Bihar, India. Email: dhananjayk00@gmail.com

Shailesh Kumar Gupta, Department of Electrical & Electronics Engineering, IIMT Engineering College, Meerut (U.P.), India. Email: geniousshailesh@gmail.com

Omveer Singh, Department of Electrical Engineering, Gautam Buddha University, Greater Noida (U.P.), India. Email: omveers@gmail.com

Akhlaque Ahmad Khan, Department of Electrical Engineering, Integral University, Lucknow, UP, India. Email: akhlaque@iul.ac.in

Ahmad Faiz Minai, Department of Electrical Engineering, Integral University, Lucknow, UP, India. Email: fzminai@gmail.com

Qamar Alam, Department of Electrical Engineering, Integral University, Lucknow, UP, India. Email: qamar@iul.ac.in

Farhad Ilahi Bakhsh, National Institute of Technology, Srinagar, India. Email: farhad@nitsri.ac.in

Kasinath Jena, Dept. of Electrical and Electronic Engineering, Arka Jain University, Jamshedpur, India. Email: kasi.jena@gmail.com

Jitendra Kumar Tandekar, Dept. of Electrical Engineering, Samrat Ashok Technological Institute Vidisha, India. Email: jitendra.tandekar@gmail.com

Niraj Kumar Dewangan, Department of Mechatronics Manipal Institute of Technology, Manipal Academy of Higher Education, Manipal, Karnataka, India. Email: nirajdewangan11@gmail.com

Vishal Rathore, Department of Electrical and Electronics Engineering, Bhopal Institute of Technology and Science, Bhopal, India. Email: vishalrathore01@gmail.com

Haroon Rehman, Department of Electrical Engineering, Aligarh Muslim University, Aligarh, UP, India. Email: hiqba003@fiu.edu

Adil Sarwar, Department of Electrical Engineering, Aligarh Muslim University, Aligarh, UP, India. Email: hiqba003@fiu.edu

Md Reyaz Hussan, Department of Electrical Engineering, Aligarh Muslim University, Aligarh, UP, India. Email: hiqba003@fiu.edu

Akhlaque Ahmad Khan, Department of Electrical Engineering, Integral University, Lucknow, UP, India. Email: akhlaque@iul.ac.in,

Ahmad Faiz Minai, Department of Electrical Engineering, Integral University, Lucknow, UP, India. Email: fzminai@gmail.com

Mohammed Aslam Husain, Department of Electrical Engineering, REC, Ambedkar Nagar, UP, India. Email: mahusain87@gmail.com

Mohammad Naseem, Department of Electrical Engineering, Integral University, Lucknow, UP, India. Email: mnansar@gmail.com

Aishwarya V., Department of Electrical & Electronics Engineering, APJ Abdul Kalam Technological University, Trivandrum, India. Email: aishwarya@tistcochin.edu.in

Zahid Afzal Thoker, Department of Electrical Engineering, National Institute of Technology, Srinagar, India. Email: thokerzahid_10phd17@nitsri.ac.in

Shameem Ahmad Lone, Department of Electrical Engineering, National Institute of Technology, Srinagar, India. Email: sadial_14@yahoo.com

Amit Ojha, Dept. Electrical Engineering, Maulana Azad National Institute of Technology, Bhopal (M.P.), India. Email: ojha.amit@gmail.com

Shailendra Jain, Dept. Electrical Engineering, Maulana Azad National Institute of Technology, Bhopal (M.P.), India. Email: sjain68@gmail.com

Zahoor Ahmad Ganie, Department of Electrical Engineering, Islamic University of Science and Technology, Kashmir, India. Email: zahoor.rifi@gmail.com

Abdul Hamid Bhat, Department of Electrical Engineering, National Institute of Technology Srinagar, India. Email: bhatdee@nitsri.net

Salman Ahmad, Department of Electrical Engineering, Islamic University of Science and Technology, Kashmir, India. Email: salmanahmad19@gmail.com

Javeed Bashir, Department of Electrical Engineering, Islamic University of Science and Technology, Kashmir, India and Indian Institute of Technology Roorkee, India. Email: javeedbashir001@gmail.com

Ahmed Sharique Anees, Department of Electrical Engineering, Islamic University of Science and Technology, Kashmir, India. Email: shariq.anees@islamicuniversity.edu.in

Faizan Fayaz Bhat, Department of Electrical Engineering, National Institute of Technology Srinagar, Hazratbal, Srinagar (J&K), India. Email: bfaizan999@gmail.com

Zahid Ahmad Tantry, Department of Electrical Engineering, National Institute of Technology Srinagar, Hazratbal, Srinagar (J&K), India. Email: zahidtantry969@gmail.com

Md Ibrahim, Department of Electrical Engineering, Integral University, Lucknow, India. Email: ibrahimalig3893@gmail.com

Tajamal Hayat Parray, Department of Electrical Engineering, National Institute of Technology, Hazratbal, Srinagar, Jammu and Kashmir, India. Email: tajamulhayat05@gmail.com

Mohammad Ali Abido, K.A.CARE Energy Research & Innovation Center at King Fahd University of Petroleum & Minerals (KFUPM), Saudi Arabia and Interdisciplinary Research Center for Sustainable Energy Systems, KFUPM, Dhahran, Saudi Arabia.

Preface

Multilevel converters have gained much attention in recent years for medium-high-voltage and high-power industrial and residential applications. The main advantages of a multilevel converter over two-level converters include less voltage stress on power semiconductors, low dv/dt, low common voltage, reduced electromagnetic interference, and low total harmonics distortion among others. Better output power quality is ensured with the increasing number of levels in the synthesized output voltage waveform. Several multilevel topologies have been reported in the literature, such as neutral point clamped (NPC), flying capacitor (FC), cascaded H-bridge (CHB), hybrid cascaded H-bridge, asymmetrical cascaded H-bridge, flying capacitor, modular multilevel converter (MMC), active neutral point clamped converter (ANPC), and packed U-cell type converter, and various topologies based on reduced device count and reduced number of sources have also been proposed in the literature.

The multilevel converters have been commercialized in many standard and customized products such as mixers, blowers, pumps, compressors, conveyors, fans, crushers, rolling mills, extruders, grinding machines, solar photovoltaic systems, wind power conversion, traction and ship propulsion, electric vehicles, variable speed drives, static compensators (STATCOMS), high-voltage direct-current (HVDC) transmission, and hydro pump storage, to name a few. However, the multilevel converter, although a proven and enabling technology, still presents numerous challenges in topologies, modulation, and control and need-based applications. Since multilevel converters offer a wide range of possibilities, research and development in the area of multilevel converter topologies, modulation, and control and in various applications are still growing in depth and width. To further improve multilevel converters energy efficiency, reliability, power density, and cost, many research groups across the world are working to broaden the application areas of multilevel converters and to make them more attractive and competitive compared to classic topologies. Multilevel converters represent a very important topic in modern power electronics, and

Preface

Multilevel converters have gained much attention in recent years for medium/high-voltage and high-power industrial and residential applications. The main advantages of a multilevel converter over two-level converters include less voltage stress on power semiconductors, low dv/dt, low common voltage, reduced electromagnetic interference, and low total harmonics distortion among others. Better output power quality is ensured with the increasing number of levels in the synthesized output voltage waveform. Several multilevel topologies have been reported in the literature, such as neutral point clamped (NPC), flying capacitor (FC), cascaded H-bridge (CHB), hybrid cascaded H-bridge, asymmetrical cascaded H-bridge, flying capacitor, modular multilevel converter (MMC), active neutral point clamped converter (ANPC), and packed U-cell-type converter, and various topologies based on reduced device count and reduced number of sources have also been proposed in the literature.

The multilevel converters have been commercialized in many standard and customized products such as mixers, blowers, pumps, compressors, conveyors, fans, crushers, rolling mills, extruders, grinding machines, solar photovoltaic systems, wind power conversion, traction and ship propulsion, electric vehicles, variable speed drives, static compensators (STATCOMS), high-voltage direct-current (HVDC) transmission, and hydro pump storage, to name a few. However, the multilevel converter, although a proven and enabling technology, still presents numerous challenges in topologies, modulation, and control and need-based applications. Since multilevel converters offer a wide range of possibilities, research and development in the area of multilevel converter topologies, modulation, and control and in various applications are still growing in depth and width. To further improve multilevel converters' energy efficiency, reliability, power density, and cost, many research groups across the world are working to broaden the application areas of multilevel converters and to make them more attractive and competitive compared to classic topologies. Multilevel converters represent a very important topic in modern power electronics, and many research groups are working on topologies, modulation techniques, and control strategies to improve the global behavior of the system.

This book, "*Multilevel Converters*", intended to provide a deep insight about multilevel converters' operation, modulation, and control strategies and the various applications of multilevel converters such as in variable speed drives, renewable energy generation, and power systems. The book will serve as a reference for academic researchers, university students, and practicing engineers who are working in the area of multilevel converters. This book contains 17 chapters on multilevel converter topologies, modulation, control, and applications and organized as detailed below.

Chapter 1 discusses the invention and applications of multilevel converters, which were created to overcome voltage limitations in semiconductor devices. The chapter's main objective is to comprehensively study multilevel inverters, including their advantages, drawbacks, and recent applications. The study begins with an examination of dual two-level inverter systems, highlighting their benefits, challenges, techniques for issue resolution, and various applications.

Chapter 2 highlights the historical development of multilevel inverters. It mentions the evolution of MLI configurations, including newer models with fewer switches that produce more voltage levels. The chapter emphasizes the industrial uses of MLI in on-grid renewable energy systems and discusses important MLI configurations such as flying capacitor MLI (FC-MLI), neutral point clamped (NPC) or diode clamped MLI (DC-MLI), and cascaded H-bridge MLI (CHB-MLI). Finally, it notes the attractiveness of multilevel technology for photovoltaic (PV) applications and provides an overview and classification of MLI topologies relevant to PV systems.

Chapter 3 provides a detailed explanation of both conventional and new multilevel converter topologies. It also covers various pulse width modulation techniques and discusses their applications in motor drives. To support the theoretical concepts, the chapter includes simulation and hardware results to validate the discussed findings.

Chapter 4 introduces a novel inverter modulation technique to generate 11 voltage levels at the inverter output. This modulation optimizes capacitor charging and discharging, reducing inrush current peaks significantly, improving reliability, and extending the inverter's lifespan, making it well-suited for solar PV applications even in challenging weather conditions. This innovative approach also reduces stress on switch voltage and current. To validate the effectiveness of this enhanced technology, comprehensive experiments are conducted using hardware-in-the-loop setups and simulation environments as well as rigorous testing in a laboratory prototype.

Chapter 5 discusses a four-level open-end winding induction motor (OEW-IM) drive controlled by a single inverter through space vector pulse width modulation (SVPWM). It employs two two-level voltage source inverters (VSIs) connected to each side of the OEW-IM. To achieve four-level inversion, the DC-link voltages of the dual-inverter system are maintained at a 2:1 ratio. The four-level OEW-IM drive faces two main challenges: 1) circulating zero sequence currents in the OEW-IM phase windings and 2) preventing overcharging of the lower DC-link voltage capacitor by the higher DC-link voltage capacitor. The latter issue is mitigated by providing independent DC power supplies to the dual-inverter system. The former challenge is addressed through a novel SVPWM scheme, which promises improved performance compared to existing SVPWM methods.

Chapter 6 discusses the use of field-oriented control (FOC) in induction motor drives, which is widely adopted for high-performance applications. FOC enhances motor drive performance by independently controlling the torque and field flux components, making the motor operation similar to a separately excited DC motor. This chapter specifically introduces a diode-clamped multi-level inverter for the closed-loop control of an induction motor drive using FOC. It covers the operation of the induction motor in both motoring and regenerating modes, powered by a diode-clamped three-phase three-level inverter circuit. The entire drive system is developed and analyzed using MATLAB/Simulink.

Chapter 7 focuses on the development of innovative topological designs for multilevel inverters that offer several key advantages, including a low component count, high voltage gain, cost-effectiveness, and minimized voltage stress on power switches. The study introduced five-level single-cell multilevel inverter (SCMLI) topologies designed to efficiently generate multilevel voltage waveforms, particularly suitable for medium- and high-power applications. Experimental testing has been conducted to validate the practical feasibility of each proposed topology. The results highlight several significant advantages, including reduced voltage stress, high voltage gain, the ability to operate at various power factors, reduced switching components, and high overall efficiency. The suggested topologies underwent both simulation in the MATLAB/Simulink environment and practical experimentation using the dSPACE-1104 controller, further reinforcing their viability and potential for practical implementation in real-world applications.

The switched-capacitor multilevel converters (SCMLCs) are gaining popularity for the integration of solar photovoltaic (SPV) systems into the grid. The current emphasis is on minimizing the number of components in SCMLCs, leading to cost savings and the creation of smaller converters. SCMLCs are particularly suitable for grid integration of low-voltage sources like SPV and fuel cells due to their ability to provide significant voltage boosting. In this context, Chapter 8 discusses a 13-level SCMLC topology for the grid-connected operation of SPV systems. This topology achieves threefold voltage boosting while utilizing a reduced component count, making it a promising solution for efficiently integrating SPV systems into the grid.

Chapter 9 discusses the multilevel inverters useful for solar photovoltaic (SPV) systems and provides a concise overview. The selection of suitable maximum power point tracking (MPPT) strategies for grid-connected SPV/hybrid systems with multilevel inverter has also been discussed in detail.

Chapter 10 discusses electric-drive-reconstructed onboard converter (EDROC), which can function in either the grid-to-vehicle (G2V) or the vehicle-to-grid (V2G) mode. It gives an extensive examination of the global implementation of power converters for electric vehicle systems and contemporary solutions for electric vehicle (EV) charging infrastructure.

In Chapter 11, the packed U-cell (PUC) inverter is highlighted as an innovative technology that achieves a seven-level voltage output using only six active switches, one isolated DC source, and a second capacitor source. However, maintaining a precise voltage ratio between the two capacitor sources is crucial. The chapter delves into the modulation and control of the PUC topology, focusing on its application in solar photovoltaic systems.

Chapter 12 proposes a reduced switch count-based multilevel converter for application in unified power quality conditioner (UPQC) in distribution power systems. The independent control of real and reactive power exchanged between the UPQC and the power networks using multilevel converters is investigated in this chapter in detail.

Chapter 13 illustrates the process of computing power losses, assessing efficiency, and conducting a harmonic analysis of a three-phase infinite-level inverter (TILI). The TILI boasts of exceptional DC-link voltage utilization with a factor of 1.23, reduced input voltage demands, minimized switching power losses, decreased voltage stress, enhanced working efficiency, and improved output power quality.

Chapter 14 discusses the modeling, analysis, and control of a doubly fed induction generator (DFIG) using direct torque control space vector pulse width modulation (DTC-SVPWM) which is employed in wind energy conversion systems, featuring a partially rated converter and facilitating a direct grid connection. The analysis covers aspects such as the quality of stator current, dynamic response, and steady-state response.

Chapter 15 discusses a novel robust sliding mode control strategy for the static var compensator (SVC) for reactive power support to reduce the mismatch in the system reactive power. To improve the performance of a sliding mode controller (SMC) and eliminate the chattering problem, a sliding mode observer (SMO) is constructed to estimate the SVC parameters effectively. The Lyapunov stability analysis is conducted to guarantee the stability of the controller. Simulink/MATLAB is used to validate the efficacy of the proposed controller through simulation studies.

For modular multilevel converter (MMC)-based applications, it is crucial to ensure trouble-free operation for 2 to 3 years. Various types of electrical failures, such as open circuits, short circuits, DC-bus short circuits, and single line-to-ground faults, can potentially disrupt the MMC's operation, leading to a voltage drop within the system. Therefore, the development of a robust fault-tolerant strategy is imperative. A fault-tolerant plan typically comprises three distinct phases: pre-fault preparation, fault isolation and localization, and post-fault restoration and control. Chapter 16 primarily focuses on the post-fault control procedures aimed at swiftly and reliably restoring voltage levels to their pre-fault conditions.

Finally, Chapter 17 explores the application of cascaded H-bridge multilevel inverter (CHB-MLI)-based active filter systems and demonstrates their dynamic performance under transient and steady-state conditions. Specifically, a nine-level CHB-MLI-based SAPF system is simulated using MATLAB/Simulink, showcasing its effectiveness in addressing power quality issues arising from non-linear loads.

This book serves as a valuable reference for individuals interested in the area of multilevel converter topologies, modulation, control, and practical application. Both researchers and practicing engineers will find this book highly adaptable and enriching for their work. Covering a wide range of subject areas and providing a wealth of information from fundamental to advanced levels, this single volume stands as a comprehensive guide for anyone engaged in this field.

We extend our heartfelt gratitude to all the contributors of this book for their invaluable contributions in crafting exceptional literature for the research community. Their dedication and expertise have resulted in the creation of high-quality content that will undoubtedly benefit and enhance research in this field.

We also wish to express our appreciation to our colleagues and friends for their insightful contributions and helpful discussions throughout the production of this edited book.

Salman Ahmad
Department of Electrical Engineering, Islamic University of Science and Technology, Awantipora J&K, India
Farhad Ilahi Bakhsh
Department of Electrical Engineering, National Institute of Technology, Srinagar J&K, India
P. Sanjeevikumar
Department of Electrical Engineering, IT and Cybernetics, University of South-Eastern Norway, Porsgrunn, Norway

1

Analysis of Dual Two-Level Converters for Multilevel Performance

Shailesh Kumar Gupta[1*] **and Omveer Singh**[2]

[1]*Department of Electrical & Electronics Engineering, IIMT Engineering College, Meerut (U.P.), India*

[2]*Department of Electrical Engineering, Gautam Buddha University, Greater Noida (U.P.), India*

Abstract

Multilevel converters were invented to circumvent the voltage restriction of semiconductor devices. The coupling of AC and DC high voltage systems was its first use. Multilevel converters are mostly used to connect many devices in series while clamping the voltages between their pins. The clamping method is determined by the variations between multiple converter architectures. Recently, the use of multilevel inverter technology has become a highly significant alternative for controlling high-power, medium-voltage energy. The goal of this paper is to study this multilevel inverter from the ground up. A review of multilevel converters will cover their benefits, drawbacks, and applications in recent years. The examination of dual two-level inverter systems is then followed by a discussion of the benefits of dual inverter systems as well as issue identification, techniques, and applications. Then, the multilevel performance of dual two-level converters is investigated in the MATLAB/SIMULINK environment. Finally, conclusion and future scope are discussed.

***Keywords*:** Multilevel inverters, dual inverter, space vector pulse width modulation, artificial nural network based pulse width modulation, harmonic spectrum, simulation

1.1 Introduction

By the end of the 1990s, industry has started to require more powerful machinery, which can now produce megawatts of power. The medium-voltage

**Corresponding author*: drskggbu@gmail.com

Salman Ahmad, Farhad Ilahi Bakhsh, and P. Sanjeevikumar (eds.) Multilevel Converters, (1–26)

network is typically coupled to controlled ac drives in the megawatt range. Currently, connecting a solitary semiconductor switch directly to medium-voltage grids is challenging. For these reasons, working with greater voltage levels has been made possible by a new family of multilevel inverters (MLIs) [1–3]. In order to create a stepped voltage waveform for larger power levels, a multilevel inverter is a power electronic circuit that combines a number of low-rated power semiconductor switches with a number of different DC sources. Batteries, fuel cells, capacitors, and solar PV panels are just a few of the energy sources that can be used to power numerous DC input sources. Semiconductors and capacitor voltage sources are used in multilevel inverters, whose output produces voltages with stepped waveforms. The power semiconductors must endure only low voltages, whereas the commutation of the switches allows the addition of capacitor voltages, which touch high voltage at the output. When compared to the negative pole of the capacitor, the output voltage of a two-level (2-L) inverter has two values (levels), a three-level (3-L) inverter has three levels, and so on.

When the three-level inverter was first introduced in [4], the name "multilevel" was coined. By raising the inverter's level count, the output voltages have more steps, creating a waveform resembling a staircase with less harmonic distortion. A large number of levels, however, complicates control and creates issues with voltage imbalance. MLIs have three alternative topologies that have been suggested: diode-clamped MLIs [4], flying capacitor MLIs [1, 5, 6], and cascaded H-bridge MLIs [1, 7, 8]. MLIs have the ability to draw input current with very low distortion, generate smaller common-mode (CM) voltage, and generate output voltages with exceptionally low distortion and lower dv/dt, all of which reduce the stress on the motor bearings. Furthermore, it is possible to eliminate CM voltages [8] and operate with a reduced switching frequency by applying sophisticated modulation techniques.

In diode-clamped converters, the clamping is done by the diode batteries, in cascade H-bridge topologies, the clamping is done by the batteries, and so on. The former use of a multilevel converter in high voltage applications to connect the DC side to the AC grid is still used on HVDC transmission lines. In addition, low-voltage applications for which multilevel is appropriate have been discovered. The output waveforms of multilevel converters are better than those of standard converters. This feature is extremely beneficial in meeting energy quality standards. Even in drive applications, multilevel converters are now a good choice. Sophisticated control algorithms can take advantage of the system's large number of levels to improve performance.

Furthermore, a 3-L inverter also produces space vectors that are similar to those generated by a double 2-L inverter, where every sector is split into four distinguished triangular areas based on factors like the angle and amplitude of the reference voltage vector, but it adds to the computational complexity. As the number of level increases, cost and complexity also increase.

Although the 2-L dual inverter and 3-L single inverter systems have the same number of switches, the 3-L single inverter system's construction required the addition of additional capacitors and diodes, which required voltage balancing. A double inverter system, which does not require any additional diodes or capacitors, can alleviate this problem. Ripples are considerably decreased when a dual 2-L inverter system is used as a 3-L inverter. Acoustic noise and EMI emissions from the motor are significantly decreased when a dual 2-L inverter system operates as a 3-L inverter. Researchers are now concentrating on dual 2-L inverter systems to provide multilayer output because of their benefits.

This paper's objective is a comprehensive investigation of this multilevel topology. Multilevel converters will be discussed, including their advantages, disadvantages, and recent uses. The advantages of dual inverter systems, as well as issue identification, approaches, and applications, are discussed after the analysis of dual two-level inverter systems. Then, in the MATLAB/SIMULINK environment, the multilevel performance of dual two-level converters is examined. Finally, future prospects are addressed.

1.2 Pros and Cons of Multilevel Converters [9–26]

2-L inverters require high power rating components, which are frequently unavailable, in order to get the required voltage or current. As a result, numerous devices must be coupled in series or parallel to generate the required voltage or current. Apart from this, 2-L inverters have poor power quality output and necessitate filtering prior to applying the load. Consequently, the filter gets bigger and the cost increases. In order to address these problems, MLIs were developed in 1975 by R.H. Baker and L.H. Bannister, who also received a patent for the cascade H-bridge inverter [9]. MLIs which are utilized in medium- and high-power applications have the following advantages over 2-L inverters:

a. By distributing voltage stress across switches at various levels, voltage derivatives (dv/dt) on switches have been decreased.
b. A decrease in dv/dt leads to a reduction in electromagnetic interference.

c. MLIs are suitable for fault-tolerant applications.
d. The multiple-level output waveform improves power quality due to low THD in output.
e. It is feasible to prevent or reduce common-mode voltages in electric drive and solar PV applications by using the suitable switching techniques.
f. It can work with fundamental or low switching frequencies. Switching at the fundamental frequency, which lowers conduction and switching losses, results in high levels of efficiency.
g. It is capable of functioning without transformers. Thus, transformerless operation is made possible.
h. Reliable control strategies and redundant switching states provide a fault-tolerant operation.
i. MLIs frequently feature a modular design. The system offers better voltage and current possibilities because of its versatility.
j. The input current has very low distortion.
k. It offers a simple user interface for adding renewable energy sources to the grid.
l. Without the need of a filter circuit, various PWM control techniques combined with a multi-level design lead to minimal overall harmonic distortion in the output waveform.
m. MLIs exhibit a self-balancing behavior without feedback control.

Despite these advantages, MLIs also exibhit few limitations which are as follows:

a. The quantity of switching devices has significantly increased. This makes the circuit more expensive and complex. Additionally, reliability has often been compromised.
b. Since MLIs have numerous switches, it is necessary to produce multiple gate pulses, which once again necessitates the usage of sophisticated PWM-based digital signal processors.
c. Some MLI topologies demand the use of multiple separate DC supplies, due to which a system's cost rises, which is a serious problem. Another issue is the lack of a DC supply.
d. MLIs have difference in the losses experienced by inner and outside switches.

Despite these drawbacks, MLIs are highly beneficial because of their high efficiency, little filtering needs, and simplicity of use with low-rating switching devices.

1.3 Applications of Multilevel Converters

The MLI is ideal for high- and medium-power applications because it can provide high-quality voltage waveforms with minimal switching frequency function [14, 27]. Split winding wind synchronous generators was connected to the grid using a cascaded H-bridges (CHB) MLI topology to make the system more efficient. In order to decrease the amount of installation space needed for MLIs and their price, researchers have recently focused on MLI designs with fewer gate drivers and switches. [28] proposed a hybrid modular multilevel converter for connecting huge wind turbines to the grid. In grid-connected systems, many MLI topologies have been combined with RESs to add the power produced by the RESs to the grid [29]. Due to the integration of MMC and HB module features, the system is a potential converter topology on the grid-side full-scale inverter of a WECS. It could increase the WECS's overall reliability. An innovative two-stage CSD multilevel system was introduced that employed fewer switches to provide a greater variety of voltage levels. The system also solved the problem of high voltage spikes under R-L loads because it has a path for reverse load currents. The system was deemed appropriate for incorporation into systems using renewable medium-voltage electricity [30]. The application of a modular cascaded MLI in a fault-tolerant distributed control system for the grid integration of wind turbines has been discussed in [31]. Through the use of the variable frequency inverted sine wave carrier modulation approach, a hybrid cascaded MLI for fuel cell applications was produced and discussed in [32]. A fuel cell-based system with an H-bridge DC to AC inverter and a three-level boost converter was also developed in [33]. [34] has suggested a hybrid cascaded MLI for fuel cell power conditioning systems. Great conversion efficiency and low switching losses are achieved while generating high voltages at higher speeds. A CHBMLI-interfaced open-loop FC-based energy system was developed in [35] which can be used where extra flexibility and power handling capability are required. An interface between a proton exchange membrane FC and a five-level (5-L) diode-clamped (DC) MLI was demostrated in [36]. This configuration was made with the goal of lowering harmonic distortion and boosting system performance. The technology described in [37] used a multilevel current-source inverter (CSI) to provide energy while also enhancing grid power quality. It was designed to deliver electricity from a FC to the grid. CHBMLI is utilized in motor drives to balance the voltage of battery cells [38, 39]. A CHBMLI converter was presented in [40] to balance the terminal voltage between the battery cells being used and to enhance the relaibility as well as for battery energy management in EV motor drives. For the purpose of removing the balancing circuits, the technology was suggested for application in battery-powered vehicles in [41] which also provided a novel modular multilevel topology containing electrochemical cells. Lots of MLI topologies are utilized in battery-powered EVs in the last few decades for increasing the reliability and minimizing the THD as well [42–44]. In order to increase efficiency, improve control, increase dependability, and reduce electromagnetic interference, it is suggested employing multilevel converters to provide IH power simultaneously at two frequencies [45]. [46] has utilized a CMLI to quicken the deployment of an IH power supply at high frequencies and with high power. Various MLIs and topologies have been used in IH application for various purposes [47–50].

MLIs are extensively utilized in a wide range of applications, including renewable energy systems (RESs), likewise in wind energy conversion systems (WECS), photovoltaic (PV) systems, fuel cells (FC), power supply (PSs), induction heating (IH), power factor compensators, electric vehicle drives (EV), motor drives, active filters, and countless other disciplines.

1.4 Advantages of Dual Two-Level Converters

a. Double 2-L 3-phase inverter topologies with an OEW that is fed from both ends with a VSI are utilized for rolling mills, electric ship propulsion, VSD applications, as well as HEVs/EVs applications. A double 2-L 3-phase inverter design with this output is quite similar to a 3-L single inverter arrangement. The following are the most prevalent advantages of twin inverter configurations over identical multi-level single inverter configurations.
b. A double inverter system can continue to run without interruption in the event of a single inverter failure since it can operate on a single inverter as well, but a single inverter system cannot. In other words, while using a twin inverter system, the drive's fault tolerance improves. As a result, the twin 2-L inverter is ideal for land traction. Because the voltage between the phases is larger in a double inverter system than in a single inverter system, the low current rating may

be utilized. As a result of this benefit, we can investigate several applications such as electric and hybrid electric cars in which the DC bus voltage is limited [51, 52].

c. Although the number of switches in both the double 2-L inverter and 3-L single inverter systems is the same, the architecture employed in the 3-L single inverter system necessitated the inclusion of diodes and a capacitor, which needed voltage balancing. This problem may be solved by adopting a double inverter system, which eliminates the need for extra capacitors and diodes [52].
d. Ripples are considerably minimized when a double 2-L inverter system is used as a 3-L inverter. When a double 2-L inverter system is used as a 3-L inverter, acoustic noise and electromagnetic emissions from the motor are greatly decreased [52].
e. The issue of power limit may be solved by using a double 2-L inverter system, which divides the multiple batteries into two equal proportions in battery-related applications and also offers half-DC voltage that increases the dependability of few applications such as electric automobiles [52]. Battery-related applications may outperform thermal engine uses since they are both cost-effective and environmentally beneficial [53].
f. In applications such as fishing boats, where power and torque variations are required from time to time due to the fact that they operate in two modes, a single 2-L inverter system cannot distribute these parameters uniformly depending on the condition. This can be resolved by a double 2-L inverter system, which can provide higher torque in fishing activities while also providing larger power in moving conditions. In such a way, the whole productivity of the system can be enhanced in both modes [53].

1.5 Problem Identification

Although the number of switches in both double 2-L inverter and 3-L single inverter systems is the same, the topology utilized in 3-L single inverter systems needed extra diodes and a capacitor that needed balancing of voltages [53]. Multi-operational integrated components, such as the ISA, that perform various functions such as powering the flywheel, generator/motor, or torque support for the ignition engine have not only a cost and heat difficulty but also a motor-inverter system design challenge. In such systems, electricity should be generated while the engine is operating at 6,000 rpm and producing 150 Nm of starting torque, i.e., it should have a high pole structure (14 poles). To achieve the needed torque, the voltage must not be larger than 42 V; hence, such systems must have field weakening, operating at a greater speed than the permitted speed [54]. The CMV generated by 3-phase PWM inverters, which are employed in a variety of applications, produces bearing current in IM drives. Electromagnetic interference is introduced into the system by the leakage current created by CMV.

To eliminate CMV, only the vectors with space voltage that do not create the CMV were used for switching [55]. Traditional 3-phase 4-pole IM armatures feature two comparable per phase winding coil groups. Because both coil groups are linked in series, the distance between them is equivalent to two pole pitches. As a result, it gives a common mode current path [56]. It is required to reduce CMV produced by PWM inverters in link with a 3-phase in 5-phase double inverter arrangement. By selecting space vectors that are not responsible for CMV, a large number of redundancies of switching states may be used to remove CMV [57].

A massive DC linkage capacitor carrying a output current at load was required for a 3-L inverter. The fluctuating vectors with space voltage generated by this output current at load are unpleasant in a 3-level inverter [58]. A large DC link capacitor in the NPC inverter flows load current. This current not only causes neutral point variations, but it also gives the output voltage waveform a DC component. OEWIM [59, 60] can alleviate this disadvantage. Between the two inverters, interphase reactors are utilized, and the system output is utilized to provide huge AC drives. However, such interphase reactors are also to blame for the system's significant acoustic noise and losses [61]. Inverter deadtime causes zero sequence voltage that causes zero sequence current. This current causes circulating current in the OEW, resulting in substantial system losses and semiconductor device current stress [62]. As a 3-L inverter, it required large DC link capacitors to circulate load current and provide a fluctuating neutral point [63].

1.6 Applications of Dual Two-Level Converters

In 2010, [64] presented OEW-VSD for electric and hybrid electric cars with double-polyphase inverter supply. In 2005, [65] suggested twin 3-phase inverters linked to an OEW AC motor hybrid vehicle with energy storage. [66] presented a charging approach for charging the auxiliary battery in a double inverter system for applications such as electric automobiles, because the architecture employed in such applications only has one charger for the primary battery. In 2008, [67] suggested a double 2-L inverter system-based generating system for series hybrid power trains and navy ship propulsion systems. [68] presented a double 7-L inverter arrangement for a fault-tolerant OEWIM drive in 2009. In 2005, [69] introduced an SVPWM switching approach for double inverter-fed OEWIM drive to lower the CMV and make it acceptable for applications requiring repeated starts and low speeds.

Various authors disclosed the results of a unique synchronized SVPWM for the operation of double inverter-fed traction motor drives using two insulated DC voltage sources and an OEW IM in 2007 [70]. A. Lega proposed a double 2-L inverter scheme in 2007 to enhance the efficiency of the system in a variety of applications such as fishing boats and ships, as well as in applications such as automotive traction systems and land traction systems, because it surpasses the peak power limit of power switches and increases the power [53].

The VSI develops a 3-phase voltage with a governable magnitude, frequency, and phase angle. According to the literature, the PWM schemes are of two types: carrier-based as well as SVPWM. The algorithm which is used to control the PWM is commonly known as SVM. The common and most popular modulation scheme due to their benefits of better DC bus utilization and easier digital execution, SVPWM generates six voltage vectors by VSI for the first step of multiphase, i.e., 3-phase which is used to control the frequency, phase angle, and amplitude of 3-phased permanent magnet or 3-phase induction machine. The primary applications of a dual inverter system are listed below after summarizing the aforementioned:

a. It improves the efficiency of system in applications such as fishing boats and ships by managing power sharing and distribution among the batteries, which balances the discharging of batteries.
b. Because it overcomes the maximum power restriction of switching devices and boosts the power up to two times the original value, the double 2-L inverter system is employed in car traction systems as well as land traction applications.
c. When employed as an OEW system, a double 2-L inverter system can be beneficial in a variety of utilities such as series hybrid electric power trains, EVs/HEVs, rolling mills, and propulsion motors.

d. For grid integration without an isolation transformer, double inverters are also utilized as a solar panel-fed active filter and floating network inverter.
e. Double inverters are commonly employed in applications that require two separate DC-link voltages, such as ship electrical systems, large-power wind energy generation systems, and hybrid automobiles with energy storage.

1.7 Multilevel Performance of Dual 2-L 3-Phase Inverter Using ANN-Based PWM [71–74]

As shown in Figure 1.1, a dual 2-L inverter comprises one 2-L inverter on either side of the load. Inverters are divided into two types: positive end converters and negative end converters. These are the letters a, b, and c, as well as a′, b′, and c′, respectively, for load terminals. The inputs to both inverters are two distinct dc sources labeled V_{dc1} and V_{dc2}. The outputs of both inverters are linked to a power supply. Winding with an open end (OEW), the voltage V_{ref} reference vector and the dual inverter's vector outputs V_{01} and V_{02} are coupled in the following way:

$$V_{ref} = V_{01} - V_{02} \tag{1.1}$$

In a dual 2-L 3-phase inverter system, where each inverter produces active voltage vector V_P and V_N for INV1 and INV2 and their application time durations T_P and T_N are represented as well, a unified SVPWM approach has been shown to describe the process.

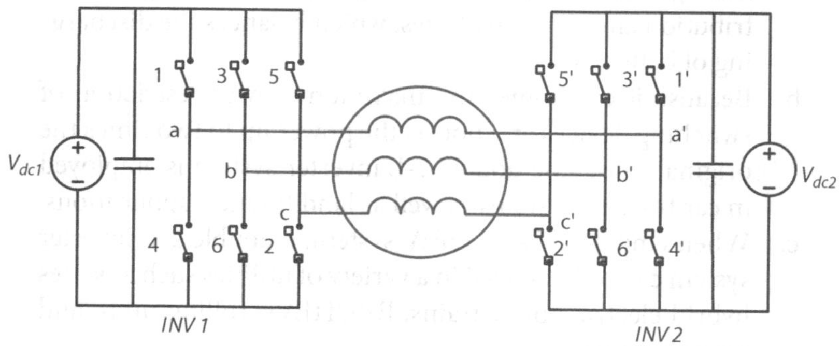

Figure 1.1 Power circuit of dual 2-L 3-phase VSI system.

$$T_P = \frac{\sqrt{3}}{V_{DC2}} V_S T_S \sin(60 - \theta) \tag{1.2}$$

$$T_N = \frac{\sqrt{3}}{V_{DC2}} V_S T_S \sin(\theta) \tag{1.3}$$

$$T_0 = T_s - T_P - T_N \tag{1.4}$$

T_0 is the instant at which zero vectors are applied, Ts is the sampling time period, and V_{ref} and V_P are the angles. The reference vector peak value Vref and the angle "α" determine the periods T_P, T_N, and T_0.

Ts is the maximum time length for T_P and T_N, and each sector is divided into three areas. If T_P > Ts, V_{ref} is shown in region 2 in Figure 1.2(b), and in area 3 if T_N > Ts (Figure 1.2(c)); otherwise, V_{ref} is shown in region 1 in Figure 1.2(a) for sector I.

Figure 1.3 depicts the vector plane with voltages for the dual inverter setup. The timing of voltage vectors in various sectors and regions is shown in Table 1.1, where $V_{1P} = V_P$ and $V_{2N} = V_N$ denote the voltage vectors used in region 1. The complementary vector in INV1 is labeled V_{1N}, whereas it is labeled V_{2P} in INV2. The time durations of V_{1P}, V_{2N}, V_{1P}, and V_{2N} are T_{1P}, T_{2N}, T_{1N}, and T_{2P}, respectively. When Vs is in area 1, no complementary voltage vectors are required; hence, T_{1N} and T_{2P} are set to zero.

Figure 1.4 shows the switching sequences for both inverters with voltage vectors for different places across a sample period, using sector I as an example. In area 1, both inverters are in the switching mode. INV1 clamps while INV2 switches when the reference vector is in region 2; when the reference vector is in region 3, INV1 switches while INV2 clamps. The

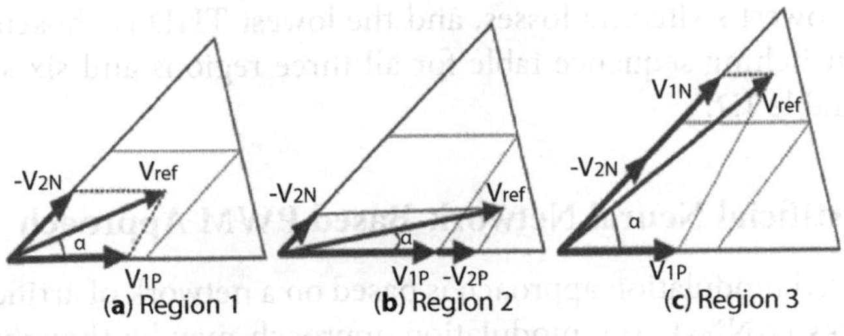

Figure 1.2 Vectors generating process for reference vector in sector I using SVPWM scheme.

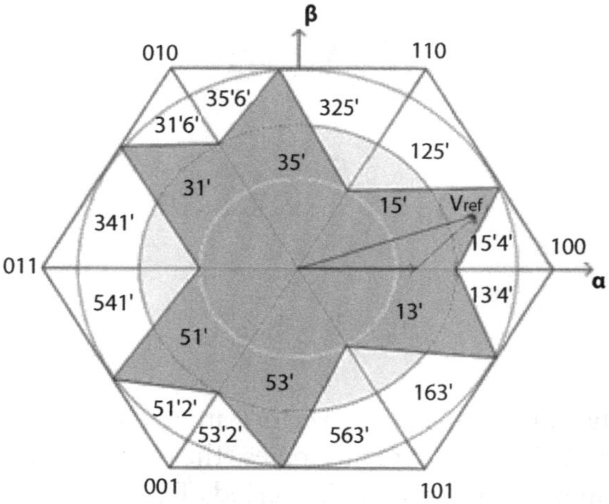

Figure 1.3 Voltage vector plane and switching vector selection for the dual 2-L 3-phase inverter system.

Table 1.1 Voltage vector time distributions for the dual 2-L 3-phase inverter.

			INV1		INV2	
Sector I–VI			V_{1P}	V_{1N}	V_{2P}	V_{2N}
Region-1	$T_P \leq T_s$	$T_N \leq T_s$	$T_{1P} = T_P$	$T_{1N} = 0$	$T_{2P} = 0$	$T_{2N} = T_N$
Region-2	$T_P > T_s$	$T_N \leq T_s$	$T_{1P} = T_s$	$T_{1N} = 0$	$T_{2P} = k$ $(t_P - T_s)$	$T_{2P} = 0$
Region-	$T_P \leq T_s$	$T_N > T_s$	$T_{1P} = T_P$	$T_{1N} = (T_N$ $- T_s)/k$		$T_{2N} = T_s$

switching sequence of the double two-level 2-L VSI for 3-L operation must also be considered. The switching sequence with the fewest switching states, the lowest switching losses, and the lowest THD is chosen in general. The switching sequence table for all three regions and six sectors is shown in Table 1.2.

1.7.1 Artificial Neural Network-Based PWM Approach

The proposed modulation approach is based on a network of artificial neural networks (ANNs). The modulation approach may be thought of as a nonlinear input–output mapping.

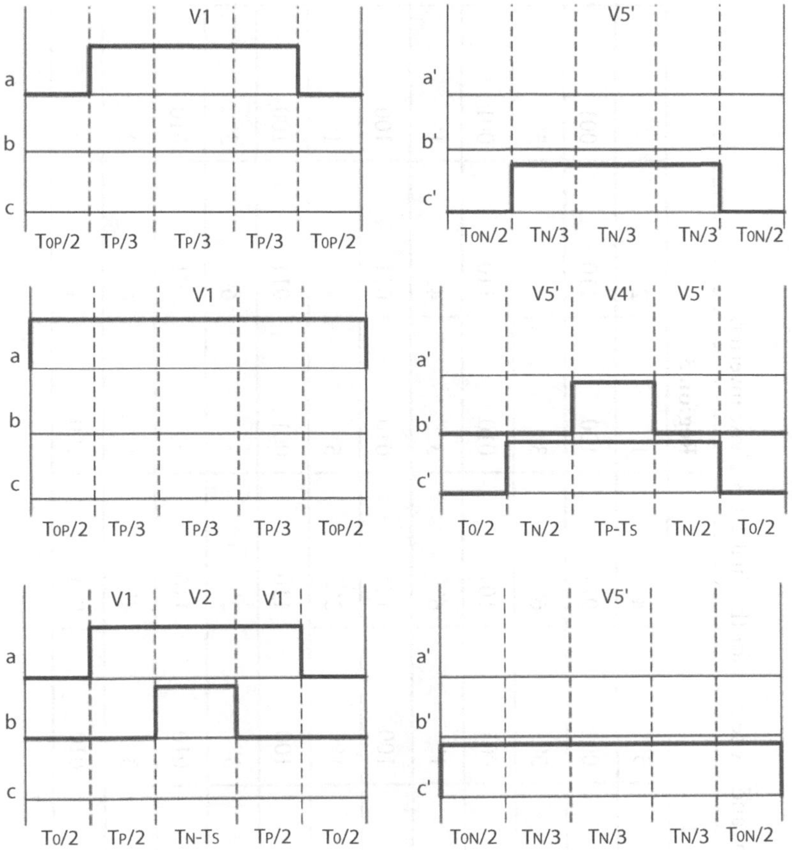

Figure 1.4 Switching waveforms in regions 1, 2, and 3 for sector I for the dual 2-L 3-phase inverter.

The pulse width modulation approach was used to construct the reference vector V_{ref} amplitude and 'α' angle for the dual 2-L 3-phase inverter and the succeeding pulse width pattern is at the inverter output. Simulation results of a standard SVM-based 3-L rectifier were used to create the input–output data patterns (and) for training the neural network. A hidden layer containing a few neurons, input and output layer neurons, and suitable transfer functions are used to create an initial neural network setup. To prevent the neuron outputs from getting saturated, small random weights are originally chosen. The output (known as forward pass) is computed using the specified input pattern (V_{ref} magnitude and 'α' angle) and compared to the intended output pattern. The weights are then changed in the backward direction using a backpropagation approach until the error (MSE)

Table 1.2 Switching sequence for all the three regions and six sectors for the dual 2-L 3-phase inverter.

	Region 1		Region 2			Region 3		
Sector I	1	5′	1	5′	4′	1	2	5′
	100	001	100	001	011	100	110	001
Sector II	3	5′	3	5′	6′	3	2	5′
	010	001	010	001	101	010	110	001
Sector III	3	1′	3	1′	6′	3	4	1′
	010	100	010	100	101	010	011	100
Sector IV	5	1′	5	1′	2′	5	4	1′
	001	100	001	100	110	001	011	100
Sector V	5	3′	5	3′	2′	5	6	3′
	001	010	001	010	110	001	101	010
Sector VI	1	3′	1	3′	4′	1	6	3′
	100	010	100	010	011	100	101	010

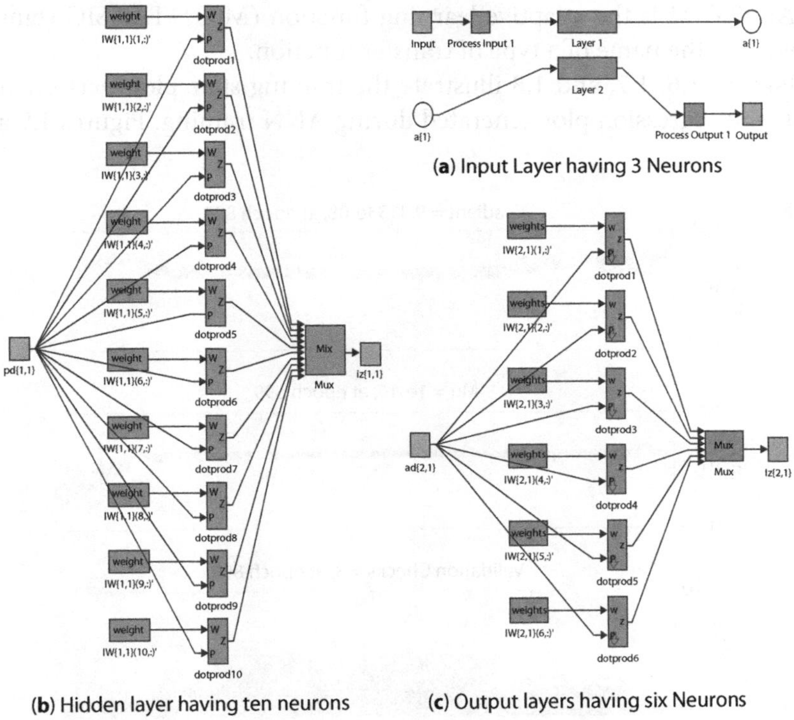

Figure 1.5 (a) Input layer having three neurons. (b) Hidden layer having 10 neurons. (c) Output layers having six neurons.

is very small and acceptable. Figure 1.5 depicts the neural network plan used in the experiment. Figure 1.5(a and b) shows the three levels of the neural network: an input layer with three neurons, a hidden layer with 10 neurons, and a hidden layer with 10 neurons. Figure 1.5(c) depicts a six-neuron output layer.

1.7.2 Simulation Results

The simulation results of the ANN PWM technique are described in this section. In the ANN-based PWM scheme, the reference signal acts as the input neuron values, while the modulating signals serve as the target values. The ANN block is built after training and simulation using the MATLAB neural network application. Figure 1.5(a–c) shows three-neuron input and output layers as well as a 10-neuron hidden layer. The circuit is a feed-forward backpropagation type, and the simulation is done with the training function "trainlm". MSE is the performance function, while

LEARNGDM is the adaptive learning function (MSE). TANSIG (tansigmoidal) is the name of a type of transfer function.

Figures 1.6, 1.7, and 1.8 illustrate the training state plot, performance plot, and regression plot generated during ANN training. Figures 1.9 and

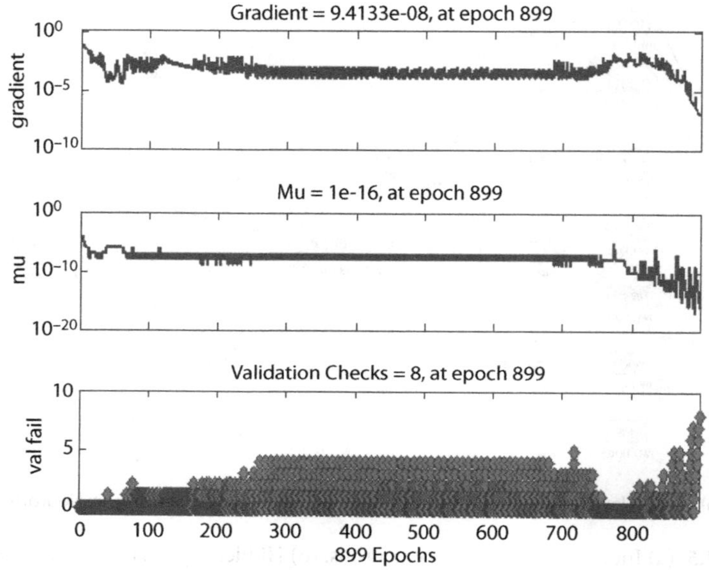

Figure 1.6 Training state plot using ANN tool.

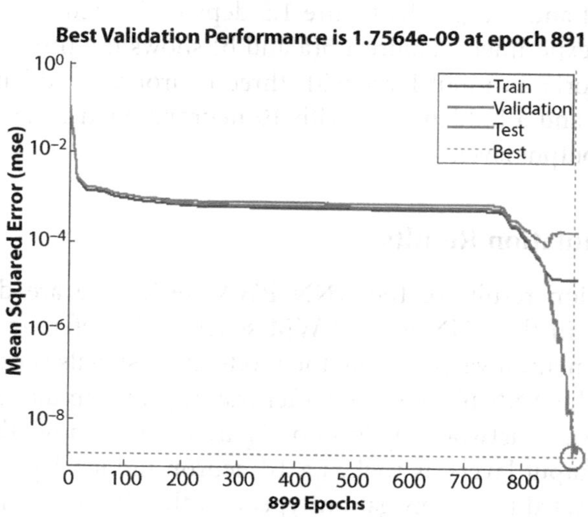

Figure 1.7 Performance plot using ANN tool.

1.10 illustrate the output voltages across load from the simulations without and with filters, respectively. Figure 1.9 represents the multilevel output. The output is sinusoidal, and it is the same as what the space vector PWM approach produces. The harmonic spectrum of load voltage for phase 'a' is shown in Figure 1.11 at 50 Hz; the fundamental has a magnitude of 0.5588 peak (0.3952 rms) and a THD of 3.19%, as indicated in the graph. The results are quite similar as that developed by SVPWM approach in [75] for multilevel performance.

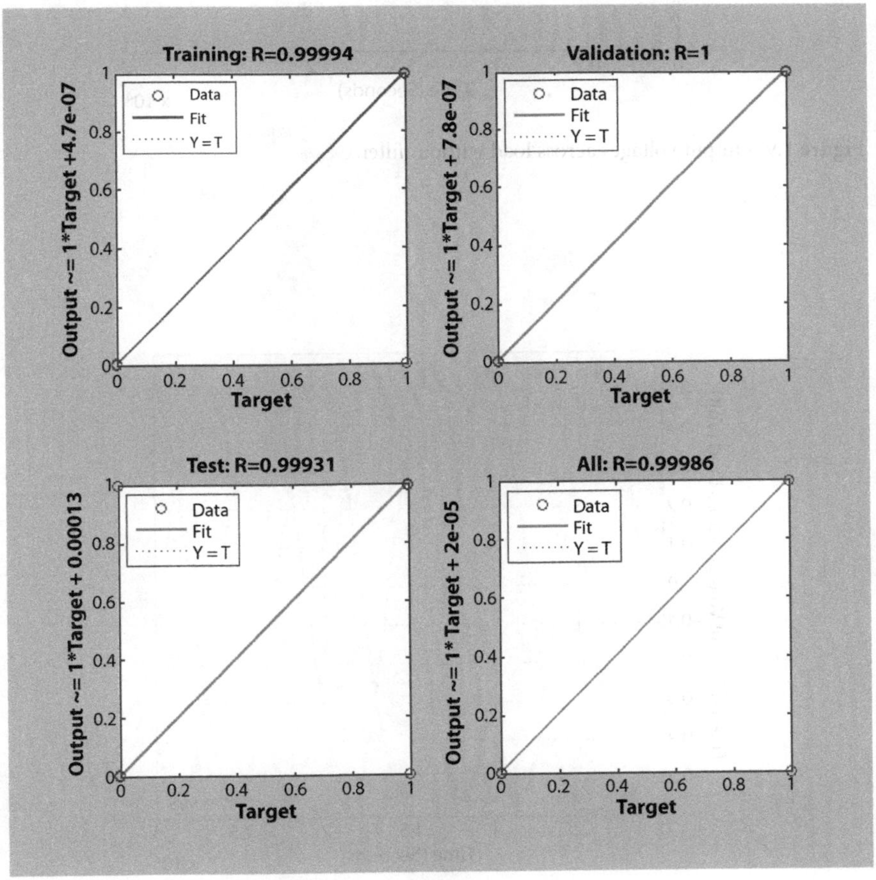

Figure 1.8 Regression plot using ANN tool.

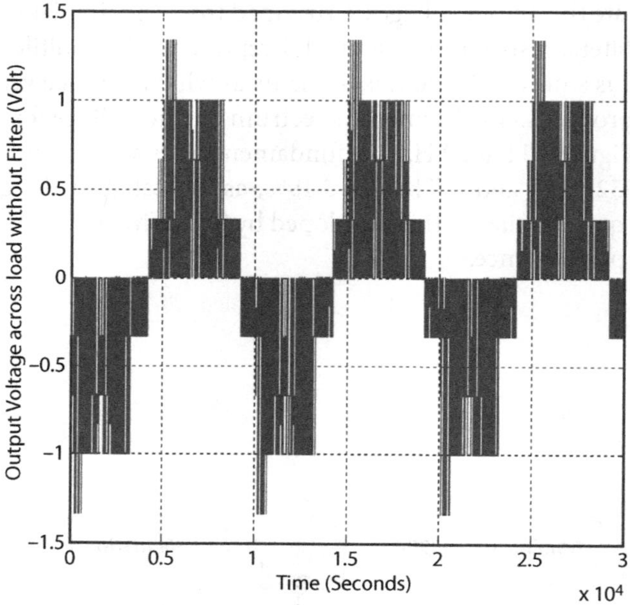

Figure 1.9 Output voltages across load without filter.

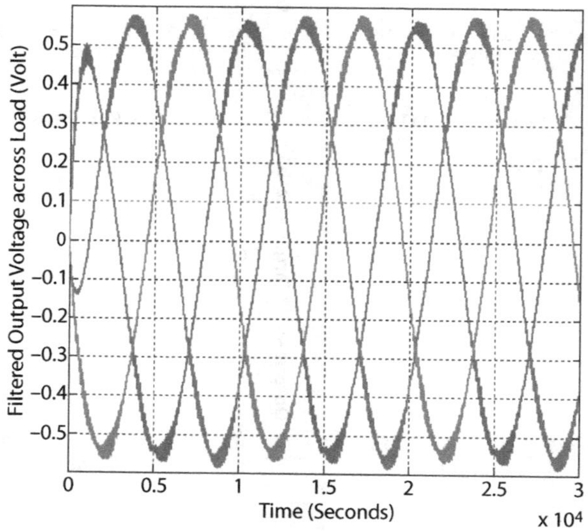

Figure 1.10 Filtered output voltage across load.

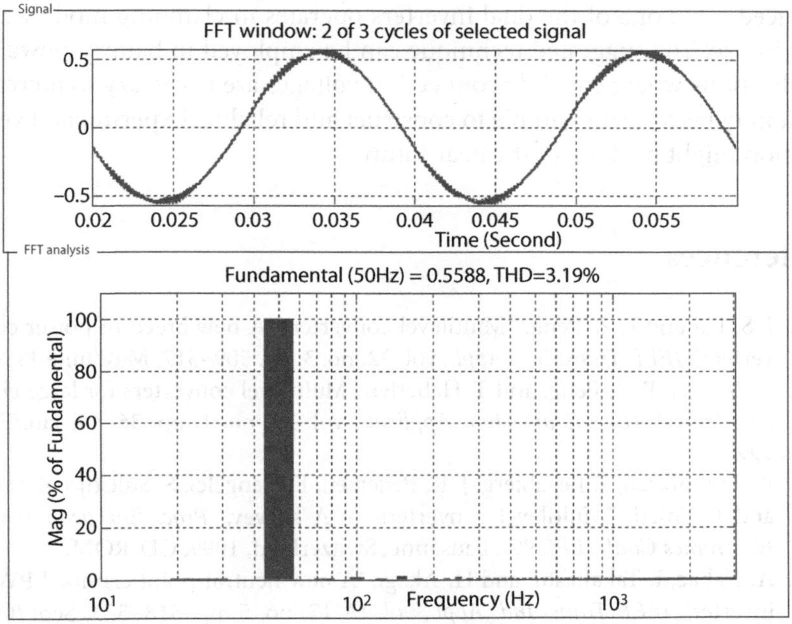

Figure 1.11 Harmonic spectrum of load voltage for particular phase 'a' using ANN-based PWM.

1.8 Conclusion

In this article, the author examines MLIs, covering their advantages, disadvantages, and uses in recent years. It is discovered that as the level increases, MLIs become more expensive and complex. Dual 2-L inverter systems are explored to get around this restriction, and then there is a discussion of their advantages, problem-solving methods, and applications. In this study, the author applies the ANN PWM technique to the analysis of a dual 2-L three-phase voltage inverter system. An Open End Winding with double 2-L 3-phase VSI is used to conduct the performance analysis of output voltage behavior based on the fundamental component of magnitude and THD since ANN technology speeds up digital implementation while also requiring less computing work. It is discovered that both inverters in region 1 are in a switching state. Since the reference vector is in region 2, INV1 acts in a clamping mode, whereas INV2 operates in a switching mode. INV1 will operate in a switching mode when the reference vector is in region 3, whereas INV2 will operate in a clamping mode. The losses are reduced since one of the dual inverters operates in clamping mode while it is in use. The suggested technique can be employed in battery-powered applications where equal dc connection voltages are necessary to increase reliability because it is simple to construct and reliable. Experimental verification might be done in the near future.

References

1. J. S. Lai and F. Z. Peng, "Multilevel converters–A new breed of power converters," *IEEE Trans. Ind. Appl.*, vol. 32, no. 3, pp. 509–517, May/June 1996.
2. L. Tolbert, F.-Z. Peng, and T. Habetler, "Multilevel converters for large electric drives," *IEEE Trans. Ind. Applicat.*, vol. 35, no. 1, pp. 36–44, Jan./Feb. 1999.
3. R. Teodorescu, F. Beaabjerg, J. K. Pedersen, E. Cengelci, S. Sulistijo, B. Woo, and P. Enjeti, "Multilevel converters — A survey," *Proc. European Power Electronics Conf. (EPE'99)*, Lausanne, Switzerland, 1999, CD-ROM.
4. A. Nabae, I. Takahashi, and H. Akagi, "A new neutral-point clamped PWM inverter," *IEEE Trans. Ind. Appl.*, vol. IA-17, no. 5, pp. 518–523, Sept./Oct. 1981.
5. T. A. Meynard and H. Foch, "Multi-level choppers for high voltage applications," *Eur. Power Electron. Drives J.*, vol. 2, no. 1, p. 41, Mar. 1992.
6. C. Hochgraf, R. Lasseter, D. Divan, and T. A. Lipo, "Comparison of multilevel inverters for static var compensation," *Conf. Rec. IEEE-IAS Annual Meeting*, pp. 921–928, Oct. 1994.
7. P. Hammond, "A new approach to enhance power quality for medium voltage ac drives," *IEEE Trans. Ind. Appl.*, vol. 33, no. 1, pp. 202–208, Jan./Feb. 1997.
8. E. Cengelci, S. U. Sulistijo, B. O. Woom, P. Enjeti, R. Teodorescu, and F. Blaabjerge, "A New Medium Voltage PWM Inverter Topology for Adjustable Speed Drives," *Conf. Rec. IEEE-IAS Annal. Meeting*, St. Louis, MO, pp. 1416–1423, Oct. 1998.
9. R. H. Baker and L. H. Bannister, "Electric power converter," U.S. Patent 3 867 643, Feb. 1975.
10. A. Nabae, I. Takahashi and H. Akagi, "A New Neutral-Point-Clamped PWM Inverter" *IEEE Trans. Ind. Appl.*, vol. IA-17, no. 5, pp. 518–523, 1981.
11. L. M. Tolbert and X. Shi, "Multilevel Power Converters" *Power Electronics Handbook*, Elsevier:, Amsterdam, The Netherlands, pp. 385–416, 2018.
12. N. Prabaharan and K. Palanisamy, "A Comprehensive Review on Reduced Switch Multilevel Inverter Topologies, Modulation Techniques and Applications", *Renew. Sustain. Energy Rev.*, vol.76, pp. 1248–1282, 2017.

13. I. Pharne and Y. Bhosale, "A Review on Multilevel Inverter Topology", *Proceedings of the 2013 International Conference on Power, Energy and Control (ICPEC)*, Dindigul, India, pp. 700–703, 6–8 February 2013.
14. N. S. Hasan, N. Rosmin, D. A. A. Osman, and A. H. Musta'amal, "Reviews On Multilevel Converter and Modulation Techniques", *Renew. Sustain. Energy Rev.*, vol. 80, pp. 163–174, 2017.
15. J. Rodriguez, S. Bernet, P. K. Steimer, and I. E. Lizama, "A Survey on Neutral-Point-Clamped Inverters", *IEEE Trans. Ind. Electron.*, vol. 57, no. 7, pp. 2219–2230, 2010.
16. C. Balamurugan, S. Natarajan and M. Arumugam, "A Review on Various Multilevel Inverter Topologies" *Glob. J. Adv. Res.*, vol. 2, no. 11, 2015.
17. M. Vijeh, M. Rezanejad, E. Samadaei and K. Bertilsson, "A General Review of Multilevel Inverters based on Main Submodules: Structural Point of View", *IEEE Trans. Power Electron.*, vol. 34, no. 10, pp. 9479–9502, 2019.
18. R. A. Krishna and L. P. Suresh, "A Brief Review on Multi Level Inverter Topologies", *Proceedings of the 2016 International Conference on Circuit, Power and Computing Technologies (ICCPCT)*, Nagercoil, India, pp. 1–6, 18–19 March 2016.
19. F. Kieferndorf, M. Basler, L. Serpa, J.-H. Fabian, A. Coccia, and G. Scheuer, "A New Medium Voltage Drive System based on ANPC-5L Technology" *Proceedings of the 2010 IEEE International Conference on Industrial Technology*, Via del Mar, Chile, pp. 643–649, 14–17 March 2010.
20. J. G. Subarnan, "Multilevel inverters: An Enabling Technology", *Hybrid-Renewable Energy Systems in Microgrids*, Elsevier: Amsterdam, The Netherlands, pp. 61–80, 2018.
21. B. P. McGrath and D. G. Holmes, "Analytical Modelling of Voltage Balance Dynamics for a Flying Capacitor Multilevel Converter" *IEEE Trans. Power Electron.*, vol. 23, no. 2, pp. 543–550, 2008.
22. T. Suneel, "Multi Level Inverters: A Review Report", *Int. J. New Technol. Sci. Eng. (IJNTSE)*, vol. 1, no.1, pp. 82–109, 2014.
23. P. C. Deepak and S. N. Rao, "Cascaded H-Bridge Multilevel Inverter Using Inverted Sine Wave PWM Technique", *International Journal of Emerging Trends in Electrical and Electronics (IJETEE)*, vol. 6, no. 1, pp. 39–44, 2013.
24. M. Malinowski, K. Gopakumar, J. Rodriguez and M. A. Pérez, "A Survey on Cascaded Multilevel Inverters", *IEEE Trans. Ind. Electron.*, vol. 57, no. 7, pp. 2197–2206, 2010.
25. J. Venkataramanaiah, Y. Suresh and A. K. Panda, "A Review on Symmetric, Asymmetric, Hybrid and Single DC Sources based Multilevel Inverter Topologies" *Renew. Sustain. Energy Rev.*, vol. 76, pp. 788–812, 2017.
26. F. Masoudina, E. Babaei, M. Sabahi and H. Alipour, "New Cascaded Multilevel Inverter with Reduced Power Electronic Components," *Iranian J. Electr. Electron. Eng.*, vol. 16, no. 1, pp. 107–113, 2020.

27. S. Alatai, M. Salem, D. Ishak, M. Kamarol, M. Jamil and A. Bughneda, "Design and Analysis of Five-Level Cascaded LLC Resonant Converter", *Proceedings of the 2020 IEEE International Conference on Power and Energy (PECon)*, Penang, Malaysia, pp. 66–70, 7–8 December 2020.
28. S. Debnath and M. Saeedifard, "A New Hybrid Modular Multilevel Converter for Grid Connection of Large Wind Turbine". *IEEE Trans. Sustain. Energy*, vol. 4, no. 4, pp. 1051–1064, 2013.
29. R. Mahalakshmi and K. S. Thampatty, "Grid connected multilevel inverter for renewable energy applications", *Procedia Technol.*, vol. 21, pp. 636–642, 2015.
30. L. Wang, Q. Wu and W. Tang, "Novel cascaded switched-diode multilevel inverter for renewable energy integration", *IEEE Trans. Energy Convers.*, vol. 32, no.4, pp. 1574–1582, 2017.
31. M. A. Parker, L. Ran and S, J. Finney, "Distributed Control of a Fault-Tolerant Modular Multilevel Inverter for Direct-Drive Wind Turbine Grid Interfacing" *IEEE Trans. Ind. Electron.*, vol. 60, no. 2, pp. 509–522, 2013.
32. R. Seyezhai and B. Mathur, "Hybrid multilevel inverter using ISPWM technique for fuel cell applications", *Int. J. Comput. Appl.*, vol. 9, no.1, pp. 41–47, 2010.
33. J. R. Rahul, A. Kirubakaran and D. Vijayakumar, "A new multilevel DC-DC boost converter for fuel cell based power system" *Proceedings of the 2012 IEEE Students' Conference on Electrical, Electronics and Computer Science*, Bhopal, India, pp. 1–5, 1–2 March 2012.
34. R. Seyezhai and B. Mathur, "Hybrid cascaded H-bridge multilevel inverter for fuel cell power conditioning systems", *Proceedings of the 2008 43rd International Universities Power Engineering Conference*, Padua, Italy, pp. 1–5, 1–4 September 2008.
35. R. Singh and J. Nakka, "Design, simulation and performance analysis of Fuel Cell based energy system with Cascaded H-Bridge Multilevel Inverter" *Proceedings of the 2016 IEEE 1st International Conference on Power Electronics, Intelligent Control and Energy Systems (ICPEICES)*, Delhi, India, pp. 1–6, 4–6 July 2016.
36. L. S. Kumar and O. S. Teja, "Harmonic analysis of diode clamped multilevel inverter base PEM fuel cell energy system", *Proceedings of the 2017 International Conference on Energy, Communication, Data Analytics and Soft Computing (ICECDS)*, Chennai, India, pp. 2801–2806, 1–2 August 2017.
37. P. Cossutta, M. P. Aguirre, A. Cao, S. Raffo and M. I. Valla, "Single-Stage Fuel Cell to Grid Interface with Multilevel Current-Source Inverters", *IEEE Trans. Ind. Electron.*, vol. 62, no. 8, pp. 5256–5264, 2015.
38. L. Maharjan, T. Yamagishi, H. Akagi and J. Asakura, "Fault-tolerant operation of a battery-energy-storage system based on a multilevel cascade PWM converter with star configuration", *IEEE Trans. Power Electron.*, vol. 25, no. 9, pp. 2386–2396, 2010.

39. L. Maharjan, T. Yamagishi and H. Akagi, "Active-Power Control of Individual Converter Cells for a Battery Energy Storage System Based on a Multilevel Cascade PWM Converter" *IEEE Trans. Power Electron.*, vol. 27, no. 3, pp. 1099–1107, 2012.
40. Z. Zheng, K. Wang, L. Xu and Y. Li, "A Hybrid Cascaded Multilevel Converter for Battery Energy Management Applied in Electric Vehicles", *IEEE Trans. Power Electron.*, vol. 29, no. 7, pp. 3537–3546, 2014.
41. M. Quraan, T. Yeo, and P. Tricoli, "Design and Control of Modular Multilevel Converters for Battery Electric Vehicles", *IEEE Trans. Power Electron.*, vol. 31, no. 1, pp. 507–517, 2016.
42. M. Quraan, P. Tricoli, S. D'Arco and L. Piegari, "Efficiency Assessment of Modular Multilevel Converters for Battery Electric Vehicles", *IEEE Trans. Power Electron.*, vo. 32, no. 3, pp. 2041–2051, 2017.
43. A. S. Gadalla, X. Yan and H. Hasabelrasul, "Performance of the battery energy storage systems based on cascaded H-bridge multilevel converter" *The J. Eng.*, vol. 2019, no. 16, pp. 779–783, 2019.
44. Z. Gao and Q. Lu, "A Hybrid Cascaded Multilevel Converter Based on Three-Level Cells for Battery Energy Management Applied in Electric Vehicles", *IEEE Trans. Power Electron.*, vol. 34, pp. 7326–7349, 2019.
45. B. Diong, S. Basireddy, K. Corzine and Y. Familiant, "Multilevel inverters with equal or unequal sources for dual-frequency induction heating", *Proceedings of the Nineteenth Annual IEEE Applied Power Electronics Conference and Exposition*, Anaheim, CA, USA, pp. 825–831, 22–26 February 2004.
46. L. Qingfeng, W. Huamin and L. Zhaoxia, "Discuss on the application of multilevel inverter in high frequency induction heating power supply", *Proceedings of the TENCON 2006-2006 IEEE Region 10 Conference*, Hong Kong, China, pp. 1–4, 14–17 November 2006.
47. B. M. Flayyih, M. Z. Ahmed and M. Ambroze, "A novel hybrid votage-current fed induction heating power supply system using multilevel neutral point clamped inverter", *Proceedings of the 2014 IEEE International Energy Conference (ENERGYCON)*, Cavtat, Croatia, pp. 189–194, 13–16 May 2014.
48. B. M. Flayyih, M. Z. Ahmed and M. Ambroze, "An Optimum Harmonic Control of induction heating power supply system using multilevel neutral point clamped inverter", *Proceedings of the 2016 IEEE Transportation Electrification Conference and Expo*, Asia-Pacific (ITEC Asia-Pacific), Busan, Korea, pp. 531–536, 1–4 June 2016.
49. Y. Yue, Q. Xu, A. Luo, P. Guo, Z. He and Y. Li, "Analysis and control of tundish induction heating power supply using modular multilevel converter", *IET Gener. Transm. Distrib.*, vol. 12, pp. 3452–3460, 2018.
50. B. M. Flayyih, M. Z. Ahmed and S. M. Veigh, "A comprehensive power analysis of induction heating power supply system using multilevel neutral point clamped inverter with optimum control algorithm", *Proceedings of the 2015*

IEEE 11th International Conference on Power Electronics and Drive Systems, Sydney, NSW, Australia, pp. 727–731, 9–12 June 2015.
51. E. Levi, M. Jones and W. Satiawan, "A multiphase dual-inverter supplied drive structure for electric and hybrid electric vehicles", *2010 IEEE Vehicle Power and Propulsion Conference*, Lille, France, 1-3 September 2010.
52. F. Patkar and Fazlli, "PWM techniques for control of dual-inverter supplied six-phase drives", Liverpool John Moores University, Thesis, 2013.
53. A. Lega, "Multilevel converters dual two-level inverter scheme", University of Bologna, Ph.D. Thesis, March 2007.
54. J. Kim, J. Jung and K. Nam, "Dual-inverter control strategy for high-speed operation of EV induction motors", *IEEE Trans. Ind. Electron.*, vol. 51, no. 2, pp. 312-320, April 2004.
55. M. R. Baiju, K. K. Mohapatra, R. S. Kanchan and K. Gopakumar, "A dual two-level inverter scheme with common mode voltage elimination for an induction motor drive", *IEEE Trans. Power Electron.*, vol. 19, no. 3, pp. 794-805, 2004.
56. K. Sivakumar, A. Das, R. Ramchand, C. Patel and K. Gopakumar "A three level voltage space vector generation for open end winding IM using single voltage source driven dual two-level inverter", *IEEE Region 10 Conference TENCON 2009*, Singapore, 23-26 January 2009.
57. N. Bodo, M. Jones and E. Levi, "PWM techniques for an open-end winding five phase drive with a single DC source supply" *IECON 2012 - 38th Annual Conference on IEEE Industrial Electronics Society*. Montreal, QC, Canada 25-28 October 2012.
58. V. T. Somasekhar, K. Gopakumar, A. Pittet and V. T. Ranganathan "A novel PWM inverter switching strategy for a dual two-level inverter fed open-end winding induction motor drive", *4 th IEEE International Conference on Power Electronics and Drive Systems*, Denpasar, Indonesia, 25 October 2001.
59. S. M.W. Ahmed, M. M. Eisa, G.M.A. Sowilam, A. B. Salem and M. Salem, "Open ends induction motor operation based on a dual inverter", in *Design and Test Workshop (IDT)*, 2009, Riyadh, Saudi Arabia, 15-17 November 2009.
60. S.M.W. Ahmed, G.M.A. Sowilam, A. B. Salem and M. Salem, "Practical implementation of a dual inverter operates open ends induction motor" *4 th International Design and Test Workshop (IDT)*, Riyadh, Saudi Arabia, 15-17 November 2009.
61. Y. Kawabata, M. Nasu, T. Nomoto, E. C. Ejiogu and T. Kawabata, "High efficiency and low acoustic noise drive system using open-winding AC motor and two space-vectormodulated inverters", *IEEE Trans. Ind. Electron.*, vol. 49, no. 4, pp. 783-789, August 2002.
62. Q. An, J. Liu, Z. Peng and L. Sun, "Dual space vector control of open-end winding permanent magnet synchronous motor drive fed by dual inverter", *IEEE Trans. Power Electron.*, vol. 31, no. 12, pp. 8329-8342, December 2016.

63. V.T. Somasekhar, M.R. Baiju and K. Gopakumar, "Dual two level inverter scheme for an open end winding induction motor drive with a single DC power supply and improved DC bus utilization", *IEE Proceedings, Electric Power Application*, vol. 151, no. 2, pp. 230-238, April 2004.
64. E. Levi, M. Jones and W. Satiawan, "A multiphase dual-inverter supplied drive structure for electric and hybrid electric vehicles", *2010 IEEE Vehicle Power and Propulsion Conference*, Lille, France, 1-3 September 2010.
65. B. A. Welchko, "A double-ended inverter system for the combined propulsion and energy management functions in hybrid vehicles with energy storage", *31st Annual Conference of IEEE Industrial Electronics Society, IECON 2005*, Raleigh, NC, USA,6-10 November 2005.
66. J. Hong, H. Lee and K. Nam, "Charging method for the secondary battery in dual-inverter drive systems for electric vehicles", *EEE Trans. Power Electron.*, vol. 30, no. 2, pp. 909-921, February 2015.
67. C. Rossi, G. Grandi, P. Corbelli, "Series hybrid power train based on the dual two-level inverter", *11th International Conference on Optimization of Electrical and Electronic Equipment*, Brasov, Romania, pp. 277-286, 22-24 May 2008.
68. G. Mondal, K. Sivakumar, R. Ramchand, K. Gopakumar and E. Levi, "A dual-sevenlevel inverter supply for an open-end winding induction motor drive", *IEEE Trans. Ind. Electron.*, vol. 56, no. 5, pp. 1665-1673, May 2009.
69. V. T. Somasekhar, S. Srinivas, and K. Gopakumar, "A space vector based PWM switching scheme for the reduction of common-mode voltages for a dual inverter fed open-end winding induction motor drive", *2005 IEEE 36th Power Electronics Specialists Conference*, Dresden, Germany, pp. 816-821, 16-17 June 2005.
70. V. Oleschuk, R. Bojoi, G. Griva and F. Profumo, "Dual inverter-fed traction drive with DC sources power balancing based on synchronized PWM *IEEE International Electric Machines & Drives Conference*, Antalya, Turkey pp. 260-265, 3-5 May 2007.
71. S. K. Gupta, O. Singh, M. A. Khan and D. K. Chauhan, "Pulse Width Modulation Technique for Multilevel Operation of Five-Phase Dual Voltage Source Inverters", *J. Eur. Syst. Autom. (JESA)*, vol. 54, issue 2, pp. 371-379, April, 2021.
72. S. K. Gupta, O. Singh and M. A. Khan, "Pulse Width Modulation Switching Schemes for Two Level Five-Phase Voltage Source Inverter", *European J. Electr. Eng.*, vol. 23, issue. 2, pp. 137-142, April 2021.
73. S. K. Gupta, "THD Analysis of Twin Two-Level 5-Phase Inverter System for Multi-Level Performance using Two Switching Schemes", *IEEE Delhi Section International Conference on Electrical, Electronics and Computer Engineering (DELCON-2022), NSUT*, New Delhi, 11-13 February, 2022.

74. S. K. Gupta, O. Singh, M. A. Khan and D. K. Chauhan, "Simulation Study of Five-Phase Dual Two-Level Inverter System using Unified Space Vector PWM", *10th IEEE International Conference on Communication Systems and Network Technologies (CSNT 2021)*, Oriental Institute of Science and Technology, Bhopal, India, June 18-19, 2021.
75. M. Chen and D. Sun, "A unified space vector pulse width modulation for dual two-level inverter system", *IEEE Trans. Power Electron.*, vol. 32, no. 2, pp. 889-893, February 2017.

2

Multilevel Inverters: Classification, Approaches, and Its Application in Photovoltaic System

Akhlaque Ahmad Khan[1*], Ahmad Faiz Minai[1], Qamar Alam[1] and Farhad Ilahi Bakhsh[2]

[1]Department of Electrical Engineering, Integral University, Lucknow, UP, India
[2]National Institute of Technology, Srinagar, India

Abstract
The term "traditional inverters" refers to voltage source inverters (VSI) that have an output voltage range of zero or ±Vdc. It necessitates a high switching frequency and multiple innovative PWM strategies with the goal of generating the voltage output with minimal distortion. When operating at higher frequency ranges, these traditional inverters have limitations due to switching losses at large voltages and large power demands. However, MLI creates an output voltage that is less susceptible to total harmonic distortion (THD) at higher frequencies. Reducing common mode voltages, high voltage capabilities, low dv/dt, practically sinusoidal outputs, and a minimal output filter are just a few of the reasons that MLIs are appropriate for demanding power applications. The MLI is built on the idea of employing several low-power switching devices and multiple DC sources to produce a staircase voltage waveform with the goal of achieving greater power. There are a large number of DC sources, like fuel cells, PV panels, batteries, and capacitors. The switching devices are activated in a wide range of manners, combinations, and techniques after the DC supplies have been connected to provide a high o/p voltage. As an approach to assisting the integration of renewable energy sources (RESs), the multi-level idea first appeared for high power and large/medium voltage purposes. MLI comprises one of the most proficient power converter topologies because, when the switches are configured correctly, it delivers staircase results with minimal harmonic distortion. The traditional MLI topology gave way to the newer, more efficient models that produce more levels with fewer switches.

**Corresponding author*: akhlaque@iul.ac.in

Salman Ahmad, Farhad Ilahi Bakhsh, and P. Sanjeevikumar (eds.) Multilevel Converters, (27–50)

MLI configurations have several industrial uses, such as in on-grid RES, for which they are necessary as a cutting-edge power converter configuration. Currently, flying capacitor MLI (FC-MLI), neutral point-clamped (NPC) or diode-clamped MLI (DC-MLI), and cascaded H-bridge MLI (CHB-MLI) are the most important and broadly applied configurations. The use of various DC sources at the input side of the inverter makes multilevel technology attractive for photovoltaic (PV) applications. This chapter provides an overview with a classification approach of multilevel inverter (MLI) topologies and their suitability for PV systems.

Keywords: NPCMLI/DCMLI, FCMLI/CCMLI, CHBMLI, MLI topology, MATLAB/Simulink, on-grid SPV system, THD

2.1 Introduction

India depends heavily on purchased fossil fuels to power its manufacturing sectors. Owing to the country's national reliance on fuel imports, any deficit resulting from an unforeseeable circumstance might result in severe resource depletion, which could then obstruct the nation's ability to build its industries and advance economically. India nevertheless has a far lower average energy use per person than most industrialized nations. Due to a strong economic recovery and the potential for fast industrialization and modernization, which might in turn propel the demand for fossil fuel use in the foreseeable future, it is anticipated to climb even more. As a result, creating alternative energy sources is necessary to lessen our reliance on fossil fuels. Additionally, the use of efficient renewable energy (RE) might prove to be a decisive factor in the manufacturing and growth of the nation. Renewable energy sources (RESs) would assist the nation in combating climate change in addition to enhancing the supply by reducing emissions of harmful greenhouse gases (GHGs) and ecological footprints *via* clean development management (CDM) [1–4]. As a result, by producing effective electricity with nearly minimal carbon emissions, RESs have attracted more attention and developed [2]. Solar PV, wind, geo-thermal, and other REs are accessible [3, 4]. Solar irradiance has a major role in the global distribution of energy. As much as 100 kilowatts (kW) of solar energy is taken in and scattered by the Earth before it touches the surface [5]. Solar energy might contribute significantly to RESs attributable to this abundance of limitless energy [6]. PV deployment has evolved quite a bit in the last 30 years, but solar power technologies remain confronted with plenty of difficulties that make them less affordable than alternative energies like fossil fuels. The enormous capital cost, erratic nature, and low degradation efficiency of solar technology are a few of these issues [7]. From a technological and scientific perspective, a new technology that might provide greater energy conversion efficiency at a lower manufacturing cost is urgently required to enable the advent of solar systems in the electricity industry. A number of studies [8, 9] have been carried out to deal with these problems and enhance the dependability, effectiveness, performance, and cost of solar-based power sources. Solar generators use PV cells to convert sunlight into electricity. Figure 2.1 illustrates the different parts of a solar energy conversion system, which include PV cells, converters, and a control module that is responsible for controlling the power produced by the PV cells.

Numerous management strategies, converter architectures, and power tracking strategies have been designed to maximize the performance of power harvesting from RESs [9]. Significant research is still being done regarding how to effectively integrate RESs into the electricity system though. Converter topologies and associated controls are given additional consideration in this respect because of their crucial function in energy conversion and output management from such reserves. Two-level inverters are extensively employed in small companies and microgrids, although they are insufficient for large-scale applications due to the inherent development of a large voltage stress, relatively inefficient behavior, and a higher operating temperature [10]. As a result, MLIs are mostly used in on-grid, big-scale RESs with the greatest power [11]. Due to their minimal switching losses, the ability to operate at high voltages, low EMI output, high efficiency, and low THD output as a result of their multiple-level output waveform, multilevel inverters are attracting more attention from researchers [12]. Consequently, innovative power converter topologies must be designed to improve electrical performance for the desired execution, control, and management of energy to allow the most feasible production of electricity from renewable resources.

Aside from the "Introduction", the chapter is outlined as follows: Section 2.2 discusses the categorization of MLIs, the construction and operation of the three main MLI topologies, and an evaluation of their benefits and drawbacks. Section 2.3 discusses reduced switch multilevel inverter topologies. In Section 2.4, topologies are tested using MATLAB/Simulink, along

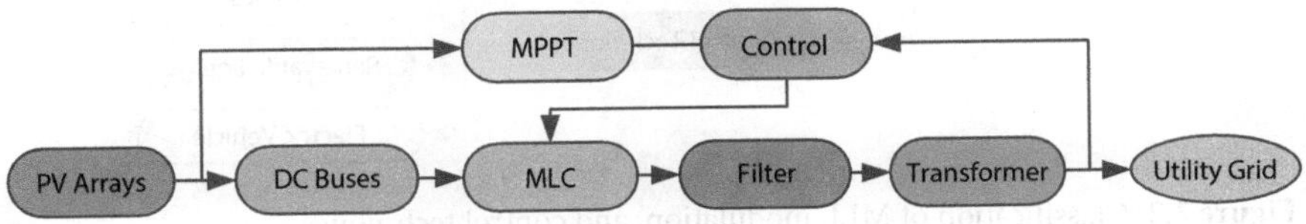

Figure 2.1 A solar power system with converters for controlling and monitoring.

with their results and THD assessment. The integration of the SPV system with MLI is demonstrated in Section 2.5, and a summary of the outcomes is presented in Section 2.6.

2.2 Multilevel Inverters (MLIs)

MLIs are being employed in electrical grids more frequently as an outcome of their capability to meet electrical performance requirements along with specific energy regulations as well as their decreased levels of THD and EMI. In comparison with traditional two-level inverters that employ the high switching frequencies of PWM, an MLI has a number of advantages [13]. MLIs are currently being considered as an industry-standard solution for systems that require exceptional power consistency and dynamic stability and range in capacities from 1 to 30 MW [14]. MLIs seem to be best suited for large-voltage settings because they can generate reduced THD o/p voltage patterns and larger voltages with a confined device rating [15]. An MLC system may interface with renewable energies, including wind, battery storage, and PV cells, in a wide range of ways [16]. The sort

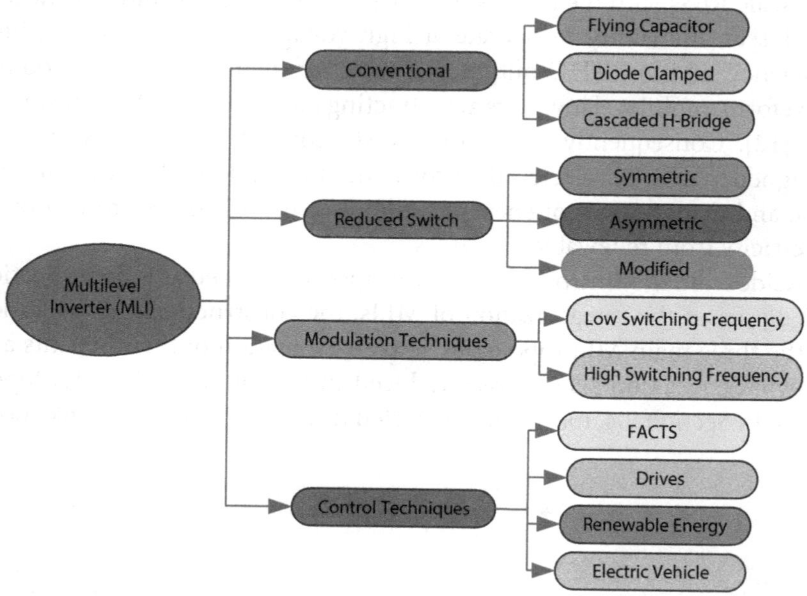

Figure 2.2 Classification of MLI, modulation, and control techniques.

of control scheme implemented to describe its functionality, effectiveness, enhance system performance, and increase accessibility is what essentially regulates the PWM of MLIs [17]. Many MLI strategies have been proposed in various studies during the last few years [18]. The three designs that are most often used in the industrial sectors are the CHB, NPC/DC, and FC architectures [19] as shown in Figure 2.2.

2.2.1 Diode-Clamped/Neutral Point-Clamped Multilevel Inverter (DCMLI/NPCMLI)

The DCMLI circuit model with seven levels is depicted in Figure 2.3. It is sometimes referred to as a neutral point-clamped MLI (NPCMLI). By plugging the semiconductor switches in series, this sort of inverter may achieve the necessary voltage rating and output levels. Clamps are connected to the inner voltage terminals through either two additional diodes or a single high-frequency capacitor. An "l" level inverter's switching components must block voltages over Vdc/ (l-1). Clamping diodes necessitate an assortment of voltage ratings for a range of inner voltage values [7].

The neutral point in Figure 2.3 is the intersection of the elements that set this circuit apart from a typical inverter. By using diodes to clamp the switch voltage, the DC bus voltages are reduced to half their original level.

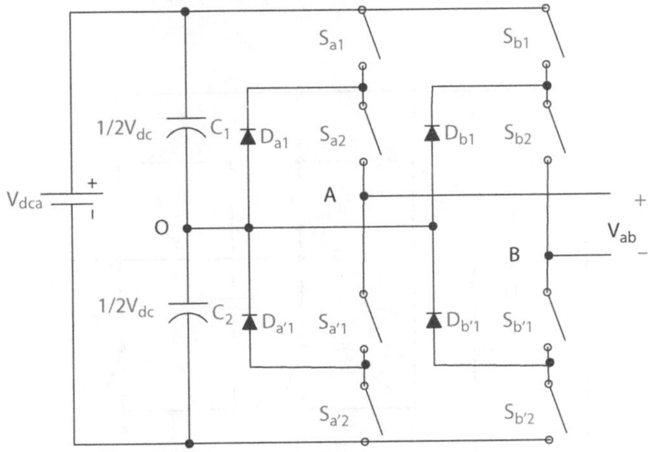

Figure 2.3 Neutral point-clamped/diode-clamped inverter.

2.2.2 Flying Capacitor/Capacitor-Clamped Multilevel Inverter (FC/CCMLI)

The schematic representation of a FC-MLI is shown in Figure 2.4. Compared to NPC-type inverters, a lot more auxiliary capacitors are required for the "l" level: (l-1) main capacitors and ((l-1) x (l-2))/2 auxiliary capacitors [7]. Each capacitor in the ladder-like structure of the DC side capacitors has a different voltage than the capacitor after it. The voltage increase across two adjacent capacitor legs influences the o/p pattern's voltage steps. A 15-level capacitor-clamped inverter's voltage formulation is more flexible than that of a DCMLI. The FCMLI does not require every ON (conducting) switch to be in a sequential series like the DC-MLI requires. Only line-line redundancies exist in the DCMLI, whereas phase redundancies exist in the FCMLI. These redundant systems provide charging and discharging for certain capacitors. Additionally, it may be used in the control system to balance the voltages at different levels.

2.2.3 Cascaded H-Bridge Multilevel Inverter (CHBMLI)

A CHBMLI consists of a series H-bridge (single-phase full bridge) inverter, as shown in Figure 2.5. The fundamental component of this MLI is a voltage synthesizer that takes input from numerous separated DC sources (SDCs) such as batteries, fuel cells, or solar panels to produce an isolated, usable DC voltage. The voltages produced by the various H-bridges have been added to get the final synthetic phase voltage. When such H-bridges' DC link voltages are exactly the same, the CHBMLI is said to be symmetric.

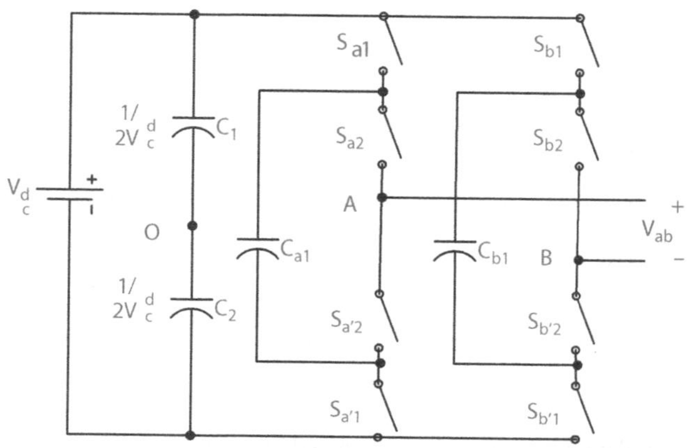

Figure 2.4 Capacitor-clamped/flying capacitor multilevel inverter.

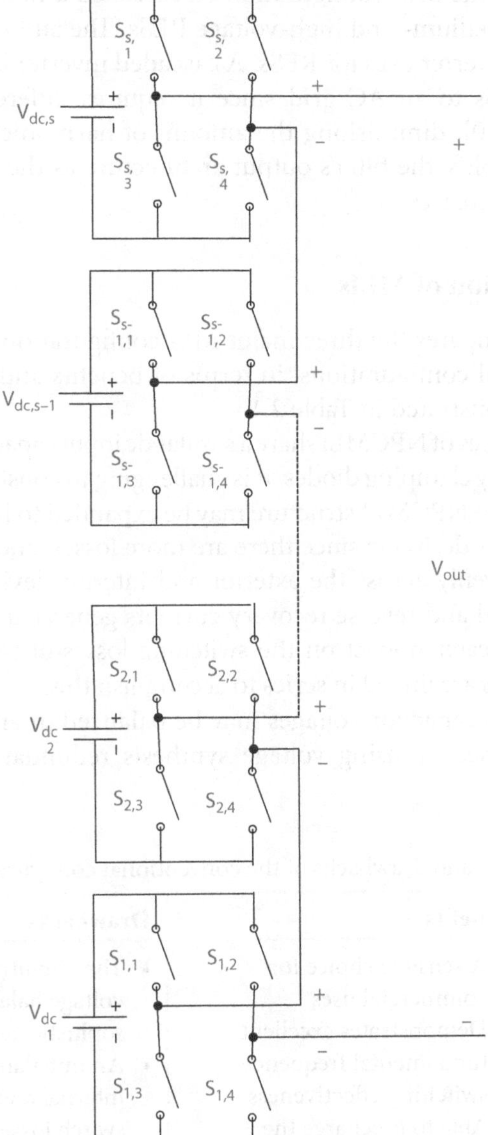

Figure 2.5 CHBMLI with k independent DC sources.

The circuit could be asymmetric, although if the HBs' DC link voltages have varying values, asymmetrical inverters are strongly advised for SPV systems when solar PV voltages are changeable due mainly to ecological parameters [20].

Within all of the MLI configurations, a cascaded arrangement has been employed for medium- and high-voltage RESs. The author [20] has listed a few cascade inverter uses for RESs. A cascaded inverter is the best option for linking RESs to an AC grid since it requires different DC sources. According to [10], diminishing the amount of harmonics in the cascade output can simplify the filter's output architecture as the total number of voltage levels increases.

2.2.4 Evolution of MLIs

This section compares the three major MLI configurations, also known as the conventional configurations, in terms of benefits and drawbacks [13, 20, 21], as demonstrated in Table 2.1.

The three phases of NPCMLI share a similar dc input capacitor, but because there are so many clamping diodes, it is challenging to construct the physical arrangement. The NPCMLI structure may be expanded to include more levels. These are less desirable since there are more losses, and these losses are distributed unevenly across the exterior and interior devices. Conduction losses introduced and reverse recovery currents generated during commutation have a greater impact on the switching losses of the other devices. Clamping diodes are linked in series to accomplish this.

The clamping capacitor voltages may be balanced in an FCMLI within a couple of cycles by using voltage synthesis redundancy. The biggest

Table 2.1 Benefits and drawbacks of the conventional configurations.

Topology	Benefits	Drawbacks
NPC/DC	• A sensible choice for commercial uses • Demonstrates excellent fundamental frequency switching effectiveness • Able to precharge the capacitors collectively • An easy technique of control • Requires less DC power sources • Perfect for systems that demand to be fault-tolerant	• The circuitry used in voltage balancing is sophisticated • An imbalance between internal and external switch losses • More clamping diodes were used as the level rose

(Continued)

Table 2.1 Benefits and drawbacks of the conventional configurations. (*Continued*)

Topology	Benefits	Drawbacks
FC/CC	• Lessens the need for DC sources • Phase redundancies adjust capacitor voltage • Perfect for systems that demand to be fault-tolerant • Power flow, both real and reactive, could be managed • Filters are not required to diminish harmonics	• The circuitry used in voltage balancing is sophisticated • Needs a lot of capacitors at high levels • High switching frequencies and losses while transmitting actual power • Exorbitant installation costs
CHB	• A basic, customizable framework • Simple to advance to new levels • Only needs unidirectional switches • Fit for use in fault-tolerant applications • The existence of independent DC sources reduces the risk of electric shock • Able as an asymmetric source topology • Possible to configure a single DC source arrangement	• Fewer output voltage levels • Necessitates additional gate driver circuits • To enhance output voltage, numerous DC sources are necessary • Restrictions on certain tasks using separate DC sources • Switches must support a blocking voltage that is the same as the input voltage value • Diminished flexibility (asymmetric source topology) • High implementation costs (asymmetric source topology) • Various voltage ratings apply to switches (asymmetric source topology)

disadvantage is the excessive clamping of capacitors. The clamping diodes of the DCMLI are more expensive than these capacitors though. For low switching frequencies, the CCs get larger, which reduces the inverter's power capacity.

A reduced switching frequency may be employed in CHBMLI, and no clamping capacitors or diodes are necessitated. There are no extraneous FCs or CDs, and each cell has the same shape. The design and manufacturing of circuits can therefore be composable. The primary flaw of CHBMLI is that each cell must have its own independent DC supply. As a result, it may be applied to systems for power production based on solar PV and fuel cells.

Power circuit evaluation on all three MLIs reveals that they all employ a similar number of semiconductor switches per phase. In contrast to the other two structures, the NPCMLI requires (l-1)*(l-2) clamping diodes, making it the most costly. We require 2*(l-1) heat sinks for the FCMLI and CMLI. NPCMLI, on the other hand, requires 2 * (l-1) + (l-1) * (l-2). In order to provide an "l"-level inverter output, the CMLI employs the fewest electronic components, which leads to the lowest computational burden. If any component in the H-bridge breaks, the inverter in a CHB can still be run at a lower power level.

2.3 Topologies for Multilevel Inverters With Reduced Switches

Despite the versatility of conventional MLIs, their higher component count is an issue. This research was conducted with the aim of improving quality without increasing costs by decreasing the number of switches, diodes, and supply voltages. Drives, RESs, and FACTs are just a few of the fields where novel reduced switching topologies have been developed. They are employed in both grid-connected and off-grid scenarios. These can be divided into the three subtypes of MLI based on reduced switch: symmetric, asymmetric, and modified. All hybrids and non-H-bridge topologies are included in the modified category. This study provides a thorough analysis of a number of recently identified topologies.

2.3.1 Symmetrical H-Bridge MLI

In order to lower both costs and the required space for installation, the author [22] proposes a novel MLI that makes use of basic switching technology. In comparison with standard MLI, switching has been significantly reduced. Through the use of vertical phase-shifted sinusoidal PWM, four switches, and two DC sources, the researcher [23] created a new cascaded H-bridge architecture. It is suitable for use in low-voltage settings. An alternative MLI arrangement with three equal-sized DC sources coupled to four switches is shown in [24] which employs multi-carrier sub-harmonic PWM. It is appropriate for high-power and high-voltage operations. It offers an alternate method to use RES over conventional design.

The author [25] has suggested a novel design that uses clock phase shift one-cycle control to minimize dimensions, reduce costs, and eliminate excessive voltage spikes in the resultant voltage. To boost both power density and power conversion efficiency, the author of [26] utilized the fundamental SPWM approach to construct a MLI that utilized a switched capacitor. In [27], the author constructed a more effective MLI employing a submodule with fewer switches by using SHE-PWM. It has possibilities in photovoltaics and fuel cells, which are two forms of renewable energy. In [28], the author constructed an improved MLI that uses fewer active switches and reduces the impact of dominating lower-order harmonics by using a SHE approach based on the modified PSO algorithm. This comes into practice in alternative energy and propulsion systems.

2.3.2 Asymmetrical H-Bridge MLI

The dc voltage sources in an asymmetric MLI architecture are unequal in magnitude. The primary goal is to create greater voltages with fewer switches. A CHB design with asymmetrical architecture and DC sources of various magnitudes has been outlined in [29]. In order to achieve a wider range of voltage outputs from a separate DC source, the author of [30] constructed a reduced switch MLI. Medium-voltage frameworks commonly employ this kind of design. The author [31] developed a modular, criss-cross-connected MLI topology for use in both symmetrical and asymmetrical modalities. This kind of design is used in high-voltage situations and RES. For use primarily in PV systems, the author of [32] constructed a novel MLI architecture based on a unipolar PWM technique. Leveraging simply the most basic switching techniques, the author [33] constructed an MLI topology with a minimally intrusive switch architecture that can be deployed in either a symmetric or an asymmetric design. For the objective of incorporating RES, this arrangement is utilized. The biggest drawback of this arrangement is that as the number of levels increases, more diodes are required; however, these limitations are outweighed by its benefits of a lower blocking voltage and fewer switches, which make it suitable for use in applications that require substantial power. A novel, sophisticated topology based on the researcher's staircase modulation approach is presented in [34]. Permanent magnet propellers and PV power distribution networks both make use of this technology. The primary unit, as described by the author [35], was designed to flip between symmetric and asymmetric configurations. This makes it useful in high- and medium-voltage applications. An optimal structure for a high-frequency switching power supply network was proposed by the author [36], employing a switched capacitor module. In grid-tied PV and EV applications, this architecture is employed.

2.3.3 Reduced Switch-Modified MLI

This particular area of topology studies focuses on systems that neglect the H-bridge. The H-bridge topology is often used in low- and medium-voltage systems. Because of the high voltage required by H-bridge switches, their usefulness is limited. It loses efficiency when put to use with higher voltages. Numerous cutting-edge topologies have been created as solutions for this restriction. The author [37] has developed an alternative modified MLI architecture that uses floating input DC sources combined in the opposite polarity. While the switch count could be reduced, flexibility and fault tolerance are compromised compared to traditional CHB MLI. There is a novel topology that has developed in [38] that can be used in both symmetric and asymmetric manners. This architecture operates at high and medium voltages. Specifically for the purpose of incorporating RES, the author [39] has created a modified MLI topology with a reduced switch number. In spite of the necessity for extra diodes and bidirectional switches, this design significantly reduces the overall amount of active switches. For superior efficacy and medium-voltage operations, the author [40] has created a distinctive modified MLI with fewer switches and redundant switch states. Using the SHE PWM approach, the author [41] created an E-type module-oriented MLI that is capable of providing a staircase output with a minimal number of switches. Due to its minimal voltage stress and reduced switch size, this design is frequently utilized in high-voltage and high-power settings. To generate PWM signals with NLC, the author of [42] proposes a novel square-type MLI with unequal DC sources and fewer components. As a result of its modulability and versatility, fewer switching devices are required to provide a broader assortment of voltage levels, while voltage stress is also reduced. High-voltage, high-power applications, like photovoltaics, are served by this type of architecture. Modularized topologies with lower voltage stress and fewer switches have been created for RE applications [43]. A semi-cascading MLI architecture was built for symmetric and asymmetric topologies in [44] and [45] by repurposing the same basic switching approach. Medium-voltage systems, such as PV, often use these types of topologies.

2.4 MATLAB/Simulink MLI Configurations

This section discusses most of the basic MLI configurations. In addition, MATLAB/Simulink is utilized to perform in-depth comparisons.

2.4.1 Simulation of a NPC Five-Level SPWM Inverter and Its Output Waveform

Figure 2.6 shows a Simulink model that represents a two-leg, five-level NPC inverter. It was the first multilevel configuration that gained significant popularity, and it is still commonly employed in industrial settings.

The capacitors in Figure 2.6's five-level NPC inverter produce the intermediate voltage level, and the voltages across the switches are just slightly higher than the input voltage. Because of capacitor voltage balancing concerns, the initial three structures of practical diode-clamped inverters have been the only ones accessible.

To get the five-level output voltage waveform seen in Figure 2.7, for the return leg of the five-level NPCMLI, SPWM switching is performed, and the base sinusoid has a phase lag of (π/180) × 240 in relation to the previous sinusoid waves of the initial leg.

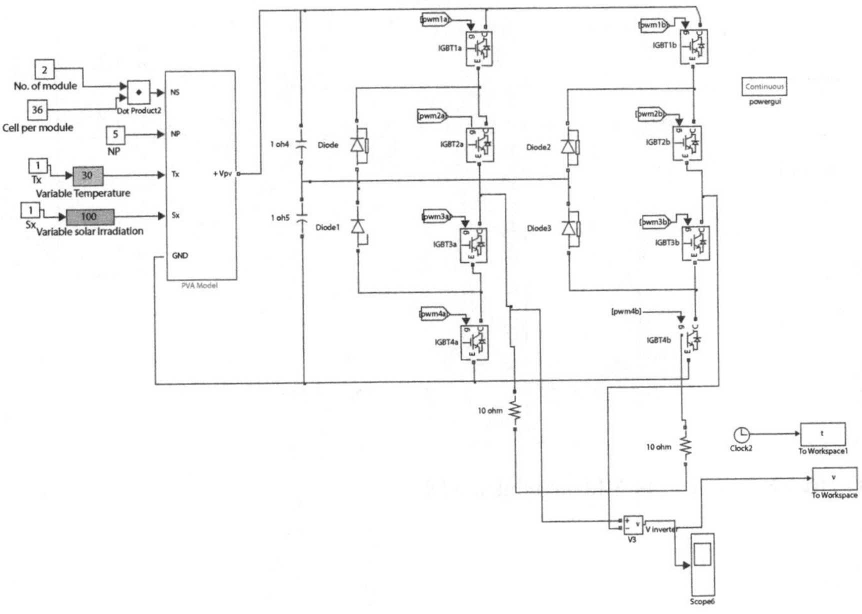

Figure 2.6 Two-leg, five-level NPCMLI Simulink model.

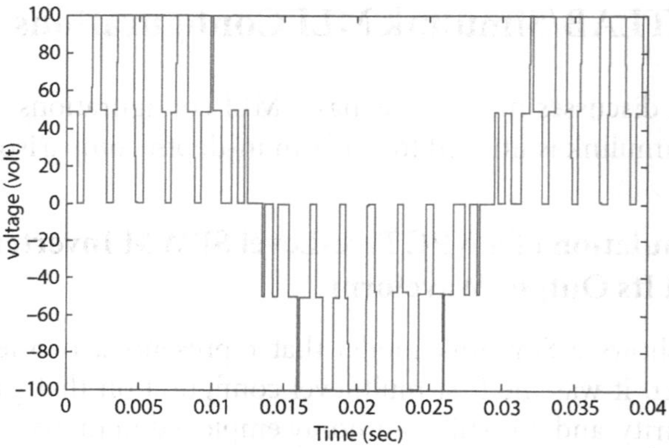

Figure 2.7 O/P waveform.

2.4.2 Simulation of a FC Five-Level SPWM Inverter and Its Output Waveform

A five-level FCMLI is represented by a Simulink model in Figure 2.8 and their output voltage in Figure 2.9.

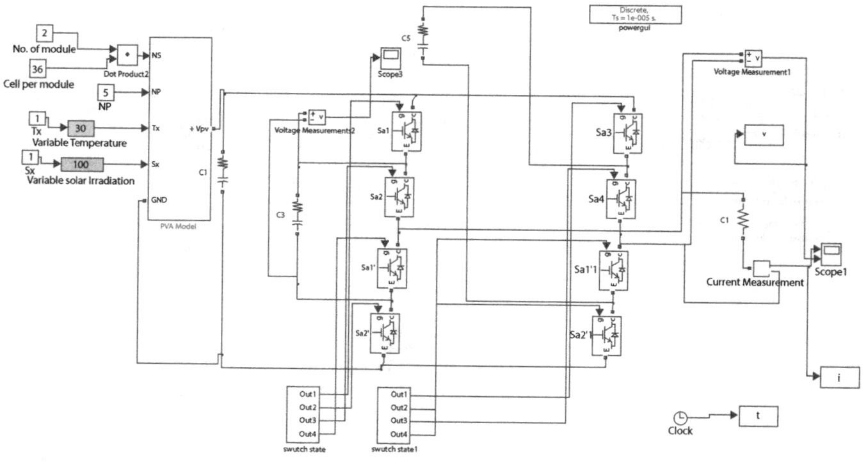

Figure 2.8 Five-level FCMLI Simulink model.

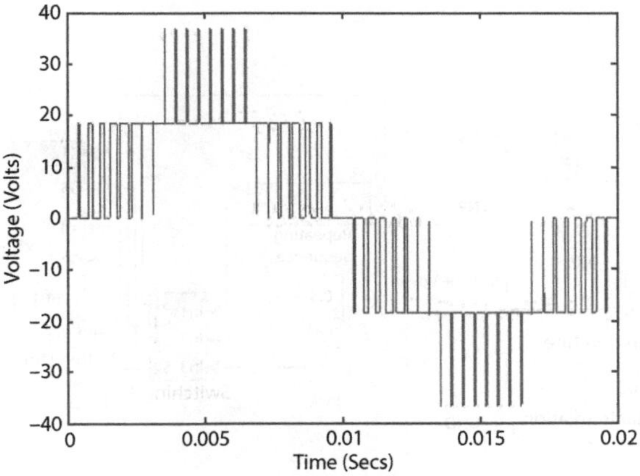

Figure 2.9 O/P waveform.

2.4.3 Simulation of a CHB Five-Level SPWM Inverter and Its Output Waveform

A five-level CHBMLI is represented by a Simulink model in Figure 2.10 and their output voltage in Figure 2.11.

2.4.4 Evolution of THD for Numerous Inverters

Table 2.2 compares the THD in the CHB, NPC/DC, FC/CC, and three-level traditional inverter (TLTI) with the five-level MLI.

A graph is drawn from Table 2.2 that compares the THD of various inverters, as shown in Figure 2.12.

Figure 2.12 exemplifies that the CHB-THD MLI's are superior to those of the remaining inverters.

2.5 Applications of MLIs in SPV Systems

The MLI's ability to produce high-quality voltage waves at lower switching frequencies [21] renders it an appropriate choice for medium- and high-voltage power requirements. There has been a significant increase in studies and advancements in MLI for applications such as PV systems to improve power quality and efficiency. One of the problems is picking the right converter for PV applications since it affects how the PV system behaves. Several studies of MLIs for solar power systems are found as a consequence in the literature [46]. In the article [47], an MLI is recommended

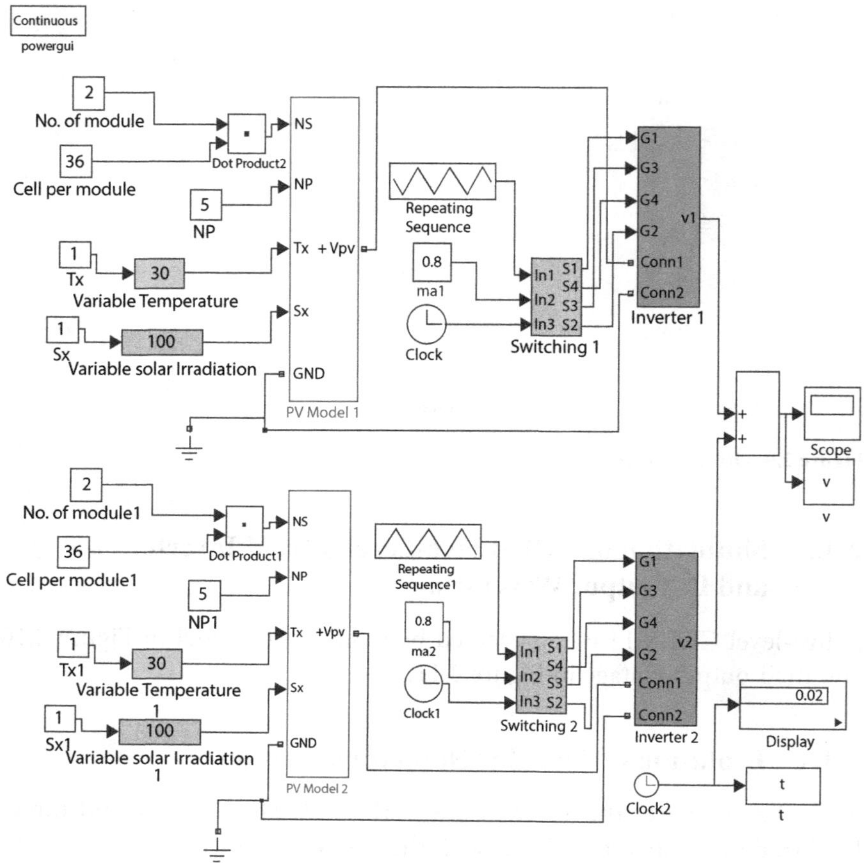

Figure 2.10 Five-level CHBMLI Simulink model.

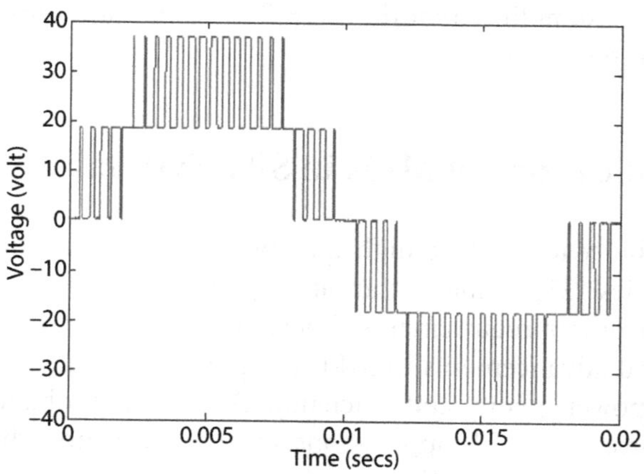

Figure 2.11 O/P waveform.

Table 2.2 Evolution of THD of numerous inverters.

Modulation index (m_a)	3-level traditional inverter	5-level NPCMLI	5-level FCMLI	5-level CHBMLI
0.8	41.46	30.77	25.81	16.79

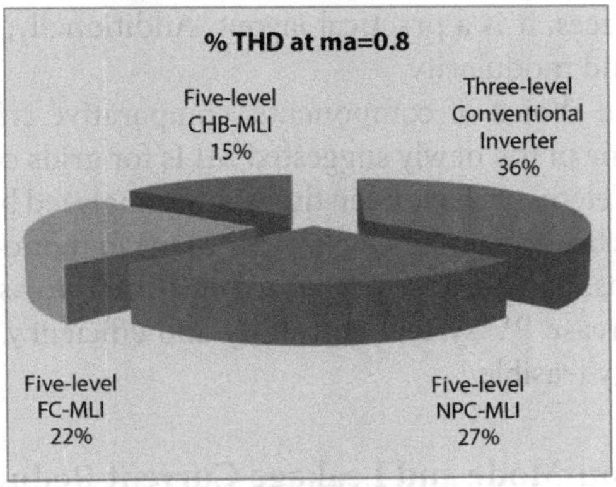

Figure 2.12 Evolution of THD of numerous inverters.

for PV systems, including comprehensive ESS. The MLI is capable of maintaining its power output in all weather conditions. In the prescribed MLI, the change from charging to discharging was flawless. The CHB inverter has the potential to be used in grid connections because to its modular design, high-quality output waves, and interface with medium-voltage services. For large-scale PV systems, [48–51] have introduced a CHBMLI with DC–DC stage separation.

The H6-bridge energy cell was used in place of the H-bridge type in the work by [46] to provide a CHB inverter for usage in PV systems. In comparison to the CHB type, the system greatly increased output voltage and current quality during partial shadowing, which increased system efficiency and grid energy injection. [49] employed a modular-modified MLI in on-grid PV installations.

2.5.1 MLIs for On-Grid PV Systems

A myriad of MLI setups have been assessed and contrasted in literary criticism in order to determine the optimal setup for PV applications that can interact at smaller and higher frequencies, achieve superior seismic waves with smaller THD, achieve the best possible EM congruence, and exhibit large efficiency [9]. A redesigned MLI configuration was introduced [13]. This structure produces a voltage waveform with seven levels and fewer harmonic components. It is legitimate to include this inverter in PV applications to generate electricity for the grid using PV modules with different V/I ratings.

In contrast to the conventional MLI arrangement, investigating [52] revealed a novel MLI configuration with fewer switches; this configuration had no need for extra parts like capacitors and inductors. Due to its asymmetrical PV sources, it is a practical layout. Additionally, it demonstrates fault tolerance and modularity.

In light of the choice of components, comparative criteria, and their applicability, some of the newly suggested MLIs for grids connected with a fewer number of elements have been thoroughly analyzed by [53]. A modular CHBMLI has been suggested by [54] for usage in mono- or three-phase on-grid situations. The purpose of this architecture is to assure greater PV module use, increase PV system versatility and efficiency, and collect the most solar energy feasible.

2.5.2 Common-Mode and Leakage Current Reduction of Transformerless MLI for SPV Systems

The on-grid PV system with a MLI is trustworthy at generating power, so a high-frequency transformer is used to create a galvanic link between the DC and massive power AC transmission lines. Due to the loss of additional components, this decreased efficiency and hiked the price. Due to their cost-effectiveness, minimal THD, and great efficiency, MLI configurations without transformers are regarded as the best choice [55]. Transformerless configurations in on-grid PV systems have the noticeable benefit of removing the problem of current leakage; nevertheless, the technological difficulty is how to lessen leakage current by keeping a constant common-mode voltage in the arrangement.

Plenty of approaches have been suggested for addressing the issue of current leakage in transformerless string converters. These methods may be broadly arranged into three distinct classes: modulating methods [56, 57], topology methods [58], and filter methods [59]. Despite the trade-off between effectiveness and current quality, the bipolar modulation technique could substantially decrease leakage current by promising a consistent common-mode voltage. In the full-bridge architecture, additional DC or AC switches are required to mitigate current leakage [59]. The filter solution suppresses leakage current by using bypassing cycles with small impedance for large-frequency, common-mode distortion; this prevents leakage current from flowing from the outside.

The author [58] has described an enhanced five-level CHBMLI to lower leakage currents in transformerless PV systems. The system was designed with fewer switches and reduced leakage current by eliminating the high-frequency transitions that happen predominantly at CMVs and terminals. The suggested system also had smaller conduction and switching losses, enabling high switching frequency performance.

2.6 Conclusion

In high-power, high- and medium-voltage environments, MLI topologies are extremely important. A number of reduced switch MLI topologies have emerged for application in areas including drives, EVs, and RESs. As in comparison with an ordinary two-level inverter, the key benefits of reduced switch MLI topologies are fewer switch number, lower switching losses, lower overall costs, a more precise output voltage, and easier control. As such, this study centers on many types of reduced switch MLI topologies, including symmetric, asymmetric, and modified MLI.

The purpose of this chapter's brief overview of MLI characteristics is to draw attention to the importance of establishing novel inverters or altered arrangements of inverters for on-grid SPV systems. MLIs have been described in detail in a number of areas, including classification scheme, benefits, drawbacks, and their capacity to improve energy conversion in contemporary SPV systems. The CHBMLI has superior THD, according to a topological analysis of MLIs. Due to the reduced number of switching devices PV applications offer greater possibilities for CHBMLI. Despite using two or more independent PV array strings to supply power to each of the cells, a CHBMLI has plenty of advantages over other possible arrangements. The latest MLIs on-grid PV systems, along with the approaches for eliminating current leakage suppression, were also addressed in this chapter.

References

1. Minai, Ahmad Faiz, Husain, Mohammed Aslam, Naseem, Mohammad and Khan, Akhlaque Ahmad. "Electricity demand modeling techniques for hybrid solar PV system" *Int. J. Emerg. Electr. Power Syst.*, vol. 22, no. 5, 2021, pp. 607-615. https://doi.org/10.1515/ijeeps-2021-0085.
2. Nema, P.; Nema, R.; Rangnekar, S. A current and future state of art development of hybrid energy system using wind and PV-solar: A review. *Renew. Sustain. Energy Rev.*, 2009, 13, 2096–2103.

3. Haque, A. Research on Solar Photovoltaic (PV) Energy Conversion System: An Overview. In *Proceedings of the Third International Conference on Computational Intelligence and Information Technology (CIIT 2013)*, Mumbai, India, 18–19 October 2013; pp. 605–611.
4. Khan, A.A.; Minai, A.F.; Pachauri, R.K.; Malik, H. Optimal Sizing, Control and Management Strategies for Hybrid Renewable Energy Systems: A Comprehensive Review. *Energies*, 2022, 15(17), 6249. https://doi.org/10.3390/en15176249.
5. IEA International Energy Agency. *World Energy Outlook*; OECD: Paris, France, 2004.
6. Panwar, N.; Kaushik, S.; Kothari, S. Role of renewable energy sources in environmental protection: A review. *Renew. Sustain. Energy Rev.*, 2011, 15, 1513–1524.
7. Khan A.A, Minai A.F, Alam Q, Mallick M.A (2014) Performance analysis of various switching scheme in multilevel inverters using MATLAB/SIMULINK. *Int. J. Curr. Eng. Technol.*, Association of American Publisher U.S.A, 4:718–724.
8. Masuko, K.; Shigematsu, M.; Hashiguchi, T.; Fujishima, D.; Kai, M.; Yoshimura, N.; Yamaguchi, T.; Ichihashi, Y.; Mishima, T.; Matsubara, N.; *et al.* Achievement of More Than 25% Conversion Efficiency With Crystalline Silicon Heterojunction Solar Cell. *IEEE J. Photovolt.*, 2014, 4, 1433–1435.
9. Hassaine, L.; OLias, E.; Quintero, J.; Salas, V. Overview of power inverter topologies and control structures for grid connected photovoltaic systems. *Renew. Sustain. Energy Rev.*, **2014**, 30, 796–807. [CrossRef]
10. Monfared, M.; Golestan, S. Control strategies for single-phase grid integration of small-scale renewable energy sources: A review. *Renew. Sustain. Energy Rev.*, **2012**, 16, 4982–4993. [CrossRef]
11. Daher, S.; Schmid, J.; Antunes, F.L. Multilevel inverter topologies for stand-alone PV systems. *IEEE Trans. Ind. Electron.*, 2008, 55, 2703–2712. [CrossRef]
12. Omer, P.; Kumar, J.; Surjan, B.S. A Review on Reduced Switch Count Multilevel Inverter Topologies. *IEEE Access 2020*, 8, 22281–22302. [CrossRef]
13. Bughneda, A.; Salem, M.; Richelli, A.; Ishak, D.; Alatai, S. Review of Multilevel Inverters for PV Energy System Applications. *Energies*, 2021, 14, 1585. https://doi.org/10.3390/en14061585
14. Rodríguez, J.; Bernet, S.; Wu, B.; Pontt, J.O.; Kouro, S. Multilevel voltage-source-converter topologies for industrial medium voltage drives. *IEEE Trans. Ind. Electron.*, 2007, 54, 2930–2945.
15. R. K. Pachauri, A. F. Minai, V. Pandey, Shashikant, V. Goyal and R. Swain, "Optimal Placement of Modules in PV Array Systems to Achieve Higher GMPP under Obscured Irradiations," *2021 International Conference on Control, Automation, Power and Signal Processing (CAPS)*, 2021, pp. 1-7, doi: 10.1109/CAPS52117.2021.9730692.
16. M. Kumar, A. F. Minai, A. A. Khan and S. Kumar, "IoT based Energy Management System for Smart Grid," *2020 International Conference on*

Advances in Computing, Communication & Materials (ICACCM), 2020, pp. 121-125, doi: 10.1109/ICACCM50413.2020.9213061.
17. A. F. Minai, T. Usmani, A. Iqbal and M. A. Mallick, "Artificial Bee Colony Based Solar PV System with Z-Source Multilevel Inverter," *2020 International Conference on Advances in Computing, Communication & Materials (ICACCM)*, 2020, pp. 187-193, doi: 10.1109/ICACCM50413.2020.921306.
18. Naseem, M., Husain, M.A., Minai, A.F. *et al.* Assessment of Meta-Heuristic and Classical Methods for GMPPT of PV System. *Trans. Electr. Electron. Mater.*, 22, 217–234 (2021). https://doi.org/10.1007/s42341-021-00306-3.
19. Hasan, N.S.; Rosmin, N.; Osman, D.A.A.; Musta'amal, A.H. Reviews on multilevel converter and modulation techniques. *Renew. Sustain. Energy Rev.*, 2017, 80, 163–174.
20. Gopi, R.R.; Sreejith, S. Converter topologies in photovoltaic applications–A review. *Renew. Sustain. Energy Rev.*, 2018, 94, 1–14.
21. A. A. Khan, A. F. Minai, L. Devi, Q. Alam and R. K. Pachauri, "Energy Demand Modelling and ANN Based Forecasting using MATLAB/Simulink," *2021 International Conference on Control, Automation, Power and Signal Processing (CAPS)*, 2021, pp. 1-6, doi: 10.1109/CAPS52117.2021.9730746.
22. Babaei, E., & Hosseini, S. H. (2009). New cascaded multilevel inverter topology with minimum number of switches. *Energy Convers. Manage.*, 50(11), 2761-2767.
23. Khosroshahi, M. T. (2014). Crisscross cascade multilevel inverter with reduction in number of components. *IET Power Electron.*, 7(12), 2914-2924.
24. Oskuee, M. R. J., Salary, E., & Najafi-Ravadanegh, S. (2014). Creative design of symmetric multilevel converter to enhance the circuit's performance. *IET Power Electron.*, 8(1), 96-102. Wang, L., Wu, Q. H., & Tang, W. (2017).
25. Novel cascaded switched-diode multilevel inverter for renewable energy integration. *IEEE Trans. Energy Convers.*, 32(4), 1574-1582.
26. He, L., & Cheng, C. (2017). A bridge modular switched-capacitor-based multilevel inverter with optimized SPWM control method and enhanced power-decoupling ability. *IEEE Trans. Ind. Electron.*, 65(8), 6140-6149.
27. Lee, S. S., Sidorov, M., Lim, C. S., Idris, N. R. N., & Heng, Y. E. (2017). Hybrid cascaded multilevel inverter (HCMLI) with improved symmetrical 4-level submodule. *IEEE Trans. Power Electron.*, 33(2), 932-935.
28. Panda, K. P., & Panda, G. (2018). Application of swarm optimisation-based modified algorithm for selective harmonic elimination in reduced switch count multilevel inverter. *IET Power Electron.*, 11(8), 1472-1482.
29. Manjrekar, M. D., Steimer, P. K., & Lipo, T. A. (2000). Hybrid multilevel power conversion system: A competitive solution for high-power applications. *IEEE Trans. Ind. Appl.*, 36(3), 834-841. Najafi, E., & Yatim, A. H. M. (2011).
30. Design and implementation of a new multilevel inverter topology. *IEEE Trans. Ind. Electron.*, 59(11), 4148-4154.
31. Khosroshahi, M. T. (2014). Crisscross cascade multilevel inverter with reduction in number of components. *IET Power Electron.*, 7(12), 2914-2924.

32. Prabaharan, N., & Palanisamy, K. (2016). Comparative analysis of symmetric and asymmetric reduced switch MLI topologies using unipolar pulse width modulation strategies. *IET Power Electron.*, 9(15), 2808-2823.
33. Alishah, R. S., Nazarpour, D., Hosseini, S. H., & Sabahi, M. (2014). Novel topologies for symmetric, asymmetric, and cascade switched-diode multilevel converter with minimum number of power electronic components. *IEEE Trans. Ind. Electron.*, 61(10), 5300-5310. Su, G. J. (2005).
34. Multilevel DC-link inverter. *IEEE Trans. Ind. Appl.*, 41(3), 848-854. Babaei, E., Laali, S., & Bayat, Z. (2014).
35. A single-phase cascaded multilevel inverter based on a new basic unit with reduced number of power switches. *IEEE Trans. Ind. Electron.*, 62(2), 922-929.
36. Liu, J., Cheng, K. W. E., & Ye, Y. (2013). A cascaded multilevel inverter based on switched-capacitor for high-frequency AC power distribution system. *IEEE Trans. Power Electron.*, 29(8), 4219-4230.
37. Gupta, K. K., & Jain, S. (2013). A novel multilevel inverter based on switched DC sources. *IEEE Trans. Ind. Electron.*, 61(7), 3269-3278.
38. Mokhberdoran, A., & Ajami, A. (2014). Symmetric and asymmetric design and implementation of new cascaded multilevel inverter topology. *IEEE Trans. Power Electron.*, 29(12), 6712-6724.
39. Odeh, C. I., Obe, E. S., & Ojo, O. (2016). Topology for cascaded multilevel inverter. *IET Power Electron.*, 9(5), 921-929.
40. Xu, S., Zhang, J., Hu, X., & Jiang, Y. (2017). A novel hybrid five-level voltage-source converter based on T-type topology for high-efficiency applications. *IEEE Trans. Ind. Appl.*, 53(5), 4730-4743.
41. Samadaei, E., Gholamian, S. A., Sheikholeslami, A., & Adabi, J. (2016). An envelope type (E-Type) module: asymmetric multilevel inverters with reduced components. *IEEE Trans. Ind. Electron.*, 63(11), 7148-7156.
42. Samadaei, E., Sheikholeslami, A., Gholamian, S. A., & Adabi, J. (2017). A square T-type (ST-Type) module for asymmetrical multilevel inverters. *IEEE Trans. Power Electron.*, 33(2), 987-996.
43. Arun, N., & Noel, M. M. (2017). Crisscross switched multilevel inverter using cascaded semi-half-bridge cells. *IET Power Electron.*, 11(1), 23-32.
44. Banaei, M. R., Oskuee, M. R. J., & Khounjahan, H. (2014). Reconfiguration of semi-cascaded multilevel inverter to improve systems performance parameters. *IET Power Electron.*, 7(5), 1106-1112.
45. Ajami, A., Oskuee, M. R. J., Mokhberdoran, A., & Van den Bossche, A. (2013). Developed cascaded multilevel inverter topology to minimise the number of circuit devices and voltage stresses of switches. *IET Power Electron.*, 7(2), 459-466.
46. Lashab, A.; Sera, D.; Hahn, F.; Camurca, L.; Terriche, Y.; Liserre, M.; Guerrero, J.M. Cascaded Multilevel PV Inverter with Improved Harmonic Performance during Power Imbalance Between Power Cells. *IEEE Trans. Ind. Appl.* 2020, 56, 2788–2798.

47. Lashab, A.; Sera, D.; Martins, J.; Guerrero, J.M. Multilevel DC-link converter-based photovoltaic system with integrated energy storage. In *Proceedings of the 2018 5th International Symposium on Environment-Friendly Energies and Applications (EFEA)*, Rome, Italy, 24–26 September 2018; pp. 1–6.
48. Zhao, W.; Choi, H.; Konstantinou, G.; Ciobotaru, M.; Agelidis, V.G. Cascaded H-bridge multilevel converter for large-scale PV grid-integration with isolated DC-DC stage. In *Proceedings of the 2012 3rd IEEE International Symposium on Power Electronics for Distributed Generation Systems (PEDG)*, Aalborg, Denmark, 25–28 June 2012; pp. 849–856.
49. Faiz Minai, A.; Khan, A.A.; Pachauri, R.K.; Malik, H.; García Márquez, F.P.; Arcos Jiménez, A. Performance Evaluation of Solar PV-Based Z-Source Cascaded Multilevel Inverter with Optimized Switching Scheme. *Electronics*, **2022**, *11*, 3706. https://doi.org/10.3390/electronics11223706.
50. A. A. Khan, A. F. Minai (2023), Introduction to Grid-Forming Inverters (GFMIs), In B. (Eds.). (2023). *Grid-Forming Power Inverters: Control and Applications* (1st Ed.). CRC Press. Page 1-14, ISBN: 9781003302520. https://doi.org/10.1201/9781003302520-1.
51. A.A. Khan, A. F. Minai (2023)"A Strategic Review: The Role of Commercially Available Tools for Planning, Modeling, Optimization, and Performance Measurement of Photovoltaic Systems", *Energy Harvesting and Syst. (EHS)*, ISSN: 2329-8766. https://doi.org/10.1515/EHS-2022-0157.
52. Shen, K.; Zhao, D.; Zhao, G.; Wang, S. Photovoltaic supplied grid-connected modular multilevel converter with active power injection and reactive power compensation capability. In *Proceedings of the IECON 2017–43rd Annual Conference of the IEEE Industrial Electronics Society*, Beijing, China, 29 October–1 November 2017; pp. 7837–7842.
53. Hamidi, M.N.; Ishak, D.; Zainuri, M.A.A.M.; Ooi, C.A. An asymmetrical multilevel inverter with optimum number of components based on new basic structure for photovoltaic renewable energy system. *Sol. Energy*, 2020, 204, 13–25.
54. Bana, P.R.; Panda, K.P.; Naayagi, R.; Siano, P.; Panda, G. Recently developed reduced switch multilevel inverter for renewable energy integration and drives application: Topologies, comprehensive analysis and comparative evaluation. *IEEE Access*, 2019, 7, 54888–54909.
55. Xiao, H.F.; Liu, X.P.; Lan, K. Zero-voltage-transition full-bridge topologies for transformerless photovoltaic grid-connected inverter. *IEEE Trans. Ind. Electron.*, 2014, 61, 5393–5401.
56. Guo, X.; Jia, X. Hardware-based cascaded topology and modulation strategy with leakage current reduction for transformerless PV systems. *IEEE Trans. Ind. Electron.*, 2016, 63, 7823–7832.
57. A. F. Minai, A. A. Khan, M. A. Siddiqui, F. I. Bakhsh, M. A. Hussain and R. K. Pachauri, "Genetic Algorithm Based SPV System with Cascaded H-Bridge Multilevel Inverter," *2023 International Conference on Power, Instrumentation,*

Energy and Control (PIECON), Aligarh, India, 2023, pp. 1-6, doi: 10.1109/PIECON56912.2023.10085864

58. Valderrama, G.E.; Guzman, G.V.; Pool-Mazún, E.I..; Martinez-Rodriguez, P.R.; Lopez-Sanchez, M.J.; Zuñiga, J.M.S. A single-phase asymmetrical T-type five-level transformerless PV inverter. *IEEE J. Emerg. Sel. Top. Power Electron.*, 2018, 6, 140–150.
59. Grigoletto, F.B. Five-level transformerless inverter for single-phase solar photovoltaic applications. *IEEE J. Emerg. Sel. Top. Power Electron.*, 2019, 8, 3411–3422.

3

Multilevel Inverter Topologies, Modulation, and Applications in Motor Drives

Zahoor Ahmad Ganie[1]*, Abdul Hamid Bhat[2] and Salman Ahmad[1]

[1]*Department of Electrical Engineering, Islamic University of Science and Technology, Kashmir, India*

[2]*Department of Electrical Engineering, National Institute of Technology Srinagar, Hazratbal, Srinagar, India*

Abstract

Multilevel inverters (MLI) are increasingly being recognized as a feasible technology in various fields because of their high-quality produced waveform, lower switching losses, fault tolerant operation, and so on. Nevertheless, the cost of a large number of components usually outweighs these benefits. While the majority of the important areas still use conventional topologies, since the last decade more attention is paid in developing reduced switch MLI topologies to further reduce the switching loss. In this chapter, conventional and new multilevel converter topologies have been explained in detail. Various pulse width modulation techniques and applications in motor drives has also been discussed. The simulation and hardware results have been given to validate the conceptual findings.

Keywords: Multilevel inverters, reduced switch count inverters, pulse width modulation techniques, selective harmonic elimination, low switching frequency, power quality

3.1 Introduction

Power electronics converters play a crucial role in the heavy reliance on electricity in modern power systems. The significance of these converters

**Corresponding author*: Zahoor.ganie@islamicuniversity.edu.in

Salman Ahmad, Farhad Ilahi Bakhsh, and P. Sanjeevikumar (eds.) Multilevel Converters, (51–74)

is further acknowledged and expected to grow further as smart grids evolve with the increasing integration of power generated from renewable sources. This integration is facilitated by the introduction of current power semiconductor devices, sensors and digital controllers. Due to the limitations of semiconductor blockage voltage in the medium voltage range, the use of multilevel inverters (MLI) [1] have come up as the favored solution for handling medium voltage and high power applications [2]. Hence, industries requiring medium voltage and high power applications have thoroughly considered different multilevel inverters (MLI). Additionally, in some cases, MLI is preferred to comply with grid codes due to its ability to minimize total harmonic distortion (THD) [3]. Consequently, the need for filters can be significantly reduced, leading to fewer electromagnetic compatibility issues. Furthermore, compared to the overall operating voltage, semiconductor devices experience significantly lower voltage stress. This enables the production of high voltage waveforms using relatively low-voltage-rated switches. In the case of a multilevel motor drive, motors coupled to MLIs experience reduced stress on their bearings due to substantially lower common mode voltage. Several multilevel topologies offer the flexibility to utilize different switching combinations to achieve specific voltage levels. These redundant states also enable the implementation of fault-tolerant procedures through programmable methods.

Selective harmonic elimination pulse width modulation (SHE-PWM) is generally preferred for eradicating harmonics from the output due to its superior control and performance compared to other techniques [4–6]. In this technique, the output voltage waveform undergoes Fourier series analysis to decompose it [7]. The amplitude and symmetry of the voltage level influence the shape and complexity of the solution space [8]. To eliminate unwanted harmonics, the SHE-PWM technique employs intricate and nonlinear transcendental equations to generate ideal switching patterns for the inverter output voltage. However, achieving convergence towards an optimal solution poses significant challenges due to the involvement of cosine and sine mathematical functions, especially when the switching instants and the levels of inverter increase. The main challenge lies in meticulously solving the SHE-PWM waveform, and the chosen solution strategy heavily depends on the formulation of the waveform [9]. Various techniques are proposed for determining the switching instants for PWM waveforms. Although numerical techniques are effective iterative methods, their success largely depends on selecting an appropriate initial guess [10]. Numerical techniques that accurately determine the initial guess converge rapidly and discover all possible solutions. Generally, the SHE equations are solved using the Newton–Raphson technique [11, 12]. Algebraic methods offer the advantage of determining all possible solutions without the need for initial guesses. These methods involve transforming transcendental equations into polynomial equations, which can be solved after conversion into triangular form to identify the optimal switching instants. On the other hand, intelligent optimization algorithms are well suited for handling transcendental equations, are computationally efficient, and do not heavily rely on initial guesses [13–18]. An objective function depending on the fundamental and lower-order harmonic components of non-linear transcendental equations is organized in these intelligent techniques [19]. The objective function is minimized to generate optimized switching patterns which, in turn, controls the magnitude of harmonics. The way the objective function is formulated significantly influences the execution of the optimization technique. The chapter is organized as follows: the conventional multilevel inverter topologies are discussed in Section 3.2; in Section 3.3, new advent multilevel inverter topologies are discussed; pulse width modulation techniques are discussed in Section 3.4; selective harmonic elimination is described in Section 3.5; in Section 3.6, results and their discussion are given; and, finally, conclusion is presented in Section 3.7.

3.2 Conventional Multilevel Inverter Topologies

Due to the special structure of MLIs, high voltages and high power levels can be achieved without the use of transformers [20]. The conventional MLI topologies are NPC inverter, FC-MLI, CHB-MLI, and active neutral point MLI. The classification of MLI topologies is given in Figure 3.1.

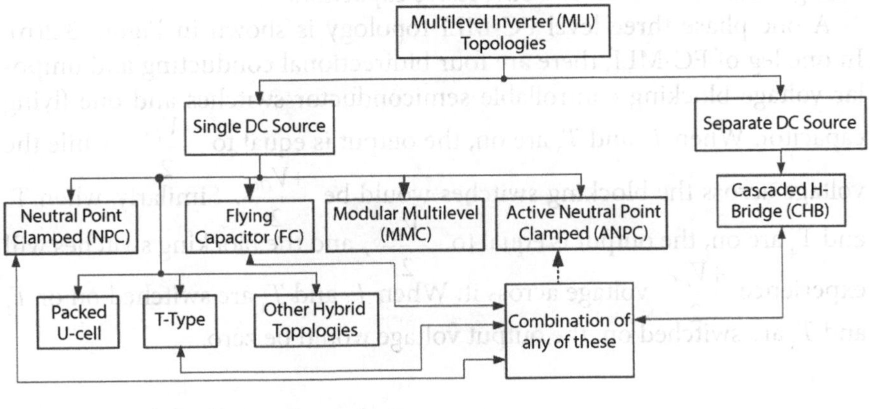

Figure 3.1 Multilevel converter topologies.

3.2.1 Neutral Point-Clamped (NPC) Inverter Topology

The NPC converter was first reported in [21] by Nabae *et al.* in 1981. A one-phase three-level NPC inverter configuration is shown in Figure 3.2(a). Clamping diodes are required to establish a connection between the virtual neutral and the midpoint of the phase-leg. Additional voltage level is generated by this neutral, hence the name three-level converter. In a three-level configuration, each leg comprises of four controlled semiconductor switches (T_1 to T_4) and two diodes (D_1, D_2). The DC link has two capacitors which bifurcate the input DC voltage into two halves. The point is referred to as zero potential point and represented by O. It produces three distinct output voltages as $\frac{+V_{DC}}{2}$, 0, and $\frac{-V_{DC}}{2}$ by turning on and off the suitable combinations of switches. When T_1 and T_2 are on, it produces $\frac{-V_{DC}}{2}$, and when T_3 and T_4 are on, it produces $\frac{+V_{DC}}{2}$. The output voltage is maintained at zero voltage when either D_1T_2 or D_2T_3 is on. All the switches in a particular leg are never turned on together to avoid a direct short circuit. The controlled switches T_1T_3 and T_2T_4 work complementary to each other.

3.2.2 Flying Capacitor (FC) MLI Topology

The FC-MLI topology [22] demands a lot of capacitances for clamping the voltage of power electronic switches to one voltage level of capacitor. The capacitors are not directly connected to any of the positive or negative bus and, hence, are named as flying or floating capacitors. For L-level converter, if the capacitors are of equal value, $(L - 1)$ main DC bus capacitors are required. The change in the levels is decided by the magnitude of the voltage change between two successive capacitors.

A one-phase three-level FC-MLI topology is shown in Figure 3.2(b). In one leg of FC-MLI, there are four bidirectional conducting and unipolar voltage blocking controllable semiconductor switches and one flying capacitor. When T_1 and T_2 are on, the output is equal to $\frac{+V_{DC}}{2}$, while the voltage across the blocking switches would be $\frac{+V_{DC}}{2}$. Similarly, when T_3 and T_4 are on, the output is equal to $\frac{-V_{DC}}{2}$, and the blocking switches will experience $\frac{+V_{DC}}{2}$ voltage across it. When T_2 and T_4 are switched on or T_1 and T_3 are switched on, the output voltage would be zero.

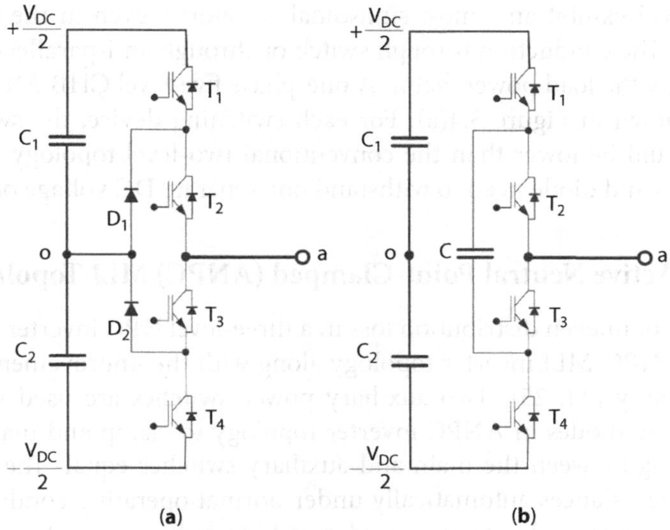

Figure 3.2 Conventional multilevel topologies: (a) one-phase three-level NPC topology and (b) one-phase three-level FC-MLI topology.

3.2.3 Cascaded H-Bridge (CHB) MLI Topology

An isolated DC source along with H-bridge forms the basic unit. A CHB-MLI topology is formed when these units are coupled in series [23]. An individual H-bridge is shown in Figure 3.3. Each H-bridge in a CHB-MLI produces three distinct output voltages: $+V_{DC}$, 0, and $-V_{DC}$. The resultant output voltage is produced by combining the voltages that are produced by different modules. For the same magnitude V_{DC} of the DC sources of different modules, the phase voltage generated will have a range from $-nV_{DC}$ to $+nV_{DC}$. To produce higher levels, a greater number of H-bridges can be interconnected in a cascaded configuration. Thus, the form of the output

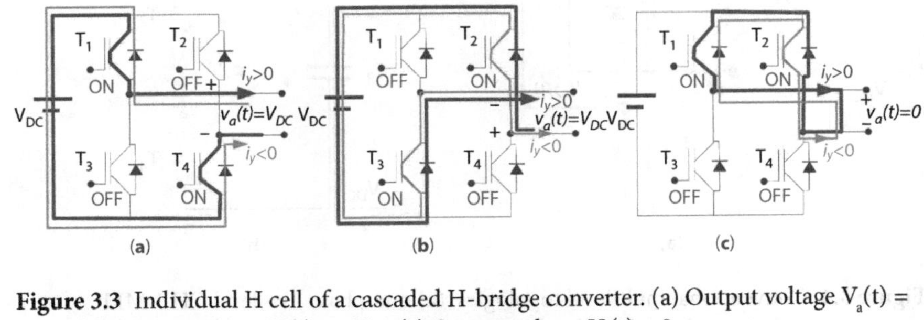

Figure 3.3 Individual H cell of a cascaded H-bridge converter. (a) Output voltage $V_a(t) = +V_{DC}$ (b) Output voltage $V_a(t) = -V_{DC}$ (c) Output voltage $V_a(t) = 0$.

voltage will exhibit an almost sinusoidal waveform, even in the absence of filters. The conduction through switch or through anti-parallel diode is decided by the load power factor. A one-phase five-level CHB-MLI topology is shown in Figure 3.4(a). For each switching device, the switching stress would be lower than the conventional two-level topology because the switch and diode need to withstand one separate DC voltage only.

3.2.4 Active Neutral Point-Clamped (ANPC) MLI Topology

The issue of uneven distribution loss in a three-level NPC inverter is taken care by ANPC-MLI inverter topology along with the improvement in the power ability [24, 25]. Two auxiliary power switches are used in place of clamping diodes in ANPC inverter topology to clamp and make voltage sharing between the main and auxiliary switches equal. The capacitor voltage balances automatically under normal operating conditions in an ANPC inverter due to increased switching redundancy. A one-phase three-level ANPC-MLI topology is shown in Figure 3.4(b). The rating of T_5 and T_6 are the same as that of main switches. For producing the output voltage of $\frac{+V_{DC}}{2}$, T_1 and T_2 are switched on. Switch T_6 is turned on in this state so that equal voltage appears across T_3 and T_4.

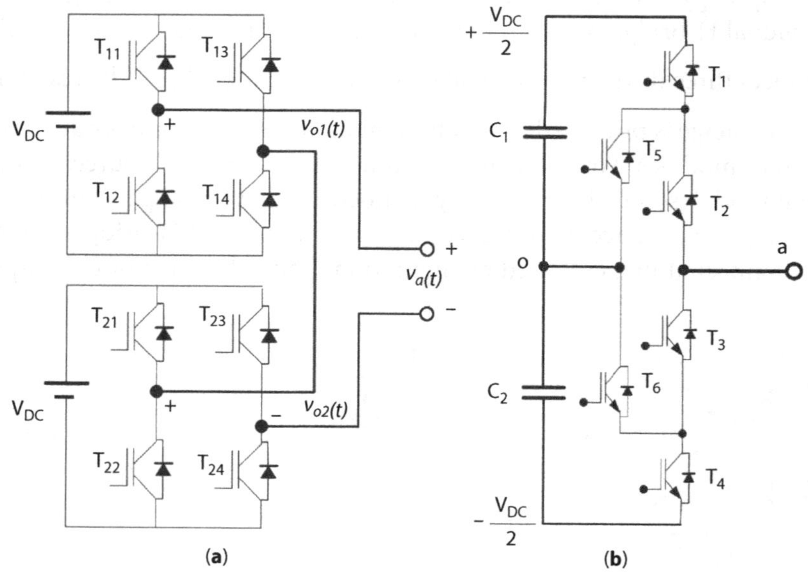

Figure 3.4 Conventional multilevel topologies: (a) one-phase five-level CHB-MLI topology and (b) one-phase three-level ANPC-MLI topology.

3.3 New Advent MLI Topologies

Numerous novel topologies have been proposed over the last few years. These configurations have been developed to incorporate a minimal number of devices, increased levels, and expanded application areas [26]. In this section, these topologies are explained in detail.

3.3.1 Modular Multilevel Inverter (MMI) Topology

The configuration of MMI topology is similar to the CHB inverter as shown in Figure 3.5(a) [27]. However, MMC inverter requires only one DC source instead of several separate DC sources as required in CHB inverter. The basic unit of MMC is called a sub-module (SM). The number of SM's can be increased or decreased to generate the appropriate output voltage. Due to the modular design of MMI topology, extremely high voltage and current carrying capability can be achieved. The working of sub-modules at low switching frequency is also possible due to the generation of a large number of voltage steps.

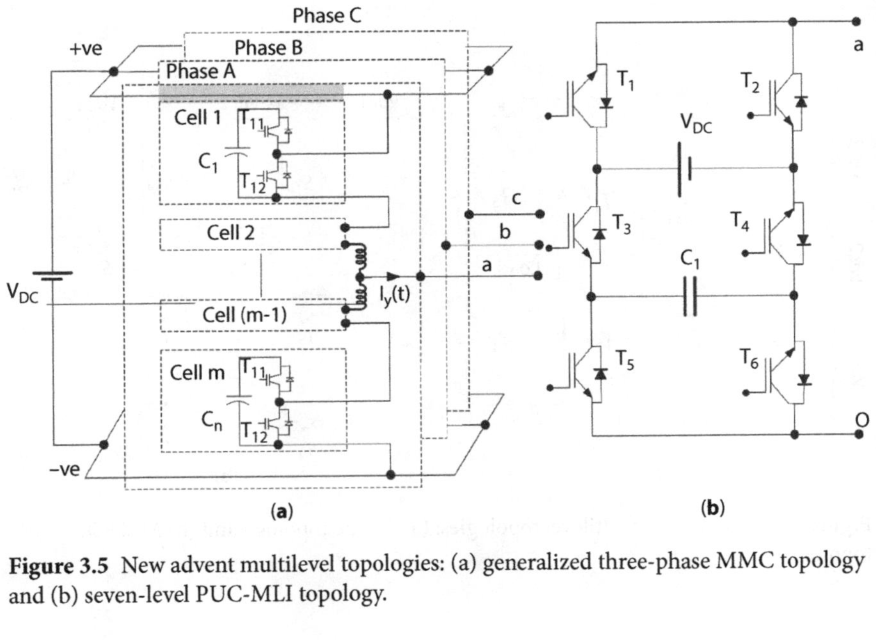

Figure 3.5 New advent multilevel topologies: (a) generalized three-phase MMC topology and (b) seven-level PUC-MLI topology.

3.3.2 Packed U-Cell (PUC) MLI Topology

The PUC-MLI topology was first reported in 2010 [28]. This topology combines the features of FC-MLC and CHB-MLC with reduction in semiconductor switches and capacitors. The basic unit is referred as packed U-cell which comprises a DC source and two power semiconductor switches. The PUC-MLI requires less gate driver circuits, has low power loss, and has low overall cost. A one-phase seven-level PUC-MLI topology is shown in Figure 3.5(b). For generating a seven-level output waveform, it is necessary to maintain the voltage across the capacitor at $(⅓)^{rd}$ of V_{DC}. The levels in the output voltage will be 0, $\pm v_c$, $\pm 2v_c$, and $\pm 3v_c$, where ($v_c = (⅓)^{rd} V_{DC}$).

3.3.3 T-Type MLI

T-type configuration was proposed by Gerardo *et al.* in 2006 [29]. Compared to conventional topologies, this topology uses significantly fewer switching devices for achieving equivalent voltage levels. A one-phase T-type topology is shown in Figure 3.6(a) having two switches S_1S_2 and S_3S_4 which are bidirectional conducting and bipolar voltage blocking and four switches T_1 to T_4 which are bidirectional conducting and unipolar voltage blocking. The input DC sources must be equal, i.e., $V_{dc1} = V_{dc2} = V_{dc3} = V_{dc}$, so that all the possible voltage levels are produced across the load.

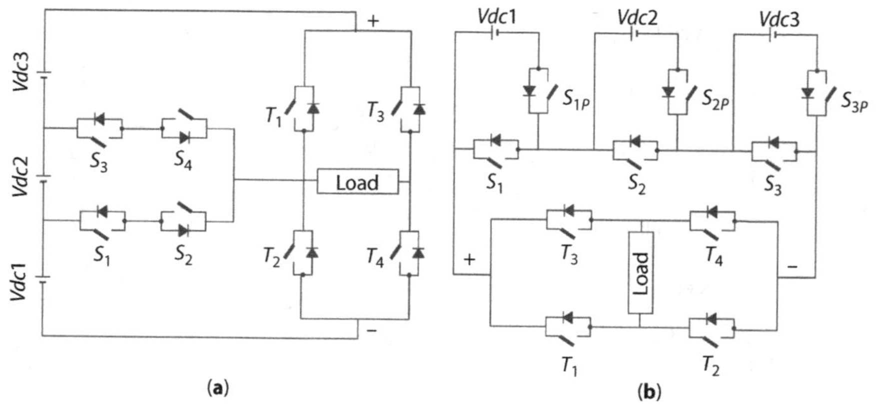

Figure 3.6 New advent multilevel topologies: (a) T-type topology and (b) MLDCL topology.

3.3.4 Multilevel DC Link Inverter (MLDCL)

MLDCL was proposed by Gui Jia Su in 2005 [30] and is shown in Figure 3.6(b). A half bridge cell along with a DC source forms a basic unit which are coupled together to form a level generator which generates the required number of levels. The polarity generator has four power electronic switches with the switches in each leg conducting in complementary fashion to generate the multilevel output waveform. This topology uses fewer switching devices as compared to conventional topologies to produce the same voltage level. For generating a particular voltage level, three power electronic switches in the level generator and two power electronic switches in the polarity generator are switched on. When DC voltage is kept in the proportion of $V_{dc1} : V_{dc2} : V_{dc3} = 1 : 2 : 3$, an output voltage with 13 distinct levels can be produced, and if they are in the proportion of $V_{dc1} : V_{dc2} : V_{dc3} = 1 : 2 : 4$, an output voltage with 15 distinct levels can be obtained. This multilevel topology does not support the subtractive combination of the input DC voltage sources.

3.3.5 Switched Series/Parallel Sources (SSPS)-Based MLI

SSPS-based MLI was first proposed by Hinago *et al.* in 2009 [31] and is shown in Figure 3.7(a). The level generator comprises power devices and individual DC voltage sources that can be switched in parallel and series. Compared to conventional topologies, this topology uses significantly fewer switching devices to produce the equivalent voltage levels. If the strength of input DC sources is the same, *viz.*, V_{dc}, then the voltage blocking capability of each power electronic device in the level generator is V_{dc}; however, for the power electronic devices in the polarity generator, it is equal to the summation of all the DC sources in the level generator.

3.3.6 Series-Connected Switched Sources (SCSS)-Based MLI

The SSCS configuration was proposed in 2008 [32, 33]. The SCSS topology is shown in Figure 3.7(b) where the DC sources are coupled in series using power electronic devices. The negative terminals of the input DC voltage sources are interconnected through power electronic devices, and the positive end of each source is also coupled to the negative side of the next source through power electronic devices, as shown in Figure 3.7(b). This interconnection of the DC sources allows this topology to generate multilevel rectified waveform which is converted into multilevel AC waveform with the help of a polarity generator. For the same strength of input DC

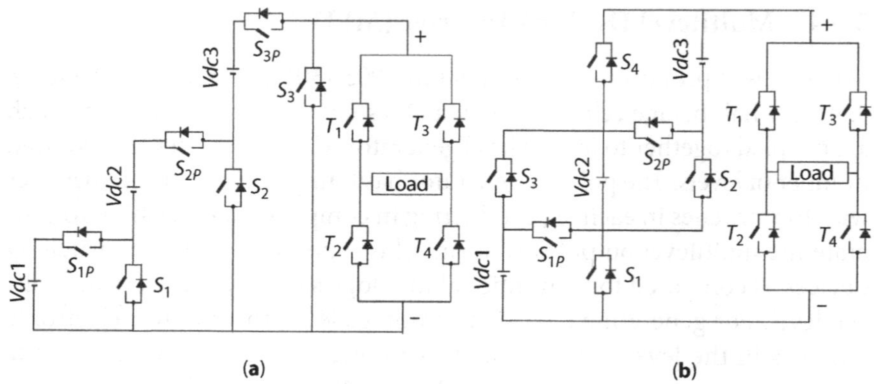

Figure 3.7 New advent multilevel topologies: (a) SSPS topology and (b) SCSS topology.

sources, the voltage blocking capability of each power electronic switch in the level generator will be different; however, for the switching devices in the polarity generator, it is equal to the summation of all the DC sources in the level generator.

3.3.7 Reversing Voltage (RV) Topology

The RV configuration was proposed by Najafi *et al.* in 2007 [34]. An RV-MLI topology is shown in Figure 3.8(a). The level generator section has three DC sources and six switches, and the polarity generator has four power electronic switches whose blocking voltage capability needs to be equal to the summation of the DC sources in the level generator. The voltage blocking capacity of the power electronic devices in the level generator is equal to V_{dc} if the strength of the input DC sources is equal to V_{dc}. Since the DC sources cannot be combined in a subtractive manner, a triode source combination cannot be used. However, for three-phase implementation, this topology uses a single DC link; therefore, DC voltage sources can be saved.

3.3.8 E-Type Topology

The E-type asymmetrical topology was first proposed by E. Samadaei *et al.* in 2016 [35]. The E-type topology shown in Figure 3.8(b) generates a 13-level output waveform. It comprises of four DC sources which are in the ratio of $V_{dc1} : V_{dc2} = 1 : 2$, two bi-directional power electronic switches, and six switching devices which are uni-directional conducting, all arranged in the shape of an envelope. The voltage blocking capacity of the switches are unequal, and this topology has redundancies in a few switching states.

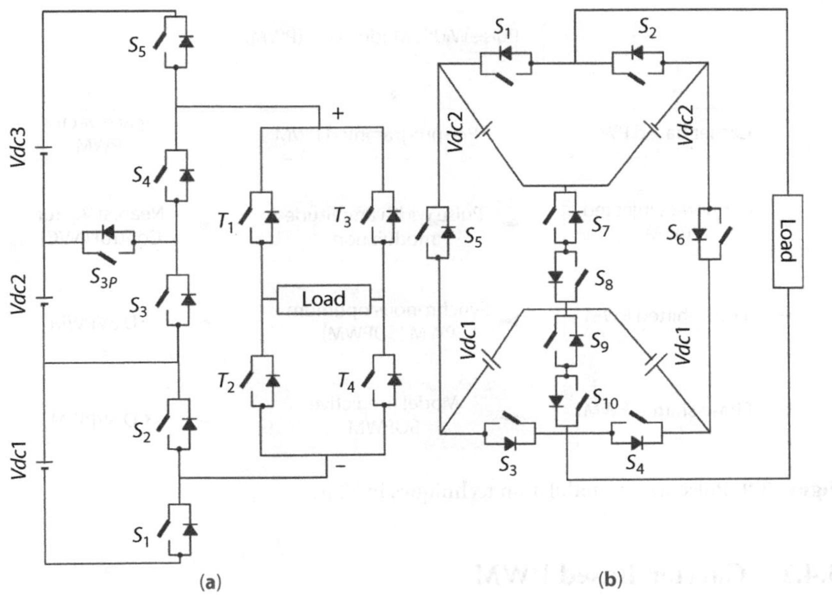

Figure 3.8 New advent multilevel topologies: (a) RV topology and (b) E-type topology.

3.4 Pulse Width Modulation Techniques

Normally, voltage source inverters (VSIs) are operated with various pulse width modulation (PWM) techniques which generates the pulsed waveform whose average (over a switching cycle) is equal to the desired magnitude [36]. Depending on the methods, the PWM is implemented in VSI. The detailed classification of various PWM schemes implemented in different VSI topologies is shown in Figure 3.9. In a broad sense, it can be categorized as high switching frequency, medium switching frequency, and low switching frequency PWM techniques. Carrier-based PWM and space vector PWM techniques are high switching frequency techniques, whereas the medium PWM technique includes hybrid PWM methods. The preprogrammed PWM schemes like selective harmonic elimination, selective harmonic mitigation, and pulse width amplitude modulation along with nearest vector control are classified as low switching frequency PWM techniques [37]. In carrier PWM, a carrier (or multiple carriers) of high frequency is compared with a modulating signal having appropriate magnitude and frequency, and based on the greater or less than logic the PWM pulses are obtained. The reference signal in some cases are obtained by injecting harmonics in it such that the DC bus utilization is increased.

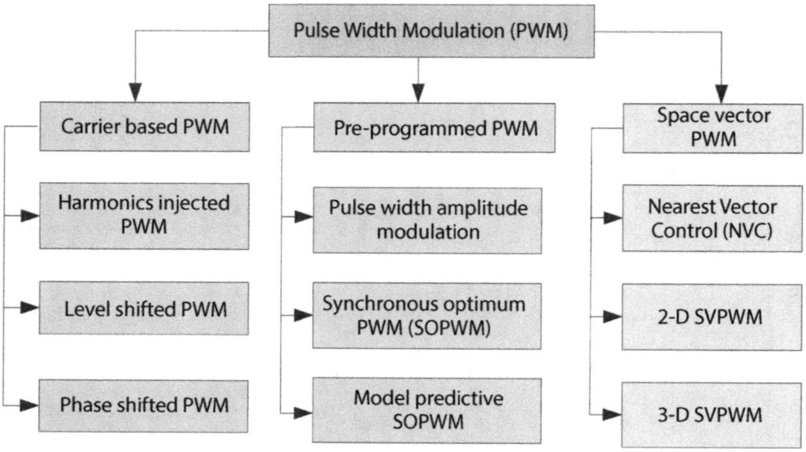

Figure 3.9 Pulse width modulation techniques in VSIs.

3.4.1 Carrier-Based PWM

In this technique, a high frequency carrier is juxtaposed with modulating signal for generating the PWM pulses. Because of the high carrier frequency, the harmonic spectra is switched to that frequency and its multiples. A three-phase VSI with CB-PWM is shown in Figure 3.10(a).

For better utilization of DC bus voltage, a third harmonic component is added in the modulating wave to get a new modulating wave which is juxtaposed with the carrier wave as shown in Figure 3.10(b). The resulting equation is given in Equation 3.1.

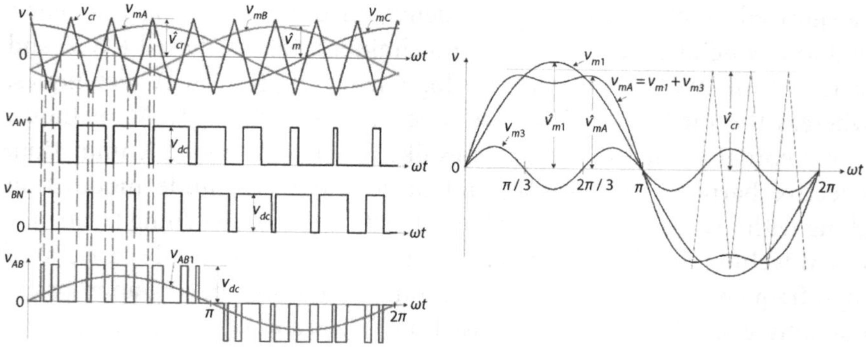

(**a**) CB-PWM scheme in three phase VSI

(**b**) Third harmonics injection PWM scheme in three phase

Figure 3.10 Pulse width modulation techniques.

$$
\begin{aligned}
V_{mA} &= v_{mA1} + v_{mA3} = m \times [\sin(\omega_0 t) + k(\sin(3\omega_0 t))] \\
V_{mB} &= v_{mB1} + v_{mB3} = m \times [\sin(\omega_0 t) + 2\pi/3) + k(\sin(3\omega_0 t))] \quad (3.1) \\
V_{mC} &= v_{mC1} + v_{mC3} = m \times [\sin(\omega_0 t) + 4\pi/3) + K(\sin(3\omega_0 t))]
\end{aligned}
$$

3.4.1.1 *Phase-Shifted PWM*

A phase-shifted carrier-based CHB MLI having three H-cells (seven-level) is shown in Figure 3.11(a). For L-level, the carrier waves required are given by $N_c = L - 1$. The phase shift in carriers is $\frac{360}{N_c}$.

3.4.1.2 *Level-Shifted PWM*

A level-shifted PWM scheme for a seven-level VSI is shown in Figure 3.11(b). The number of carriers is $N_c = L - 1$, i.e., for seven levels, six carriers are taken which are shifted in magnitude but in phase with each other.

3.4.2 Space Vector PWM

The space vector PWM is the most popularly used digital control technique for VSI. For a three-phase two-level VSI, the location of space vectors is

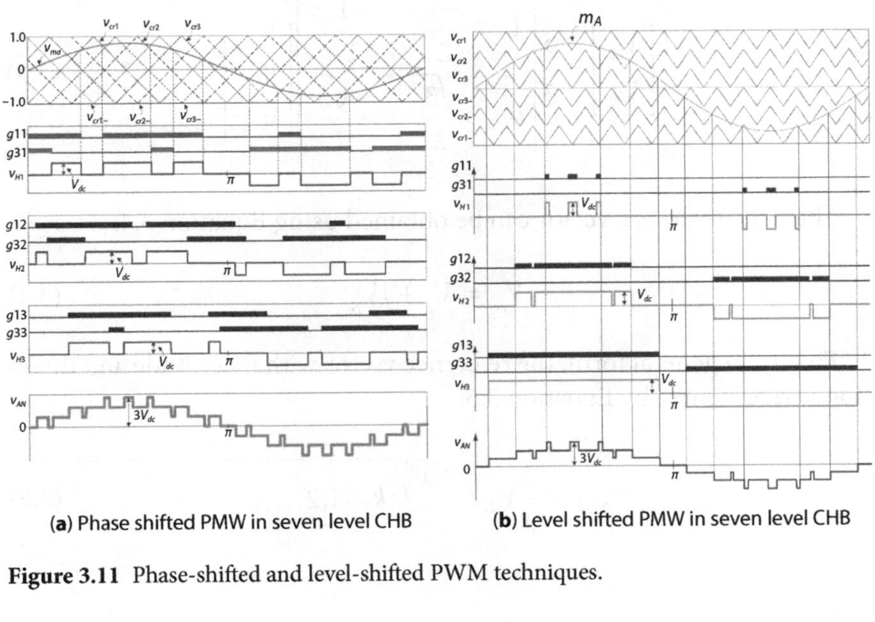

Figure 3.11 Phase-shifted and level-shifted PWM techniques.

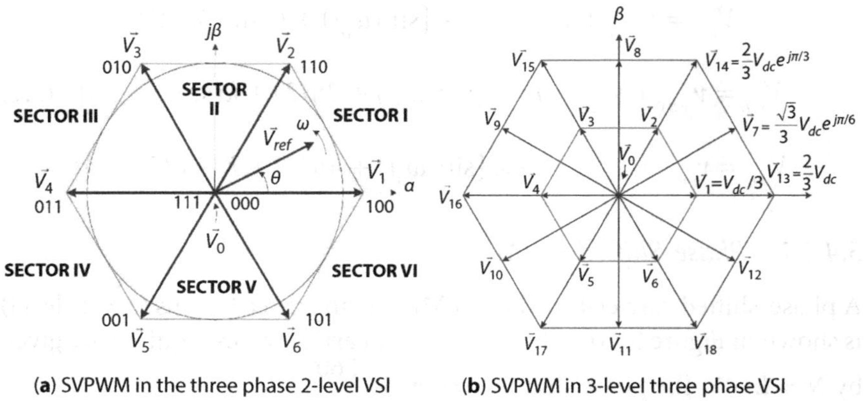

Figure 3.12 Space vector pulse width modulation in VSI.

displayed in Figure 3.12(a). Similarly, the positions of space vectors for a three-phase three-level output are shown in Figure 3.12(b). Let instantaneous phase load voltages be balanced and given by Equation 3.2.

$$v_{An} + v_{Bn} + v_{Cn} = 0 \tag{3.2}$$

Therefore, the three-phase variable can be converted into phase variables using Equation 3.3.

$$\begin{bmatrix} V_\alpha \\ V_\beta \end{bmatrix} = \frac{2}{3} \begin{bmatrix} 1 & \frac{-1}{2} & \frac{-1}{2} \\ 0 & \frac{\sqrt{3}}{2} & \frac{-\sqrt{3}}{2} \end{bmatrix} \begin{bmatrix} v_{An} \\ v_{Bn} \\ v_{Cn} \end{bmatrix} \tag{3.3}$$

Thus, the reference vector can be obtained using Equation 3.4.

$$V_{ref} = V_\alpha + jV_\beta \tag{3.4}$$

Finally, in general form, the reference vector with magnitude and direction is represented by Equation 3.5.

$$V_{ref,k} = \frac{2}{3} V_{dc} e^{j(k-1)\frac{\pi}{3}}; k = 1,2\ldots,6 \tag{3.5}$$

3.5 Selective Harmonic Elimination Technique

To control the harmonics in SHE-PWM, the switching patterns for the output waveform are determined by using Fourier series. In general form, the Fourier series analysis is given by Equation 3.6.

$$U(t) = \frac{1}{2}a_0 + \sum_{k=1}^{\infty} a_k \cos(k\omega t) + \sum_{k=1}^{\infty} b_k \sin(k\omega t) \tag{3.6}$$

a_0, a_k, and b_k are Fourier coefficients and their values are determined from Equations 3.7 and 3.8.

$$a_k = \frac{2}{T}\int_{-\frac{T}{2}}^{\frac{T}{2}} U(t)cos(k\omega t) \tag{3.7}$$

$$b_k = \frac{2}{T}\int_{-\frac{T}{2}}^{\frac{T}{2}} U(t)sin(k\omega t) \tag{3.8}$$

Due to the quarter wave symmetry possessed by the waveforms [38], the output voltage is simplified and is given by Equation 3.9.

$$U(t) = \sum_{k=1}^{\infty} b_k \times \sin(kwt) \tag{3.9}$$

b_k represents the amplitude of kth harmonic. Any harmonic of a particular order can be eliminated by equating the appropriate b_k term to zero. The fundamental component magnitude and harmonics are determined using Equation 3.10.

$$b_k = \frac{4V_s}{k\pi}\left[\sum_{j=1}^{N} (-1)^l \times \cos(k\alpha_j)\right] \tag{3.10}$$

V_s is the input DC voltage, and the number of switching angles is represented by N. The transitions that take place in a given time can be used to formulate the SHE-PWM problem for multilevel waveforms. Parameter l determines the transition of the waveform and is given by Equation 3.11.

$$l = \begin{cases} 0\ldots for..increasing..edge \\ 1\ldots for..falling..edge \end{cases} \tag{3.11}$$

To create a symmetrical and precise waveform, the constraint on the switching angles to $0 \le \alpha_1 \le \alpha_2 \le .... \le \alpha_N \le \pi/2$ has to be followed. The modulation index (m_a) determines the fundamental frequency component V_1 as given by Equation 3.12.

$$V_1 = \frac{4m_a V_s}{\pi} \tag{3.12}$$

The main limitations of SHE-PWM are as follows: (i) it is mainly an offline method, and therefore the dynamic response is not good and (ii) the first non-eliminated harmonic component has a considerable magnitude which is of a lower order and difficult to filter.

The trajectories for an 11-level CHB-MLI operating at a fundamental switching frequency are shown in Figure 3.13(a). The first four dominant

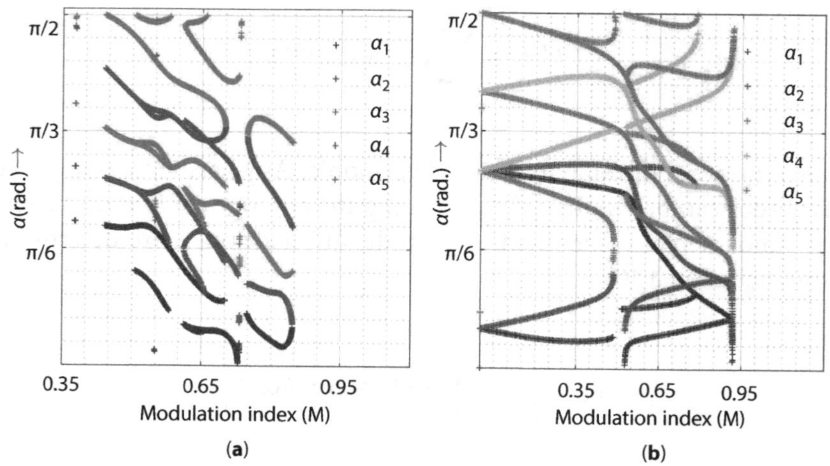

Figure 3.13 Switching angle trajectories: (a) 11-level CHB multilevel inverter for five switching angles and (b) three-level NPC inverter for five switching angles.

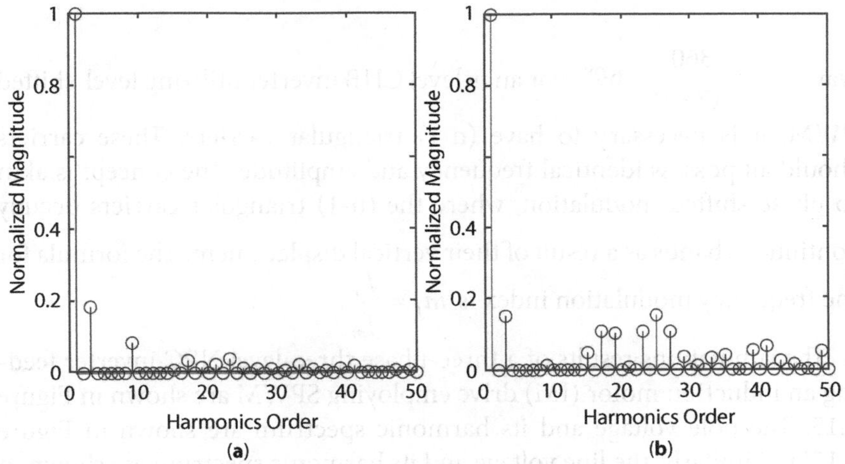

Figure 3.14 Harmonics profile: (a) 11-level CHB and (b) five switching angles NPC.

harmonics are considered for elimination and simultaneously controlling the fundamental component based on the modulation index. Similarly, the trajectories for a three-level NPC inverter operating at five times the fundamental switching frequency are shown in Figure 3.13(b). In this case, the first four dominant harmonics are also considered for elimination and simultaneously control the fundamental component based on the modulation index. In the entire range of modulation index, multiple solutions, unique solutions, or no solutions exist. In the case of multiple solutions, the solution which gives minimum %THD is chosen as the optimal solution. The harmonic spectrum of the pole voltage of an 11-level CHB-MLI with five switching angles is shown in Figure 3.14(a). Similarly, the harmonic spectrum of the pole voltage of a three-level NPC inverter with five switching angles is shown in Figure 3.14(b).

3.6 Results and Discussion

The MLIs with n levels of voltage requires (n – 1) carrier waves for the phased-shifted modulation scheme. In phase-shifted PWM, all the carrier waves have the same frequency and peak-to-peak magnitude. However, there exists a phase shift between any two consecutive carriers which is calculated as $\phi_{cr} = \dfrac{360^o}{(n-1)}$. All carrier waves in a seven-level CHB (n = 7), which we employed in our work, are 60° away from the nearby carrier waves $\phi_{cr} = \dfrac{360^o}{(7-1)} = 60^o$. For an n-level CHB inverter utilizing level-shifted PWM, it is necessary to have (n-1) triangular carriers. These carriers should all possess identical frequency and amplitude. The concept is akin to phase-shifted modulation, where the (n-1) triangular carriers occupy continuous bands as a result of their vertical displacement. The formula for the frequency modulation index is $m_f = \dfrac{f_{cr}}{f_m}$.

The simulations results of a three-phase three-level NPC inverter feeding an induction motor (IM) drive employing SPWM are shown in Figure 3.15. The pole voltage and its harmonic spectrum are shown in Figure 3.15(a). Similarly, the line voltage and its harmonic spectrum are shown in Figure 3.15(b) and the line current drawn by the IM drive and its harmonic spectrum are shown in Figure 3.15(c). The performance parameters of the IM drive are given in Figure 3.16. The rotor current, the rotor speed, and the electromagnetic torque developed by the induction motor drive are shown in Figure 3.16. The %THD of the pole voltage obtained is 54.30% which is on the higher side due to the fact it contain triplen harmonics which have not been considered for elimination, whereas, the %THD of the line voltage obtained is 36.14% which is much lower than pole voltage due to the absence of triplen harmonics in the line voltage. The %THD of

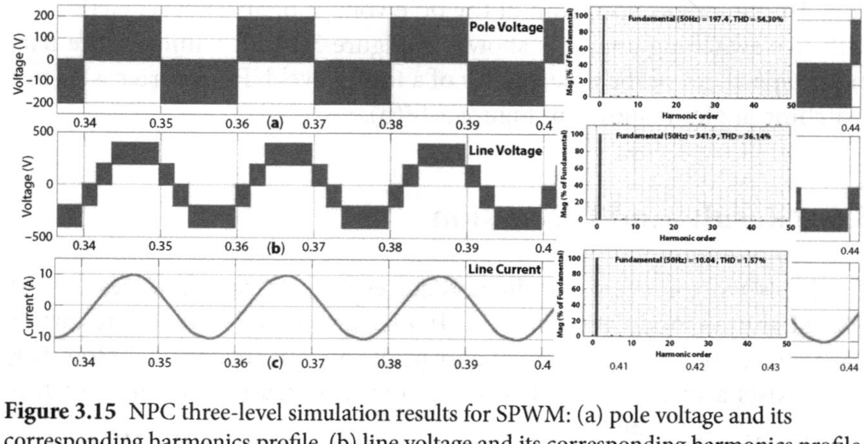

Figure 3.15 NPC three-level simulation results for SPWM: (a) pole voltage and its corresponding harmonics profile, (b) line voltage and its corresponding harmonics profile, and (c) line current drawn by the IM drive and its corresponding harmonics profile.

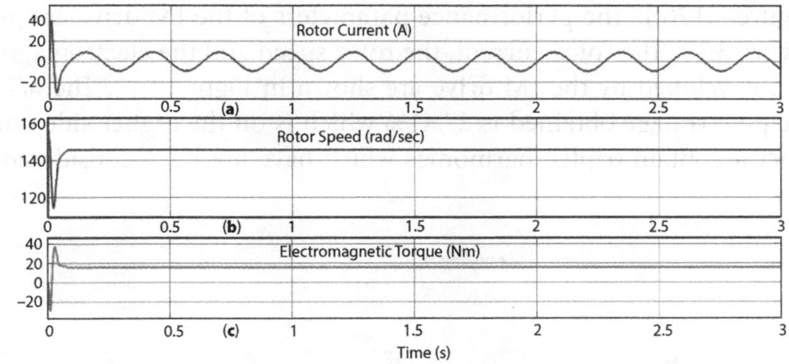

Figure 3.16 Three-phase three-level NPC inverter-fed IM drive performance parameter waveforms: (a) rotor current, (b) rotor speed, and (c) electromagnetic torque.

the stator current drawn by the induction motor drive is 1.57% which is quite low due to inductive effect of the motor drive.

The simulations results of a three-phase five-level CHB inverter feeding an IM drive employing SPWM are shown in Figure 3.17. The pole voltage and its harmonic spectrum are shown in Figure 3.17(a), similarly, the line voltage and its harmonic spectrum are shown in Figure 3.17(b) and the line current drawn by the IM drive and its harmonic spectrum are shown

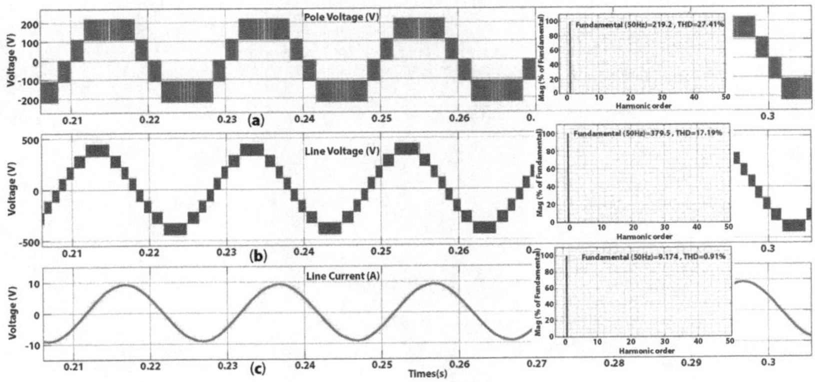

Figure 3.17 CHB five-level simulation results for SPWM: (a) pole voltage and its corresponding harmonics profile, (b) line voltage and its corresponding harmonics profile, and (c) line current drawn by the IM drive and its corresponding harmonics profile.

in Figure 3.17(c). The performance parameters of the IM drive are given in Figure 3.18. The rotor current, the rotor speed and the electromagnetic torque developed by the IM drive are shown in Figure 3.18. The %THD of the pole voltage obtained is 27.41% which is on the higher side due to the fact it contain triplen harmonics which have not been considered for

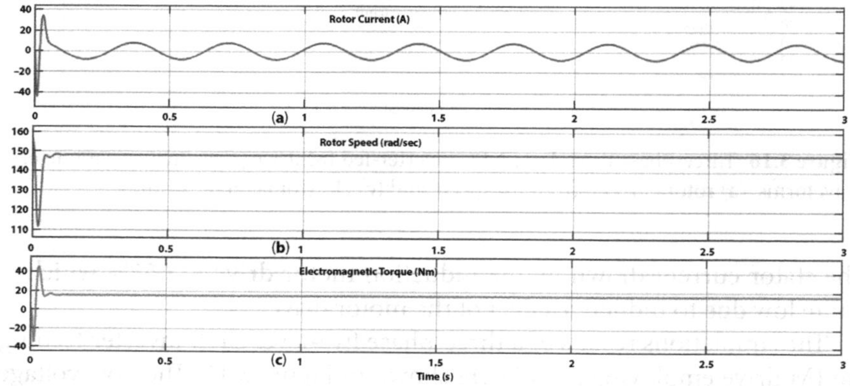

Figure 3.18 Three-phase five-level CHB inverter-fed IM drive performance parameter waveforms: (a) rotor current, (b) rotor speed, and (c) electromagnetic torque.

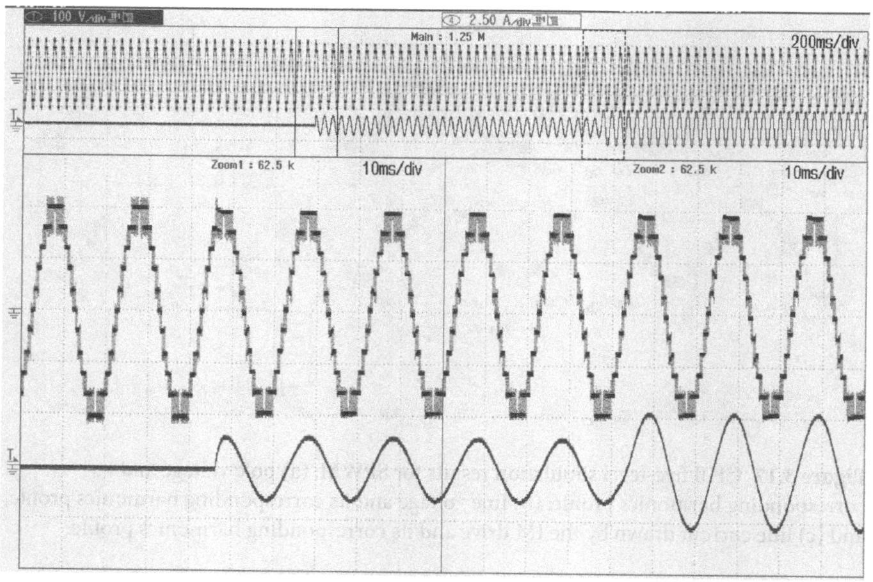

Figure 3.19 Hardware results of phase-shifted 11-level SPWM waveform: pole voltage and dynamic load current.

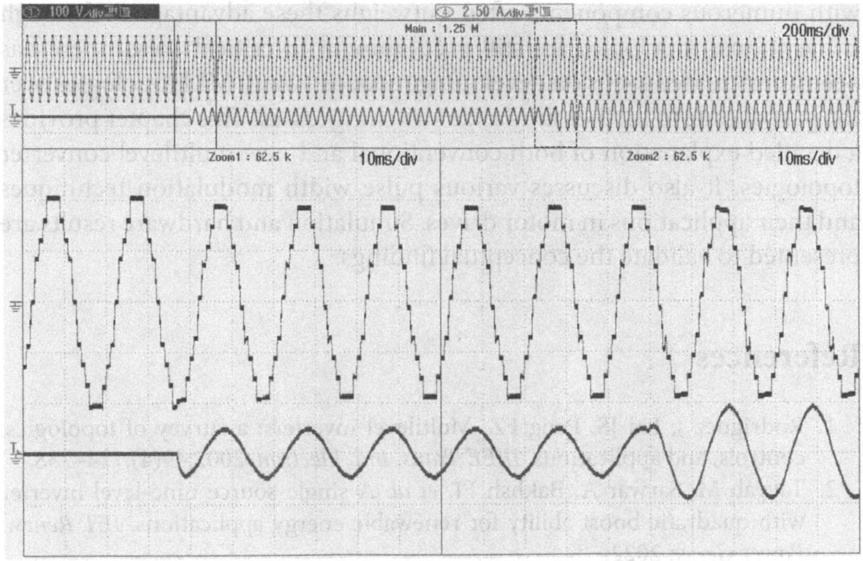

Figure 3.20 Hardware results of an 11-level CHB-MLI with SHE PWM and dynamic load.

elimination, whereas, the %THD of the line voltage obtained is 17.19% which is much lower than pole voltage due to absence of triplen harmonics in the line voltage. The %THD of the stator current drawn by the IM drive is 0.91% which is quite low due to inductive effect of the motor drive. The hardware results of a three-phase 11-level CHB-MLI with phase-shifted SPWM technique feeding an IM drive are shown in Figure 3.19. The pole voltage applied to the IM drive and the stator current drawn for dynamically changing load are shown in Figure 3.19.

The hardware results of a three-phase 11-level CHB-MLI with SHE-PWM technique feeding an IM drive are shown in Figure 3.20. The pole voltage applied to the IM drive and the stator current drawn for dynamically changing load are shown in Figure 3.20. Initially, the load on the motor is zero, hence the stator current drawn is zero, then the load on the motor is increased to 50% of the rated load and finally the motor is run at full load condition.

3.7 Conclusion

MLIs have emerged as a promising technology for various applications due to their ability to produce high-quality waveforms, minimize switching losses, ensure fault tolerance, and more. Nevertheless, the cost associated with numerous components often outweighs these advantages. Although conventional topologies are still predominant in critical areas, there has been an increased focus on developing reduced switch MLI topologies over the past decade to further minimize switching losses. This chapter provides a detailed explanation of both conventional and new multilevel converter topologies. It also discusses various pulse width modulation techniques and their applications in motor drives. Simulation and hardware results are presented to validate the conceptual findings.

References

1. Rodriguez J, Lai JS, Peng FZ. Multilevel inverters: a survey of topologies, controls, and applications. *IEEE Trans. Ind. Electron.* 2002;49(4):724–738.
2. Tayyab M, Sarwar A, Bakhsh FI, *et al.* A single-source nine-level inverter with quadratic boost ability for renewable energy applications. *IET Renew. Power Gener.* 2022.
3. Liu Y, Hong H, Huang AQ. Real-time calculation of switching angles minimizing THD for multilevel inverters with step modulation. *IEEE Trans. Ind. Electron.* 2008;56(2):285–293.
4. Ahmad S, Ganie ZA, Ashraf I, *et al.* Harmonics minimization in 3-level inverter waveform and its FPGA realization. In: *2018 3rd International Innovative Applications of Computational Intelligence on Power, Energy and Controls with their Impact on Humanity (CIPECH)*. IEEE; 2018. p. 1–5.
5. Ahmad Ganie Z, Hamid Bhat A, Ahmad S. A novel adaptive invasive weed optimization technique and least square regression for harmonics minimization in standalone PV applications. *Int. J. Circuit Theory Appl.* 2023, 51(4);p. 1642–1664.
6. Ahmad S, Uddin R, Ganie ZA, *et al.* Low Frequency Operation and dsPIC Micro-controller Implementation for Multilevel Quasi Z Source Inverter in Photovoltaic Application. *Distrib. Gener. Altern. Energy J.* 2022;p. 929–958.
7. Ahmad S, Meraj M, Iqbal A, *et al.* Selective harmonics elimination in multilevel inverter by a derivative-free iterative method under varying voltage condition. *ISA Trans.* 2019;92:241–256.
8. Ahmad S, Khan I, Iqbal A, *et al.* A novel pulse width amplitude modulation for elimination of multiple harmonics in asymmetrical multilevel inverter. In: *2021 IEEE Texas Power and Energy Conference (TPEC)*. IEEE; 2021. p. 1–6.
9. Wells JR, Nee BM, Chapman PL, *et al.* Selective harmonic control: A general problem formulation and selected solutions. *IEEE Trans. Power Electron.* 2005;20(6):1337–1345.

10. Sajid I, Sarwar A, Tariq M, *et al.* Runge Kutta optimization-based selective harmonic elimination in an H-bridge multilevel inverter. *IET Power Electron.* 2023.
11. Sun J, Beineke S, Grotstollen H. Optimal PWM based on real-time solution of harmonic elimination equations. *IEEE Trans. Power Electron.* 1996;11(4):612–621.
12. Al-Hitmi M, Ahmad S, Iqbal A, *et al.* Selective harmonic elimination in a wide modulation range using modified Newton–Raphson and pattern generation methods for a multilevel inverter. *Energies*. 2018;11(2):458.
13. Ozpineci B, Tolbert LM, Chiasson JN. Harmonic optimization of multilevel converters using genetic algorithms. In: *2004 IEEE 35th Annual Power Electronics Specialists Conference (IEEE Cat. No. 04CH37551).* vol. 5. IEEE; 2004. p. 3911–3916.
14. Ahmad S, Iqbal A. Switching angles computations using PSO in selective harmonics minimization PWM. In: *Metaheuristic and Evolutionary Computation: Algorithms and Applications.* Springer; 2021. p. 437–461.
15. Ahmad S, Iqbal A, Ashraf I, *et al.* Harmonics Minimization in Multilevel Inverter by Continuous Mode ACO Technique. In: *Innovations in Electrical and Electronic Engineering.* Springer; 2021. p. 95–104.
16. Ganie ZA, Bhat AH, Ahmad S. Harmonics Minimization in Wide Modulation Range Using Improved Differential Evolution Optimization Technique and Hardware Validation. In: *2022 2nd Asian Conference on Innovation in Technology (ASIANCON).* IEEE; 2022. p. 1–6.
17. Ganie ZA, Bhat AH, Ahmad S. Low Switching Frequency Operation of a Two-Level Voltage Source Inverter Using Imperialist Competitive Algorithm and Hardware Realization. In: *2022 1st International Conference on Sustainable Technology for Power and Energy Systems (STPES).* IEEE; 2022. p. 1–6.
18. Ganie ZA, Bhat AH, Ahmad S. Low switching frequency operation of multilevel converters for high-power applications. *Intelligent Control of Medium and High Power Converters.* 2023;p. 173.
19. Ganie ZA, Bhat AH, Ahmad S. (in press), Wide Band Solution and Sensitivity Analysis for Lower Order Harmonics Control in Multilevel Converter. *Int. J. Power Electron.* 2023.
20. Lai JS, Peng FZ. Multilevel converters-a new breed of power converters. *IEEE Trans. Ind. Appl.* 1996;32(3):509–517.
21. Nabae A, Takahashi I, Akagi H. A new neutral-point-clamped PWM inverter. *IEEE Trans. Ind. Appl.* 1981;(5):518–523.
22. Yoscovich I, Glovinsky T, Sella G, *et al.* Multi-level inverter with flying capacitor topology. *Google Patents*; 2016. US Patent 9,318,974.
23. Prasad K, Kumar GR, Kiran TV, *et al.* Comparison of different topologies of cascaded H-Bridge multilevel inverter. In: *2013 International Conference on Computer Communication and Informatics.* IEEE; 2013. p. 1–6.

24. Barbosa P, Steimer P, Steinke J, *et al.* Active-neutral-point-clamped (ANPC) multilevel converter technology. In: *2005 European Conference on Power Electronics and Applications*. IEEE; 2005. p. 10–pp.
25. Zaid M, Sarwar A, Quayyoom L, *et al.* Performance analysis and control of a novel 7-level active neutral point clamped (ANPC) topology. *IET Power Electron.* 2023.
26. Hussan MR, Sarwar A, Tayyab M, *et al.* A single-source switched-capacitor based 13-level sextuple boost inverter with reduced component count and low THD. *IET Gener. Transm. Distrib.* 2023.
27. Debnath S, Qin J, Bahrani B, *et al.* Operation, control, and applications of the modular multilevel converter: A review. *IEEE Trans. Power Electron.* 2014;30(1):37–53.
28. Ounejjar Y, Al-Haddad K, Gregoire LA. Packed U cells multilevel converter topology: theoretical study and experimental validation. *IEEE Trans. Ind. Electron.* 2010;58(4):1294–1306.
29. Ceglia G, Guzman V, Sanchez C, *et al.* A new simplified multilevel inverter topology for DC–AC conversion. *IEEE Trans. Power Electron.* 2006;21(5):1311–1319.
30. Su GJ. Multilevel DC-link inverter. *IEEE Trans. Ind. Appl.* 2005;41(3):848–854.
31. Hinago Y, Koizumi H. A single-phase multilevel inverter using switched series/parallel dc voltage sources. *IEEE Trans. Ind. Electron.* 2009;57(8):2643–2650.
32. Lee S, Kang F. A new structure of H-bridge multilevel inverter. In: *Proc. KIPE Conf;* 2008. p. 388–390.
33. Choi WK, Kang Fs. H-bridge based multilevel inverter using PWM switching function. In: *INTELEC 2009-31st International Telecommunications Energy Conference*. IEEE; 2009. p. 1–5.
34. Najafi E, Yatim A, Samosir A. A new topology-Reversing Voltage (RV)-for multi level inverters. In: *2008 IEEE 2nd International Power and Energy Conference*. IEEE; 2008. p. 604–608.
35. Samadaei E, Gholamian SA, Sheikholeslami A, *et al.* An envelope type (E-Type) module: asymmetric multilevel inverters with reduced components. *IEEE Trans. Ind. Electron.* 2016;63(11):7148–7156.
36. Holmes DG, Lipo TA. *Pulse width modulation for power converters: principles and practice*. vol. 18. John Wiley & Sons; 2003.
37. Enjeti PN, Ziogas PD, Lindsay JF. Programmed PWM techniques to eliminate harmonics-A critical evaluation. In: *Conference Record of the 1988 IEEE Industry Applications Society Annual Meeting*. IEEE; 1988. p. 418– 430.
38. Fei W, Ruan X, Wu B. A generalized formulation of quarter-wave symmetry SHE-PWM problems for multilevel inverters. *IEEE Trans. Power Electron.* 2009;24(7):1758–1766.

4

Multilevel Inverter Operation With Reduced Capacitor Inrush Currents for Solar Photo-Voltaic Applications

Mohammad Ali[1]*, Muhammad Khalid[1,2,3] and Mohammad Ali Abido[1,3]

[1]*K.A. CARE Energy Research & Innovation Center at King Fahd University of Petroleum & Minerals (KFUPM), Dahran, Saudi Arabia*

[2]*Electrical Engineering Department, King Fahd University of Petroleum & Minerals (KFUPM), Dahran, Saudi Arabiaa*

[3]*Interdisciplinary Research Center for Sustainable Energy Systems, KFUPM, Dahran, Saudi Arabia*

Abstract

Power electronic converters facilitate the integration of renewable energy sources into both households and the electrical grid. Researchers are exploring the use of multilevel inverters, given their advantages, particularly in low-power applications. This study focuses on re-investigating recent nine-level switched-capacitor multilevel inverters for their 11-level operation, aiming to enhance their reliability, especially in high-temperature conditions. This expanded operational capability allows for the generation of more voltage levels compared to previous technology and reduces inrush currents in the capacitors. However, when operating at nine levels, there is an unavoidable issue of high inrush currents during steady-state operation, which result in the overheating of capacitors due to the inherent ESR (electrolytic series resistance). This, in turn, leads to early evaporation of the electrolyte, causing variations in capacitance and an ESR increase with time. In order to mitigate these undesirable inrush currents, researchers often rely on adding a low-value inductance in the capacitor's charging circuit. Nevertheless, this solution comes with the trade-off of increased losses and a higher ringing effect of the switches In this chapter, a novel inverter modulation technique is employed to generate 11 voltage levels of the inverter's output. This modulation utilizes the unused states within the inverter operation theory to ensure the

**Corresponding author*: mohd.ali@ieee.org

Salman Ahmad, Farhad Ilahi Bakhsh, and P. Sanjeevikumar (eds.) Multilevel Converters, (75–98)

capacitor's consistent charging and discharging. Consequently, this approach significantly suppresses the steady-state inrush currents by many times and leads to enhancement of the inverter's reliability and lifespan. These improvements make these inverters particularly appropriate for solar PV applications, even in challenging weather. Furthermore, this also leads in lower switch voltage and conduction stresses. To validate the effectiveness of the enhanced topology and modulation strategy, comprehensive experiments are conducted in simulation, hardware in loop and on experimental prototype.

Keywords: Multilevel inverter, solar-PV, switched capacitors, low stress, clean energy, reliable operation

4.1 Introduction

Buildings with high energy efficiency are powered by renewable energy sources (RES) like the sun and wind. Power electronic converters are used to balance the production and consumption of energy, according to [7]. Furthermore, according to [17], improved power quality becomes a crucial component of modern distributed energy systems. In order to achieve efficient power conversion, inverters are essential. Different multilevel inverter (MLI) topologies are being investigated for low power applications. Currently, due to their advantages, multilevel converters are more focused on medium voltage high power applications [21, 23, 30]. The other innate characteristics include producing waveforms that are almost sinusoidal, shrinking the size of the filter, and decreasing harmonics. Nowadays, SCMLIs (switched-capacitor multilevel inverters) are researched in order to address applications like microinverters and fetch the gains of the multilevel inverter properties. The SCMLIs provide the advantages of multilevel inverters while utilizing fewer parts [1, 8, 18, 20].

Switched-source (SSMLI), switched-diode (SDMLI), and SCMLI are types of multilevel inverters with fewer components. For further information on the connected issues, see [11, 31, 33], respectively, and the combination can also be found in [32]. More than one DC source is employed in switched-source inverters and helps to level creation at the inverter's output. Diodes control the direction of current flow in the load in a switched-diode inverter. Switched-capacitor designs, on the other hand, use numerous capacitors in conjunction with one DC source to achieve an appropriate output level. Switched-capacitor multilevel inverters (SCMLI) provide various advantages, including less reliance on semiconductor components like IGBTs and diodes, as well as simpler control. One significant advantage is the utilization of a single DC source. These structures can even boost voltage, as shown in [5, 13, 16, 24, 35]. The notion of single-stage voltage boosting is currently gaining popularity in the field of converters [36] and inverters [6, 10].

The T-type inverters like S³CM, CSCMLI, C³-SCMLI, and UXE inverter have all been introduced in the literature. Lee presented the first topology in [14], which used 12 switches and two capacitors to enable a nine-level output with double voltage boosting relative to the DC source. Sathik *et al.* introduced the CSCMLI topology in [26], obtaining a switch reduction. Sathik *et al.* suggested the third topology in [27], which used only 10 switches, two diodes, and two capacitors. Ali *et al.* in [3] provide a more current architecture that functions as a 13-level switched-source inverter but can be tuned for nine-level operation using the same modulation method as the earlier topologies. It should be noted, however, that all of these nine-level modulation schemes rely on parallel charging of the capacitors. This technique results in significant capacitor inrush currents, which can cause premature capacitor degradation and increased heating owing to equivalent series resistance (ESR). Furthermore, the switches in these inverters subject to increased current and voltage strain, resulting in source current spikes and heightened junction temperatures, necessitating the use of larger heat sinks.

This chapter digs into the previously described inverters' 11-level operation and illustrates its advantages above nine-level operation, devising it a substantially superior and reliable converter. Figure 4.1 depicts an 11-level operation arrangement in solar-PV applications. The concerns with capacitor inrush currents and their associated impacts that were discovered in the nine-level operation have been properly addressed in this operation. With the exception of the one outlined in [27], which replaces diodes with IGBT switches, the topologies for the 11-level have the same structural arrangement as their nine-level counterparts. The inverter modulation is intended to charge the switched capacitors *via* load using the redundant states. The advantage of 11-level operation is lowering of capacitor steady-state

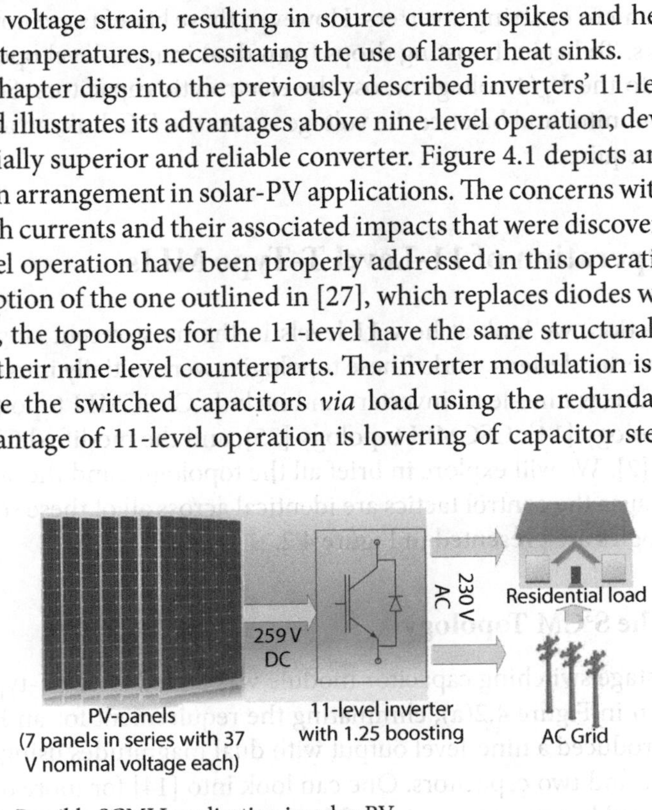

Figure 4.1 Possible SCMLI application in solar-PV.

inrush currents, which results in increased lifespan and improved inverter dependability. Furthermore, the 11-level process has the following advantages over the nine-level modulation:

- Achieves a considerable amount of reduction in capacitor ripple voltages while maintaining the same capacitance, allowing for the utilization of smaller capacitors
- Substantially alleviates switch current stresses, enhancing switch performance with minimal conduction losses
- Lowers the required switch blocking voltages, enabling the use of switches rated for lower voltage levels
- Decreases source current spikes by a large factor
- Enables dependable switch operation in high-temperature conditions by reducing junction temperatures, consequently resulting in small-sized heat sinks

The aforementioned benefits make the inverters suitable for use in solar-PV micro or string inverters. However, these benefits are not without drawbacks. To begin, boosting drops from dual in nine-level operation to 1.25 due to the $V_{dc}/4$ voltage across the electrolytic capacitors. Second, in order to monitor and balance the voltage across the DC-link, an additional sensor is required.

4.2 Operation of 11-Level T-Type MLIs

In this section, we look at how 11 levels in the most recent single-phase T-type inverter designs work. These topologies were initially introduced in the literature as nine-level inverters and include C^3-SCMLI topology [27], S^3CM topology [14], CSCMLI topology [25], and the modified C^3-SCMLI topology [2]. We will explore in brief all the topologies and the last one in detail because the control tactics are identical across all of these setups. All the topologies are presented in Figure 4.2.

4.2.1 The S^3CM Topology

A single-stage switching capacitor module was added to the T-type structure shown in Figure 4.2(a), eliminating the requirement for an H-bridge stage. It produced a nine-level output with dual magnitudes using just one DC source and two capacitors. One can look into [14] for more details on the nine-level boosting operation of this circuit.

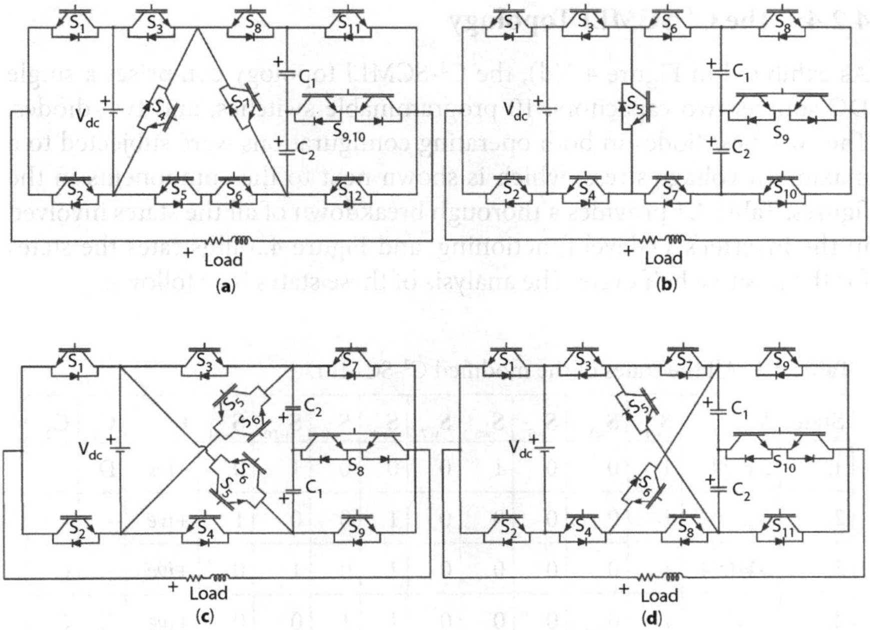

Figure 4.2 Multilevel inverter topologies: (a) S³CM [14], (b) CSCMLI [26], (c) UXE inverter [3, 4], and (d) modified C³-SCMLI topology [2].

4.2.2 CSCMLI Topology

Another example is the CSCMLI topology and described in [26]. In contrast to the S3CM architecture stated before, this configuration interestingly uses one less switch. One can refer to [26] for a thorough analysis of this circuit which is nine-level dual boosting operation. Figure 4.2(b) shows the structure.

4.2.3 The UXE Topology

In [3], a unique layout with a criss-cross topology that makes use of 12 switches and three power sources was introduced. Three DC sources set at V_dc, $V_dc/4$, and $V_dc/4$ were utilized in this design's switched-source inverters, which were employed to operate at 13 levels. As detailed in [4], which concentrated on this topology's nine-level dual boosting functionality, its further analysis and exploration include the incorporation of capacitors. Figure 4.2(c) shows the circuit arrangement.

4.2.4 The C^3-SCMLI Topology

As exhibited in Figure 4.2(d), the C^3-SCMLI topology comprises a single DC source, two capacitors, 10 programmable switches, and two diodes. The switches/diodes in both operating configurations were subjected to a maximum voltage stress, which is shown next to the components in the figures. Table 4.1 provides a thorough breakdown of all the states involved in the inverter's 11-level functioning, and Figure 4.3 illustrates the states for the positive half cycle. The analysis of these states is as follows:

Table 4.1 All the states of the modified C^3-SCMLI.

State	V_o	S_{w1}	S_{w3}	S_{w5}	S_{w6}	S_{w7}	S_{w8}	S_{w9}	S_{w10}	S_{w11}	I_o	C_1	C_2
1.	$5V_{dc}/4$	1	0	0	1	0	0	0	1	0	+*ive*	D	-
2.	V_{dc}	1	0	0	0	0	1	0	0	1	+*ive*	-	-
3.	$3Vdc/4$	1	0	0	0	0	1	0	1	0	+*ive*	-	C
4.	$V_{dc}/2$	1	0	0	0	0	1	1	0	0	+*ive*	C	C
5.	$V_{dc}/2$	1	1	0	0	1	0	0	0	1	+*ive*	D	D
4'.	$V_{dc}/2$	1	0	0	0	0	1	1	0	0	−*ive*	D	D
5'.	$V_{dc}/2$	1	1	0	0	1	0	0	0	1	−*ivc*	C	C
6.	$V_{dc}/4$	1	1	0	0	1	0	0	1	0	+*ive*	D	-
7.	0	1	1	1	0	0	0	0	0	1	+*ive*	-	-
8.	0	0	0	0	1	0	0	1	0	0	+*ive*	-	-
9.	$-V_{dc}/4$	0	0	0	0	0	1	0	1	0	+*ive*	-	C
10.	$-V_{dc}/2$	0	0	0	0	0	1	1	0	0	+*ive*	C	C
11.	$-V_{dc}/2$	0	1	0	0	1	0	0	0	1	+*ive*	D	D
10'.	$-V_{dc}/2$	0	0	0	0	0	1	1	0	0	−*ive*	D	D
11'.	$-V_{dc}/2$	0	1	0	0	1	0	0	0	1	−*ive*	C	C
12.	$-3V_{dc}/4$	0	1	0	0	1	0	0	1	0	+*ive*	D	-
13.	$-V_{dc}$	0	1	0	0	1	0	1	0	0	+*ive*	-	-
14.	$-5V_{dc}/4$	0	1	1	0	0	0	0	1	0	+*ive*	-	C

Table 4.1 at positions 6 and 6'.

Figure 4.2(d) shows the 11-level topology under discussion, which consists of a single DC source, two capacitors, and 12 IGBT switches. This arrangement is a modified variation of the topology described in [27]. The figures show the highest voltage stress that the switches and diodes could withstand when operating in either mode. The 11-level operation of the inverter is covered in Table 4.1, and Figure 4.3 shows the equivalent states for the positive cycle. We then give a succinct explanation of the positive states.

4.2.4.1 $0.25V_{dc}$

When the voltage reaches $0.25V_{dc}$, capacitor C_1 will discharge (illustrated in Figure 4.3(c)) and recharge with the flow of opposite direction current. Conversely, the charge on capacitor C_2 at this voltage level, $0.25V_{dc}$, will remain unchanged. These conditions are detailed in

4.2.4.2 $0.5V_{dc}$

The $0.5V_{dc}$ level contains redundant states, presenting two conditions shown in Figure 4.3(c and d). When there is a positive current flow, if the combined voltage of capacitors C_1 and C_2 exceeds $V_{dc}/2$, then the state shown in Figure 4.3(c) is chosen; if it is less than $V_{dc}/2$, then Figure 4.3(d) is selected. Conversely, with a negative current flow, if the capacitor link voltage is higher than $V_{dc}/2$, Figure 4.3(d) is chosen, and if it is less than $V_{dc}/2$, then Figure 4.3(c) is the preferred state.

4.2.4.3 $0.75V_{dc}$

In the case of the $0.75V_{dc}$ state, the circuit path depicted in Figure 4.3(e) is selected. The behavior of capacitor C_2 will be influenced by the current's direction, as indicated in Table 4.1 at positions 3 and 3'.

4.2.4.4 V_{dc}

Under these circumstances, as illustrated in Figure 4.3(f), the voltage across the capacitor will remain unaltered regardless of the current's direction. This observation is reflected in Table 4.1 at positions 2 and 2'.

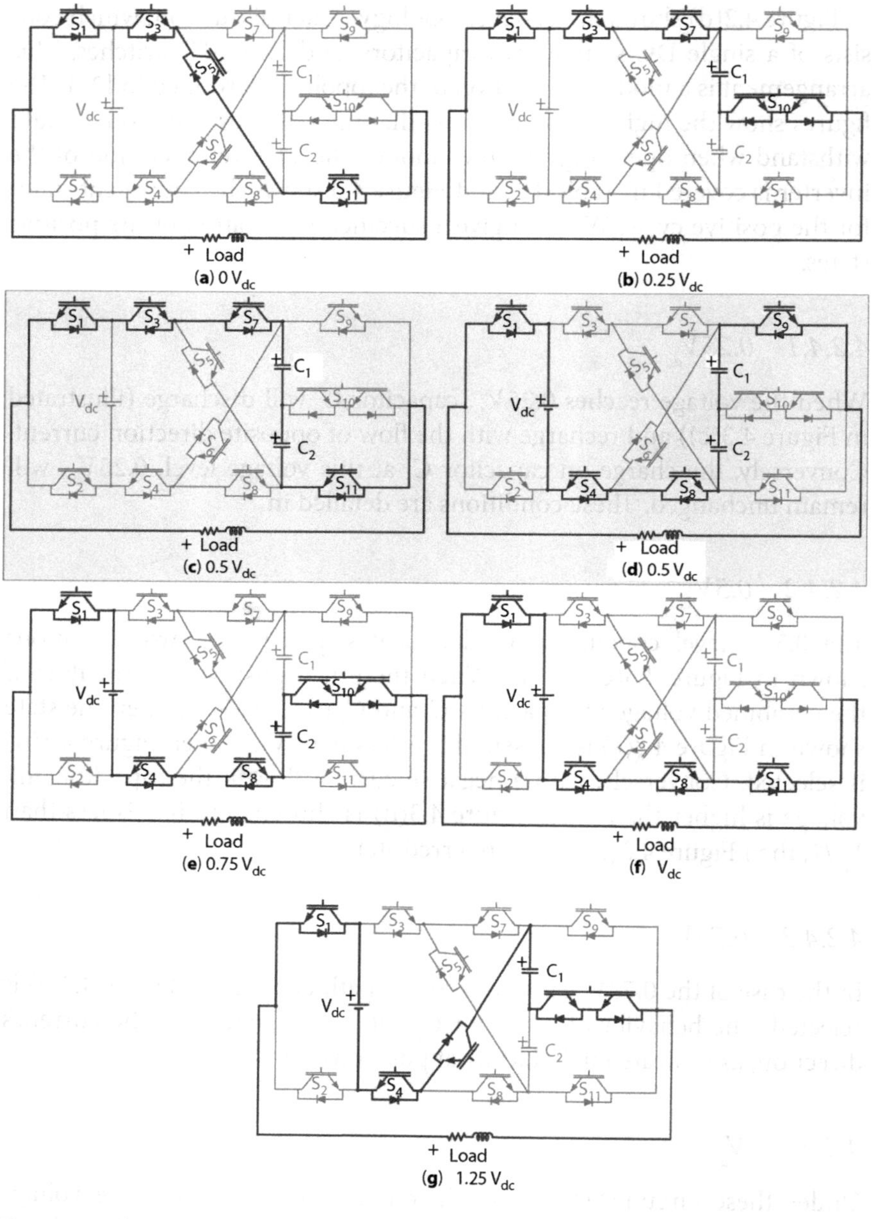

Figure 4.3 Eleven-level operation's positive states [2].

4.2.4.5 $1.25V_{dc}$

At the $1.25V_{dc}$ level, as depicted in Figure 4.3(g), capacitor C_1 will discharge when there is positive current and charge when there is negative current, as indicated in Figure 4.3(g). These behaviors are reflected in Table 4.1 at positions 1 and 1'.

4.3 Voltage Balance Algorithm of the Switched Capacitors

This section delves into the modulation and algorithm for maintaining capacitor voltage balance in the context of operating T-type inverters with 11 levels. The discussion focuses on the C³-SCMLI topology for illustrative purposes, but it is important to note that the principles discussed can be applied to all the considered inverters.

4.3.1 Modulation Strategy

The inverter modulation is achieved using the level-shifted sine-PWM technique, where the reference and carrier signals are as follows:

$$v_{ref} = 1.25M\sin(\omega_{ref}t) \tag{4.1}$$

In this context, v_{ref} represents the desired voltage reference, M represents the reference voltage's modulation index, and ω_{ref} corresponds to the angular frequency, equating to $2nf_{ref}$

$$u_n \begin{cases} \frac{1}{4}(n-1)+\frac{1}{2}(t-(k-1)T_s) \text{ for } (k-1)T_s \le t \le \frac{2k-1}{2}T_s \\ \frac{1}{4}n+\frac{1}{2}\left(t-\frac{2k-1}{2}T_s\right)\frac{1}{T_s} \text{ for } \frac{2k-1}{2}T_s < t < kT_s \end{cases} \tag{4.2}$$

The variable k denotes the number of samples within one cycle of the fundamental frequency, where $k = 1$ corresponds to the first cycle of the fundamental frequency (f_{ref}). For the 11-level operation, 10 different carriers can be generated by substituting the values of n with ±1, ±2, ±3, ±4, and ±5 into (2), as illustrated in Figure 4.4.

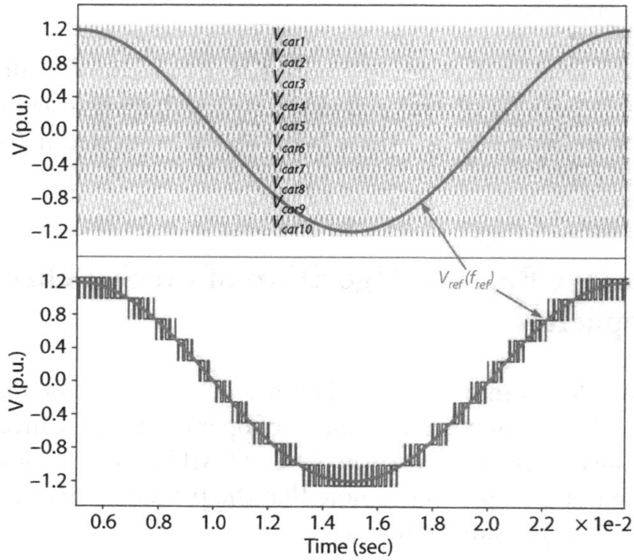

Figure 4.4 Level-shifted sine-PWM at f_s=5 kHz.

4.3.2 Voltage Balancing Algorithm

The inverter states are correctly fetched at specific voltage levels to ensure the proper operation of the inverter by using an algorithm designed based on load current polarity. The redundancies at voltage levels of $+V_{dc}/2/-V_{dc}/2$ are used to maintain the capacitor bank voltage at $V_{dc}/2$ while automatically keeping the voltages at $V_{dc}/4$.

According to Table 4.1, state 4 is chosen to charge the capacitor bank if the state is $+V_{dc}/2$, the current direction is positive, and the measured voltage is below $V_{dc}/2$. In contrast, state 5 is the chosen capacitors' discharge through load if the voltage is greater than $V_{dc}/2$, as shown in Figure 4.5. With the same reasoning, with negative current flow, states 5' and 4' from Table 4.1 are chosen for carefully controlling the capacitors' charging and discharging.

4.4 Structural and Cost Comparison

In this section, we conduct a comparative analysis of the 11-level inverter we have presented with the recent 11-level inverters. The basis for this comparison includes factors such as the number of power sources, switches,

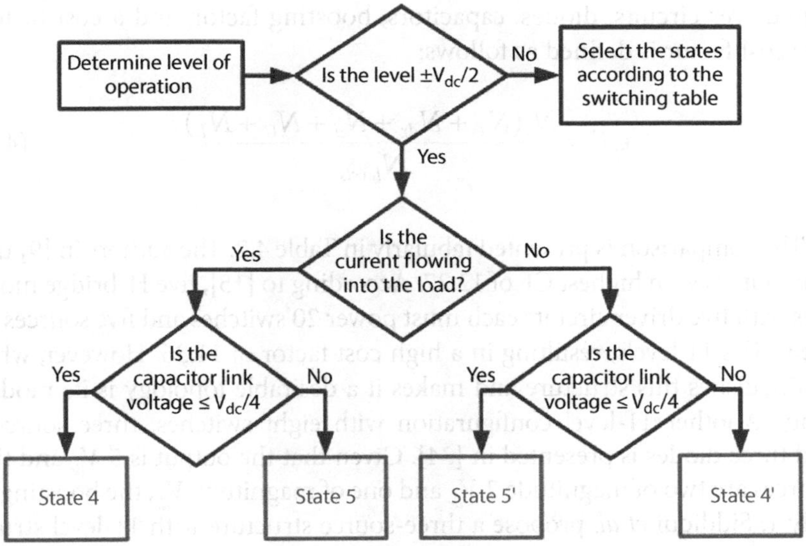

Figure 4.5 Capacitor voltage balancing algorithm.

Table 4.2 Comparison of the presented 11-level inverter with recent topologies.

MLI	N_{level}	N_s	N_T	N_D	N_{dr}	***No***	***BC***	***CF***
[9]	11	5	12	0	11	0	1	12.27
CHB [15]	11	5	20	0	5	0	1	11.36
[34]	11	3	8	3	6	0	1	4.64
[29]	11	3	8	0	7	0	1	4.09
[12]	11	1	14	0	8	4	5	2.36
[22]	11	1	9	0	7	5	1	1.91
[28]	11	1	8	0	8	5	1	1.91
P9	9	1	12	0	11	2	2	2.77
P11	11	1	12	0	11	2	1.25	2.27

CF, cost factor; N_{level}, number of levels; N_S, number of sources; N_T, number of switches; N_D, number of diodes; N_C, number of capacitors; BC, boosting capability.

gate driver circuits, diodes, capacitors, boosting factor, and a cost factor. The cost factor is defined as follows:

$$CF = \frac{N_s(N_T + N_{dr} + N_d + N_C + N_L)}{N_{levels}} \tag{4.3}$$

The comparison is presented tabularly in Table 4.2. The authors in [9] use five sources with highest CF of 12.27. According to [15], five H-bridge modules with five driver circuits each must power 20 switches and five sources in the CHB's 11 levels, resulting in a high cost factor of 11.36. However, what distinguishes this structure and makes it a desirable topology is its modularity. Another 11-level configuration with eight switches, three sources, and three diodes is presented in [34]. Given that the output is 5 V_{dc} and the sources are two of magnitude 2 V_{dc} and one of magnitude V_{dc}, the boosting is only 1. Siddiqui *et al.* propose a three-source structure with 11-level structure with eight switches and seven gate driver circuits in his citation [29]. A single-source MLI with 14 switches that are driven by eight driver circuits and four capacitors is introduced by Karimi *et al.* in [12]. Although this design may increase the output boosting to five, it uses more devices than the discussed topology. In [22], a single-source structure with nine switches and seven driving circuits is described. However, the circuit needs five capacitors to operate at 11 levels. In [28], a one-source circuit, with five capacitors and eight switches that are controlled by eight driver circuits, is covered. Unlike these topologies, the 11-level structure that is being presented uses 12 switches, two capacitors, and a single source. As a result, the low cost factor in 11-level operation is improved to 2.27 from 2.72 in nine-level operation.

4.5 Components Analysis Under Steady State

The performance of the switch and the circuit analysis are shown in this section. The causes of the capacitor's low inrush currents during 11-level operation and high inrush currents during nine-level operation are explained. The performance of the switch at lower voltage loads is also given.

4.5.1 Analysis of the Circuit

Figure 4.6(a) illustrates how capacitors charge up while the inverter is operating in parallel with the DC source. [14, 26, 27] provide information on

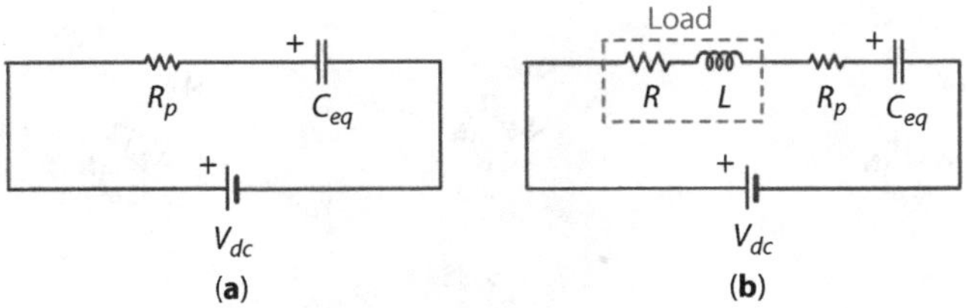

Figure 4.6 Charging paths for (a) charging through the DC-source and (b) charging using the redundant states and through load.

various T-type nine-level topologies. The resistance R_p includes the wires' and capacitors' ESR. Wire resistance and inductances are hardly noticeable. The circuit equivalent capacitance is calculated as follows:

$$C_{eq} = \frac{C_1 C_2}{C_1 + C_2} \tag{4.4}$$

When the charging condition begins, the capacitor voltage in the circuit is stated as:

$$v_{C_{eq}}(t) = V_{dc} - [V_{dc} - V_{Ceq}]e^{\frac{-t}{R_p C_{eq}}} \tag{4.5}$$

Thus, the following can be used to express the current:

$$i_{C_{eq}}(t)\frac{V_{dc} - V_{C_{eq}}}{R_p C_{eq}} e^{\frac{-t}{R_p C_{eq}}} \tag{4.6}$$

It will provide high values for low values of t and significant inrush currents as a result.

The capacitors in the MLIs' 11-level operation will be load-charged in the second circuit (Figure 4.6(b)). Therefore, it is simple to determine that by increasing the load's resistance and inductance, the time constant would grow and the ability to modify the current will be limited. Thus, when operating at 11 levels, the inrush current is significantly reduced. The following issues also appear in the circuit as a result of the reduced capacitor inrush currents:

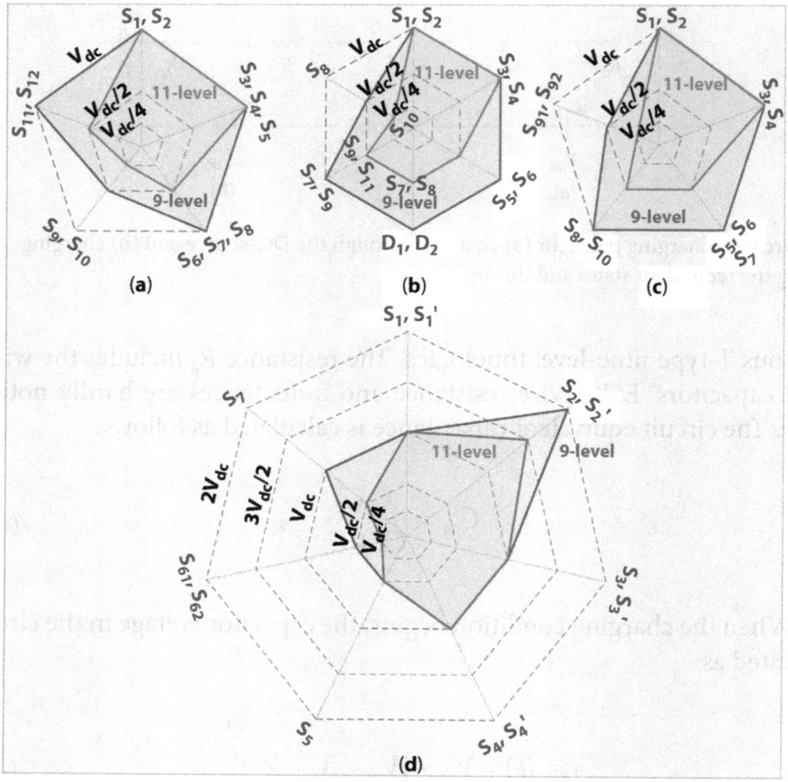

Figure 4.7 Blocking voltages of (a) S³CM, (b) C³-SCMLI, (c) CSCMLI, and (d) UXE inverters.

i. Reduced switch current stresses by many times such that the switch operates with the fewest possible conduction losses
ii. Reduced DC source current peaks, making it possible for it to be a solar-PV source
iii. Low device temperatures brought on by lower i^2R losses enable switches to operate reliably in high-temperature environments

These elements will be demonstrated and covered in the results section that follows. The blocking voltage of the switches/diodes of various SCMLIs shown in Figure 4.2, in their 9-level and 11-level operations, are shown in Figure 4.7.

4.6 HIL Results

The MLIs presented in the preceding discussion were comparatively analyzed, focusing on various aspects. One specific C³-SCMLI topology was chosen for result validation. The controller architecture was created on the

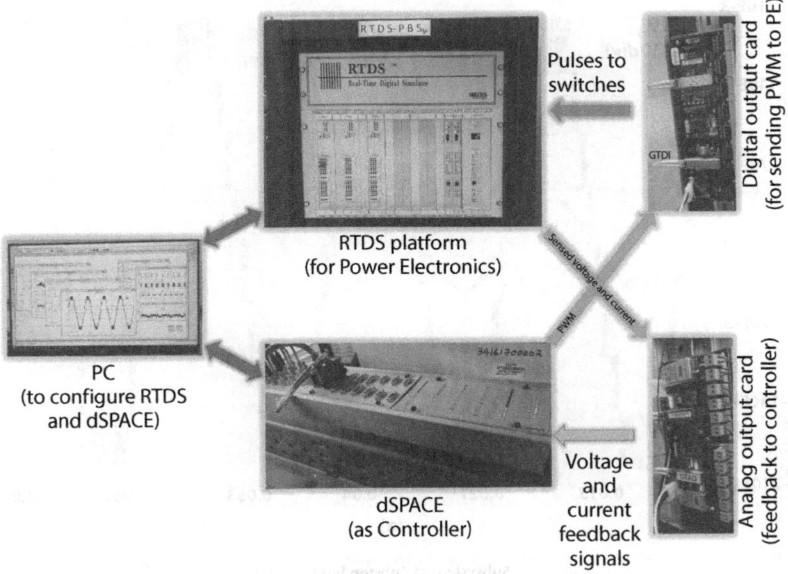

Figure 4.8 RTDS-based HIL setup for the modified C³-SCMLI.

Table 4.3 Parameters of the HIL and experimental setup.

Parameter	HIL	Experimental
V_{dc}	100 V	48 V
V_{c1}	25 V	12.5 V
V_{c2}	25 V	12.5 V
C_1, C_2	4,700 μF	4,700 μF, 63 V
Switch	IGBT	IGBT
RL load	50 Ω, 40 mH	40 Ω, 7.5 mH
Controller	dSPACE 1103	dSPACE 1103

dSPACE platform, while the power electronics circuits were created on the RTDS/RSCAD, allowing for real-time circuit validation, as shown in Figure 4.8. The RTDS signals are processed by DSPACE, which then produces PWM signals that are transmitted back to RTDS as control actions.

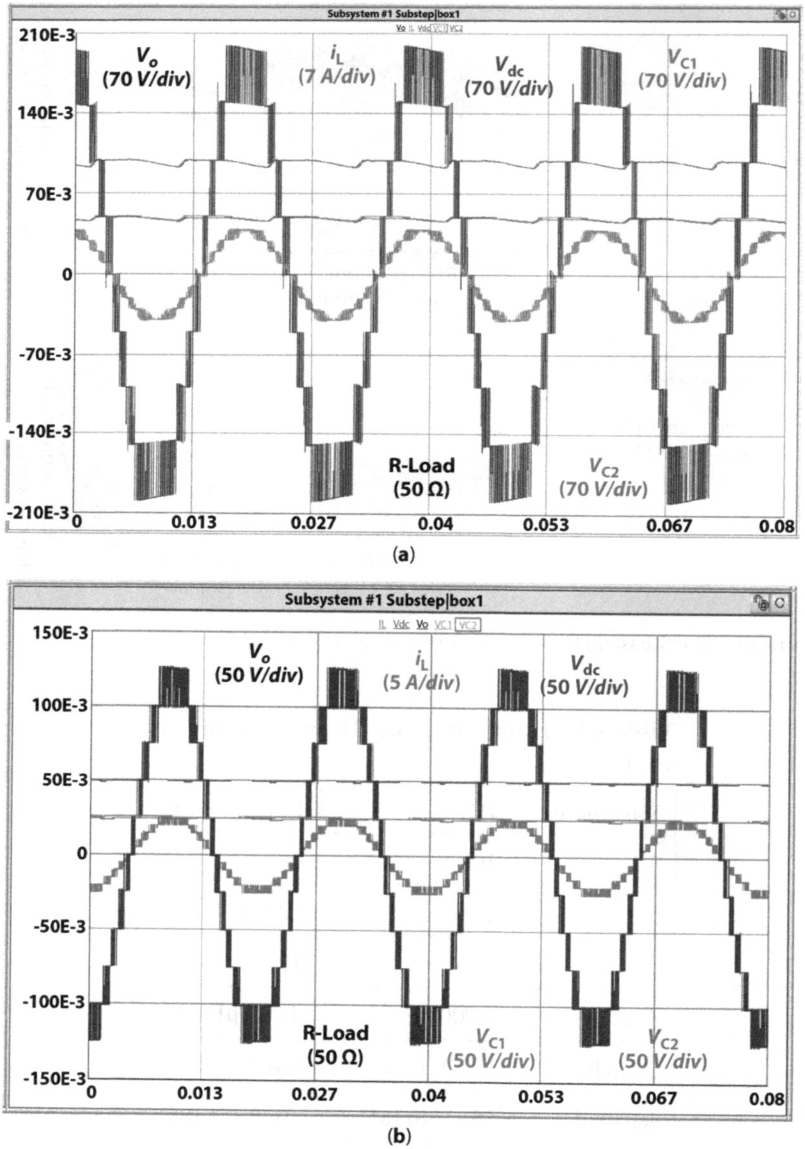

Figure 4.9 Inverter results for (a) nine-level and (b) 11-level operation, for R-load.

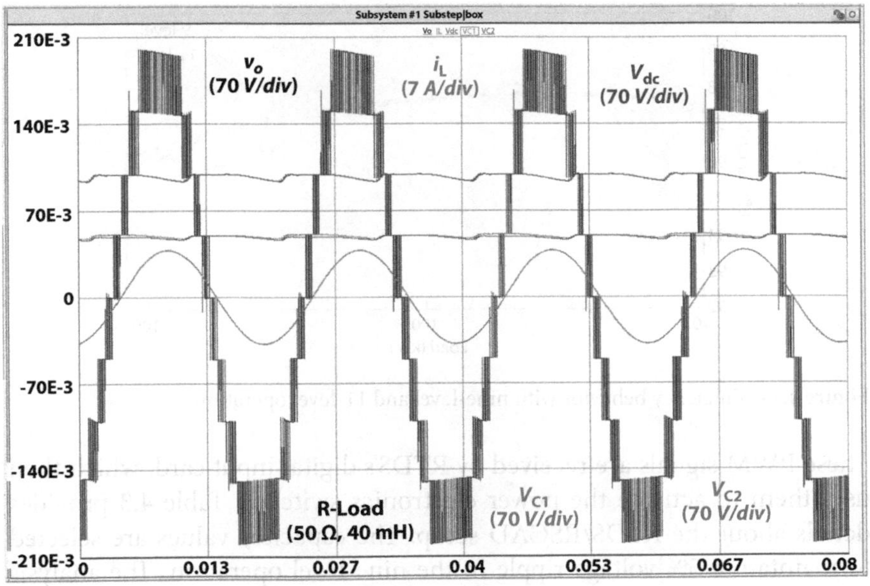

(a)

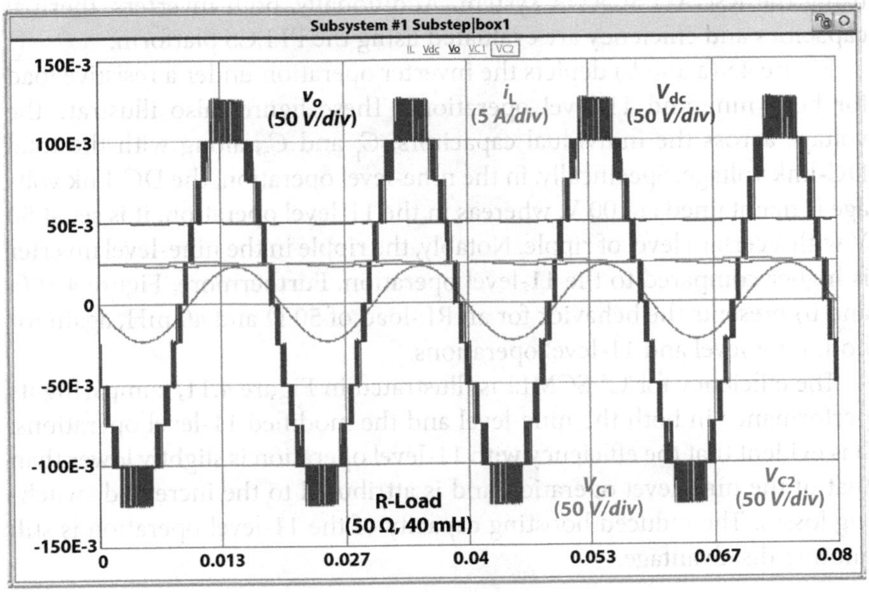

(b)

Figure 4.10 Inverter results for (a) nine-level and (b) 11-level operation, for RL-load.

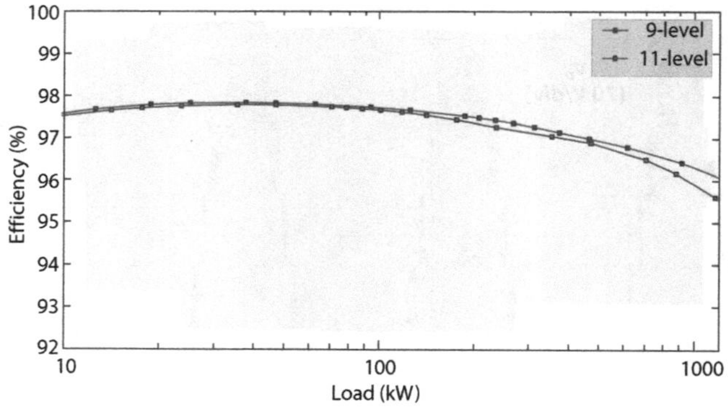

Figure 4.11 Efficiency behavior with nine-level and 11-level operation.

These PWM signals are received by RTDS's digital input card, which then uses them to activate the power electronics switches. Table 4.3 provides details about the RTDS/RSCAD setup. The capacitor values are selected to maintain a 5% voltage ripple in the nine-level operation. The analysis is conducted under steady-state conditions, and the results are captured using the RSCAD SCADA system. Additionally, both inverters' thermal capacities and efficiency are evaluated using the PLECS platform.

Figure 4.9(a and b) depicts the inverter operation under a resistive load for both nine and 11-level operations. These figures also illustrate the voltage across the individual capacitors, C_1 and C_2, along with the total DC-link voltage. Specifically, in the nine-level operation, the DC-link voltage is maintained at 100 V, whereas in the 11-level operation, it is set at 50 V with a certain level of ripple. Notably, the ripple in the nine-level inverter is higher compared to the 11-level operation. Furthermore, Figure 4.10(a and b) presents the behavior for an RL-load of 50 Ω and 40 mH, again for both nine-level and 11-level operations.

The efficiency for C^3-SCMLI is illustrated in Figure 4.11, comparing its performance in both the nine-level and the modified 11-level operations. It is evident that the efficiency with 11-level operation is slightly lower than that of the nine-level operation and is attributed to the increased switching losses. The reduced boosting capacity of the 11-level operation is still another disadvantage.

4.7 Experimental Validation

This section focuses on the experimental validation of both the nine-level and the modified 11-level inverters. The experimental prototype

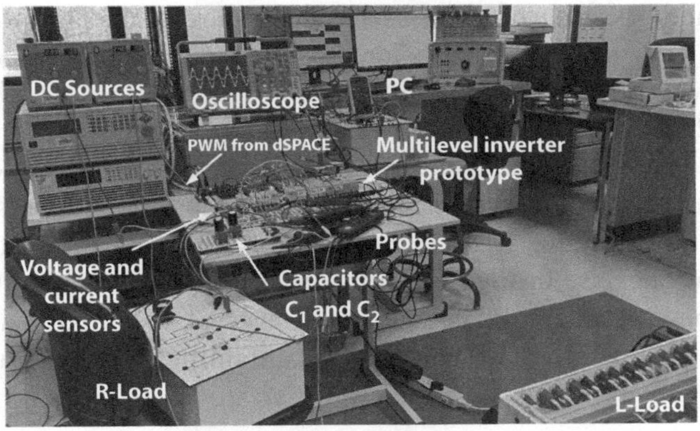

Figure 4.12 Experimental setup of the 11-level MLI.

is showcased in Figure 4.12. To create the inverter module, IGBT Plugy modules with power breadboards rated at 800 V and 20 A, obtained from PEModule® [19], were employed. Detailed specifications of the setup can be found in Table 4.3. The generation of PWM signals for the inverter was accomplished using a dSPACE 1103 unit. It is worth noting that the IGBT modules from PEModule® include integrated gate driver circuits.

The experimental results are presented in Figure 4.13. Figure 4.13(a) displays the results for the nine-level inverter operation with R-load.

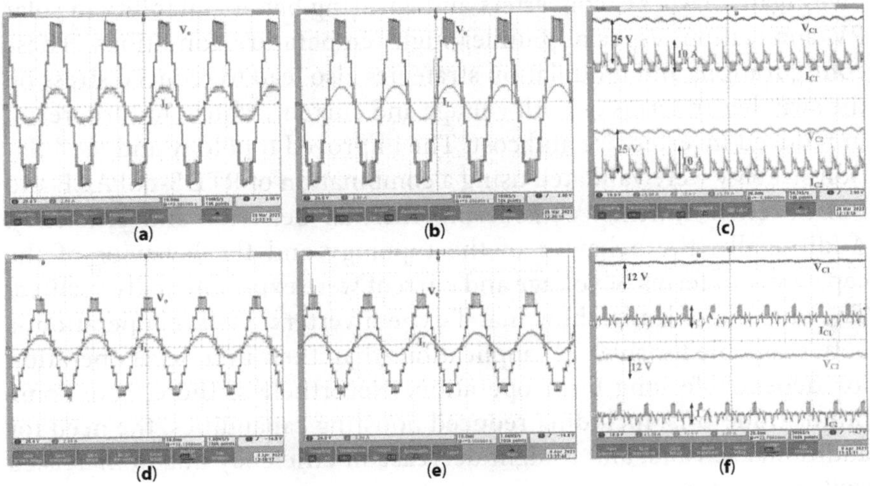

Figure 4.13 Nine-level and 11-level results of the discussed MLIs with (a) R-load, (b) RL-load, (c) capacitor voltages and currents for nine-level operation and (d) R-load, (e) RL-load, and (f) capacitor voltages and currents for 11-level operation.

The voltage boosting of 2 is observable. Figure 4.13(c) showcases the capacitor voltages and currents for capacitors C_1 and C_2. The results for the 11-level configuration are presented in Figure 4.13(d–f). The boosting of 1.25 in relation to the 48 V DC source is prominently visible. As shown in Figure 4.13(c and f), it is evident that the voltage ripple is lower for the same capacitance value. Similarly, with capacitor currents between the nine-level (Figure 4.13(c)) and 11-level (Figure 4.13(f)), there is a significant decrease in the capacitor inrush currents by at least a factor of 10. These results validate the advantages of the modified 11-level topology, particularly with respect to maintaining capacitor parameters at low values.

4.8 Conclusion

The 11-level functioning of a nine-level switched-capacitor MLI was demonstrated in this chapter. This extended operation allowed the inverters to generate more voltage levels while significantly reducing capacitor inrush currents. Achieving 11-level inverter operation relied on leveraging redundant states and ensuring that the capacitors were consistently charging and discharging through the load. As was observed with the adoption of 11-level operation, inrush current peaks decreased by a factor of 10, enhancing the reliability and lifespan of the investigated DC–AC converters and ensuring better suitability in solar PV application, especially under high-temperature conditions. These configurations and modulation strategies also lead to reduced stress on the switches in terms of both voltage and current, which, in turn, result in reduced inverter size and cost. The improved topology and modulation strategy were validated using a combination of RTDS-dSPACE and PLECS environments, considering thermal behavior and efficiency. Additionally, the behavior of the topology and the behavior of the capacitors in terms of voltage and current were experimentally verified. The outcomes clearly show that T-type inverters' 11-level operation is better suitable for solar PV applications than their nine-level operation for dependable long-term operation. Nevertheless, there were some minor trade-offs, including reduced boosting capabilities, the need for additional sensors, and a slight decrease in efficiency due to increased switching losses.

References

1. Ahmed, M., Sheir, A., Orabi, M., 2018. Asymmetric cascaded half-bridge multi-level inverter without polarity changer. *Alex. Eng. J.* 57, 2415–2426. doi:https://doi.org/10.1016/j.aej.2017.08.018.
2. Ali, M., Hossain, I., Al-ismail, F.S., Abido, M.A., Khalid, M., 2023a. Capacitor ripple reduction in T-type multilevel inverter operation for solar PV-applications. *Alex. Eng. J.* 77, 613–624. URL: https://doi.org/10.1016/j.aej.2023.06.088, doi:10.1016/j.aej.2023.06.088.
3. Ali, M., Tariq, M., Lodi, K.A., 2020. Selective Harmonic Elimination of a New UXE-Type Inverter based on Ant-Colony Optimization, in: *PEDES*, pp. 1–6.
4. Ali, M., Tariq, M., Sarwar, A., Alamri, B., 2022a. 13- , 11- and 9-Level Boosted Operation of a Single-Source Asymmetrical Inverter with Hybrid PWM Scheme. *IEEE Trans. Ind. Electron.* 69, 12817–12828. doi:10.1109/TIE.2022.3144576.
5. Ali, M., Tayyab, M., Sarwar, A., Khalid, M., 2022b. A Single-Source Multilevel Inverter with Boosted Output Voltage for Solar PV Applications, in: *IEEE GlobConet*, pp. 1–6.
6. Ali, M., Tayyab, M., Sarwar, A., Khalid, M., 2023b. A Low Switch Count 13-Level Switched-Capacitor Inverter with Hexad Voltage-Boosting for Renewable Energy Integration. *IEEE Access* 11, 36300–36308. doi:10.1109/ACCESS.2023.3265467.
7. Ghani, Z.A., Hannan, M.A., Mohamed, A., 2013. Simulation model linked PV inverter implementation utilizing dSPACE DS1104 controller. *Energy Build.* 57, 65–73. URL: http://dx.doi.org/10.1016/j.enbuild.2012.10.045, doi:10.1016/j.enbuild.2012.10.045.
8. Gupta, K.K., Ranjan, A., Bhatnagar, P., Sahu, L.K., Jain, S., 2016. Multilevel Inverter Topologies With Reduced Device Count: A Review. *IEEE Trans. Power Electron.* 31, 135–151. doi:10.1109/TPEL.2015.2405012.
9. Hassan, A.M.M., Yang, X., Ali, A.I.M., 2019. A Study of Level- Shifted PWM Single-phase 11-Level Multilevel Inverter, in: *21st International Middle East Power Systems Conference (MEPCON)*, IEEE. pp. 170–176.
10. Hassan, M., Ali, M., Tayyab, M., Sarwar, A., Tariq, M., Abido, M.A., Khalid, M., 2023. Self-Balanced Quadruple-Boost Nine-Level Switched-Capacitor Inverter for Solar PV Applications. *Arab. J. Sci. Eng.*
11. Hosseinzadeh, M.A., Sarebanzadeh, M., Rivera, M., Babaei, E., Wheeler, P., 2020. A Reduced Single-Phase Switched-Diode Cascaded Multilevel Inverter. *IEEE Trans. Emerg. Sel. Topics Power Electron.* doi:10.1109/jestpe.2020.3010793.
12. Karimi, M., Kargar, P., Varesi, K., 2019. A Novel High-Gain Switched-Capacitor Based 11- Level Inverter Topology, in: *International Power System Conference*, pp. 404–409. doi:10.1109/PSC49016.2019.9081558.

13. Khan, S.A., Upadhyay, D., Ali, M., Tariq, M., Sarwar, A., Chakrabortty, R.K., Ryan, M.J., Alamri, B., Alahmadi, A., 2021. M-Type and CD-Type Carrier Based PWM Methods and BAT Algorithm-Based SHE and SHM for Compact Nine-Level Switched Capacitor Inveter. *IEEE Access*, 1–19.
14. Lee, S.S., 2018. Single-Stage Switched-Capacitor Module (S3CM) Topology for Cascaded Multilevel Inverter. *IEEE Trans. Power Electron.* 33, 8204–8207. doi:10.1109/TPEL.2018.2805685.
15. Malinowski, M., Gopakumar, K., Rodriguez, J., Perez, M.A., 2010. A survey on cascaded multilevel inverters. *IEEE Trans. Ind. Electron.* 57, 2197–2206. doi:10.1109/TIE.2009.2030767.
16. Murshid, S., Tayyab, M., Sarwar, A., Tariq, M., Al-Durra, A., Tomar, A., 2022. Self-Balanced Twenty Five Level Switched Capacitor Multi-level Inverter with Reduced Switch Count And Voltage Boosting Capability. *IEEE Trans. Ind. Appl.* doi:10.1109/TIA.2021.3136802.
17. Nadeem, T.B., Siddiqui, M., Khalid, M., Asif, M., 2023. Distributed energy systems: A review of classification, technologies, applications, and policies:. *Energy Strategy Rev.* 48, 101096. URL: https://doi.org/10.1016/j.esr.2023.101096, doi:10.1016/j.esr.2023.101096.
18. Omer, P., Kumar, J., Surjan, B.S., 2020. A Review on Reduced Switch Count Multilevel Inverter Topologies. *IEEE Access* 8, 22281–22302. doi:10.1109/ACCESS.2020.2969551.
19 PEModule, 2023. *PLUGY Power Electronics Modules*. URL: https://www.pemodule.com/r-d-products. 2023-04-09.
20. Prabaharan, N., Palanisamy, K., 2017. A comprehensive review on reduced switch multilevel inverter topologies, modulation techniques and applications. *Renew. Sust. Energ. Rev.* 76, 1248–1282. URL: http://dx.doi.org/10.1016/j.rser.2017.03.121, doi:10.1016/j.rser.2017.03.121.
21. Rodriguez, J., Bernet, S., Steimer, P.K., Lizama, I.E., 2010. A survey on neutral-point-clamped inverters. *IEEE Trans. Ind. Electron.* 57, 2219–2230. doi:10.1109/TIE.2009.2032430.
22 Rooholahi, B., 2020. A New Transformerless Single-Phase Eleven-Level Inverter with Reduction of Switches Based on Model Predictive Control Method, in: *IEEE GPECOM*, pp. 102–107.
23. Salem, A., Abido, M.A., Blaabjerg, F., 2022. Common-Mode Voltage Mitigation of Dual T-Type Inverter Drives Using Fast MPC Approach. *IEEE Trans. Ind. Electron.* 69, 7663- 7674. doi:10.1109/TIE.2021.3102437.
24. Sandeep, N., Ali, J.S.M., Yaragatti, U.R., Vijayakumar, K., 2019. A Self-Balancing Five-Level Boosting Inverter with Reduced Components. *IEEE Trans. Power Electron.* 34, 6020–6024. doi:10.1109/TPEL.2018.2889785.
25. Sathik, M.J., Bhatnagar, K., Sandeep, N., Blaabjerg, F., 2020a. An Improved Seven-Level PUC Inverter Topology with Voltage Boosting. *IEEE Trans. Circuits Syst. II: Express Br.* 67, 127–131. doi:10.1109/TCSII.2019.2902908.

26. Sathik, M.J., Krishnasamy, V., 2019. Compact Switched Capacitor Multilevel Inverter (CSCMLI) with Self-Voltage Balancing and Boosting Ability. *IEEE Trans. Power Electron.* 34, 4009–4013. doi:10.1109/TPEL.2018.2871378.
27. Sathik, M.J., Sandeep, N., Almakhles, D., Blaabjerg, F., 2020b. Cross Connected Compact Switched-Capacitor Multilevel Inverter (C3-SCMLI) Topology with Reduced Switch Count. *IEEE Trans. Circuits Syst. II: Express Br.* 67, 3287–3291. doi:10.1109/tcsii.2020.2988155.
28. Shahrukh, M., Usmani, A.A., Shrivastava, P., Kidwai, S., 2018. New Single Phase 11 Level Inverter Topology usng Multilevel PWM Switching with Lesser Components Count, in: *IEEE Texas Power and Energy Conference*, pp. 1–6.
29. Siddique, M.D., Mekhilef, S., Shah, N.M., Memon, M.A., 2019. Optimal Design of a New Cascaded Multilevel Inverter Topology With Reduced Switch Count. *IEEE Access* 7, 24498–24510. doi:10.1109/ACCESS.2019.2890872.
30. Sun, P., Wickramasinghe, H.R., Khalid, M., Konstantinou, G., 2022a. Ac/dc fault handling and expanded dc power flow expression in hybrid multi-converter dc grids. *Int. J. Electr. Power Energy Syst.* 141, 107989. doi:https://doi.org/10.1016/j.ijepes.2022.107989.
31. Sun, R., Wang, X., Ye, Y., 2022b. Seventeen-Level Inverter Based on Switched-Capacitor and Flying-Capacitor-Fed T-type Unit. *IEEE Access* 10, 33561–33570.
32. Taghvaie, A., Adabi, J., Rezanejad, M., 2017. A Multilevel Inverter Structure based on Combination of Switched-Capacitors and DC Sources. *IEEE Trans. Industr. Inform.* 13, 2162–2171. doi:10.1109/TII.2017.2710265.
33. Thakre, K., Mohanty, K.B., Chatterjee, A., 2018. Reduction of circuit devices in symmetrical voltage source multilevel inverter based on series connection of basic unit cells. *Alex. Eng. J.* 57, 2703–2712. doi:https://doi.org/10.1016/j.aej.2017.11.001.
34. Tjokro, C., Pratomo, L.H., 2018. Design and Simulation of an Asymmetrical 11-Level Inverter for Photovoltaic Applications. *Proceedings - 2018 5th International Conference on Information Technology, Computer and Electrical Engineering, ICITACEE 2018*, 93–98. doi:10.1109/ICITACEE.2018.8576943.
35. Upadhyay, D., Ali, M., Tariq, M., Khan, S.A., Alamri, B., Alahmadi, A., 2022. Thirteen-level UXE-type inverter with 12-band hysteresis current control employing PSO based PI controller. *IEEE Access* doi:10.1109/access.2022.3146355.
36. Zaid, M., Khan, S., Mahmood, A., Ali, M., Sarwar, A., Khaild, M., 2023. A New High Gain Boost Converter with Common Ground for Solar-PV Application and Low Ripple Input Current. *Arab. J. Sci. Eng.* doi:10.1007/s13369-023-07814-9.

26. Sathik, M.J., [illegible] 2018. [illegible] Compact [illegible] Stabilized Inverter (CSCMLI) [illegible] Reduced [illegible] Ability. IEEE Trans. Power Electron. [illegible]
Sathik, M.J., [illegible] Almakhles, D.J., [illegible] 2020. Cross Connected Compact [illegible] Capacitor [illegible] Topology with [illegible] IEEE Trans. [illegible] II: Express 67, 1287–[illegible]
28. Mahdavi, M., [illegible] 2018. New Single Phase [illegible] Inverter [illegible] with Lesser Components [illegible] In: IEEE [illegible] Conf. [illegible] pp. 1–6.
29. [illegible] M.D., [illegible] M.M., [illegible] 2018. Optimal Design of a [illegible] Inverter Topology [illegible] Reduced Switch Count. IEEE Access 6, [illegible] 10.1109/[illegible]
30. Sun, P., [illegible] Khan, M., [illegible] 2022. [illegible] fault handling [illegible] power [illegible] hybrid multi-converter design. [illegible] Energy [illegible] doi.org/10.1016/[illegible]
31. [illegible] R., Wang, [illegible] 2022b. [illegible] Inverter Based on Switched-Capacitor [illegible] Capacitor [illegible] IEEE Access 10, 35561–35570.
32. Taghvaie, A., [illegible] M., 2017. [illegible] Inverter Structure Based on Combination of Switched-Capacitors and [illegible] IEEE Trans. Industr. Inform. 13, [illegible]
33. [illegible] K., Mohanty, [illegible] 2017. [illegible] circuit topology as symmetrical and asymmetrical [illegible] connection of basic unit cells. [illegible] 2743–[illegible] 10.1016/[illegible] 2017.11.001.
34. [illegible] C., [illegible] 2018. Design and [illegible] A Symmetrical 11 Level Inverter [illegible] Approach [illegible] 2018 5th International Conference on Information Technology [illegible] and Electrical Engineering (ICITACEE) [illegible] 2018.8576[illegible]
35. Upadhyay, [illegible] M., [illegible] Khan, [illegible] Ahmad, A., 2022. [illegible] hybrid hysteresis current control [illegible] doi.org/10.1109/ACCESS.2022.3140[illegible]
36. Ali, M., Khan, [illegible] Ali, [illegible] M., 2023. A New High [illegible] with [illegible] Solar PV Application [illegible] Input [illegible] doi: 10.1007/s11589-023-0[illegible]

5

Single Inverter Switched SVPWM Scheme for Four-Level Open-End Winding Induction Motor Drive

Suresh Lakhimsetty[1]*, Hareesh Myneni[2] and Obbu Chandra Sekhar[3]

[1]Department of Electrical Engineering, Sardar Vallabhbhai National Institute of Technology Surat, Surat, India
[2]Department of Electrical Engineering, National Institute of Technology Srinagar, Srinagar, India
[3]Department of Electrical Engineering, National Institute of Technology Delhi, New Delhi, India

Abstract

This chapter presents a four-level open-end winding induction motor (OEW-IM) drive controlled by a single inverter switched space vector pulse width modulation (SVPWM) scheme. Two-level voltage source inverters (VSIs) are connected to either side of the OEW-IM. To accomplish a four-level inversion, the dual-inverter system's DC-link voltages are kept at a 2:1 ratio. The four-level OEW-IM drive has two issues: (1) circulation of zero sequence currents in the OEW-IM phase windings and (2) overcharging the lower DC-link voltage capacitor by its counterpart higher DC-link voltage capacitor. The latter is avoided by supplying the dual-inverter (DI) system with independent DC power supplies. The former issue is addressed in the literature by developing decoupled SVPWM (switching both inverters) and bias SVPWM (switching one inverter while clamping the other). The current chapter presents an SVPWM scheme that increases the four-level OEW-IM drive performance even further compared to existing SVPWMs. The suggested SVPWM method simulates and validates the four-level OEW-IM drive using experimental data. The experimental findings demonstrate that the capacitor is not overcharged and the harmonic performance is improved.

***Keywords*:** Capacitor overcharging, multilevel inversion, open-end winding

**Corresponding author*: suresh@eed.svnit.ac.in

Salman Ahmad, Farhad Ilahi Bakhsh, and P. Sanjeevikumar (eds.) Multilevel Converters, (99–116)

5.1 Introduction

Two-level VSI-driven induction motor drives are employed in various industrial and transportation applications as they offer advantages such as low cost and robustness. Also, it is easier to manufacture and maintain two-level inverters [1].

On the other hand, the general requirement of lowering harmonic distortion, reduction of switching stress on power semiconductor devices (by reducing dv/dt), enhancing efficiency by lowering switching frequency, and avoidance of insulation stress to strengthen the reliability of the motor are better served by MLIs compared to the conventional two-level VSI [2–4]. Therefore, obtaining multilevel inversion using two-level VSIs as the basic building blocks is advantageous. Such a power circuit configuration would inherit ruggedness (i.e., reliability) from the two-level VSI and all of the advantages pertinent to the multilevel inverters.

The OEW-IM drives are conceived to achieve this objective [5]. The OEW configuration is attained by removing the star connection for Y-connected or the end-terminals of a Δ-connected stator winding of an electric machine in Figure 5.1(a). Thus, six terminals of the motor windings are brought out. On either side, they are fed with two conventional two-level VSIs. Such a power topology is termed the "dual-inverter (DI) fed open-end winding induction motor drive", as shown in Figure 5.1(b). As said previously, the main feature of the OEW topology is that it uses the two-level VSIs to achieve multilevel inversion, and the total DIs dc-link

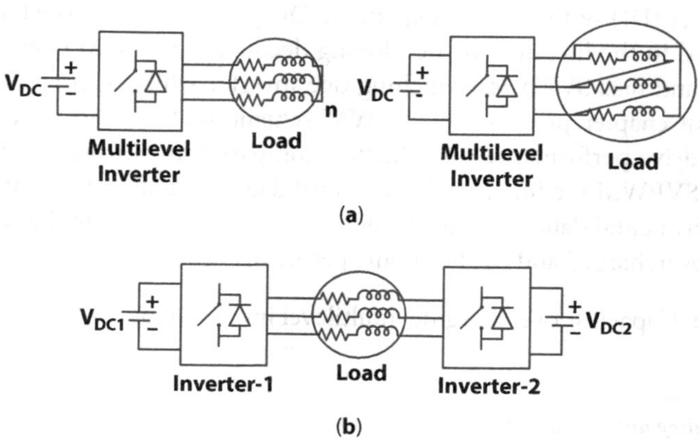

Figure 5.1 Inverter topologies based on supply techniques: (a) multilevel inverter with single-ended load and (b) dual-inverter with open-ended load.

voltage is shared between the two inverters, but the switching devices block the same DC-link voltage.

Interestingly, the OEWIMD is very flexible and offers several possibilities regarding the selection of VSI topologies, the distribution of input dc voltage levels, and the switching strategies adopted to control VSIs.

The first three-level OEWIM, which is similar to the NPC converter, was proposed by H. Stemmler *et al.* [5]. The authors suggested two topologies, shown in Figure 5.2, to realize three-level inversion. The main features of these topologies are the absence of flying capacitors and clamping diodes and the requirement of two power supplies compared to the FC, NPC, and CHB MLIs, respectively. Another advantage of this topology is that the DI system can provide 64 space-vector combinations, as each two-level VSI can assume eight independent states. These 64 switching-states are scattered over 19 space-vector locations. In contrast, the conventional three-level VSIs provide only 27 switching states and 19 vectors. Therefore, because of its redundancy, it offers more flexibility in devising PWM algorithms to control it.

KA Corzine *et al.* suggested different methods to obtain multilevel inversion for the circuit topology shown in Figure 5.3 [6]. It is shown that if the two power supplies of the DI system are asymmetric, it can obtain more than three levels. If the $V_{DC_2} = \frac{1}{2} V_{DC_1}$, in the circuit topology shown

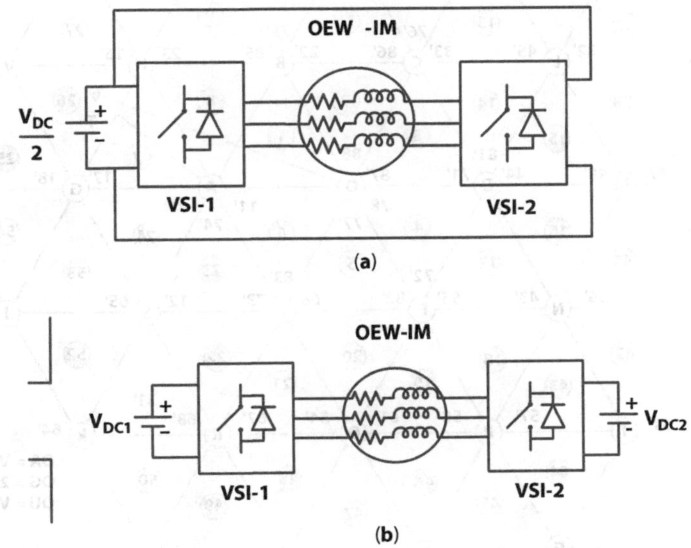

Figure 5.2 Open-end winding topologies for realizing three-level operation with symmetric power supplies: (a) single dc-link and (b) isolated dc-links.

in Figure 5.3, some of the voltage space-vectors overlap, and the voltage pattern is identical to that of a four-level inverter. Figure 5.3 shows the circuit diagram for the four-level OEW-IMD with asymmetric dc-link voltages. The corresponding space-vector diagram of Figure 5.3 is shown in

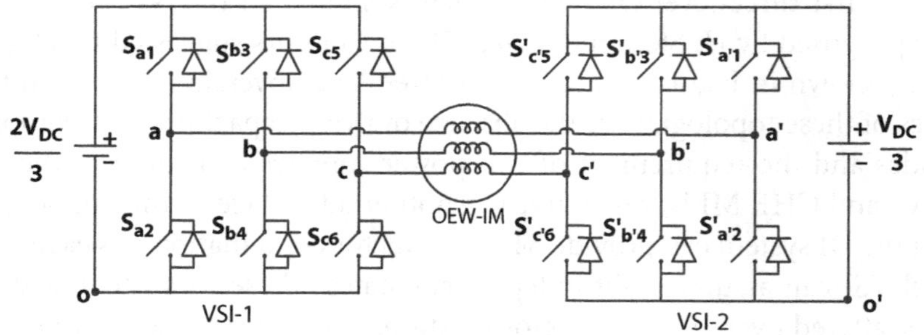

Figure 5.3 Four-level OEWIMD with dc-link voltages (2:1 ratio).

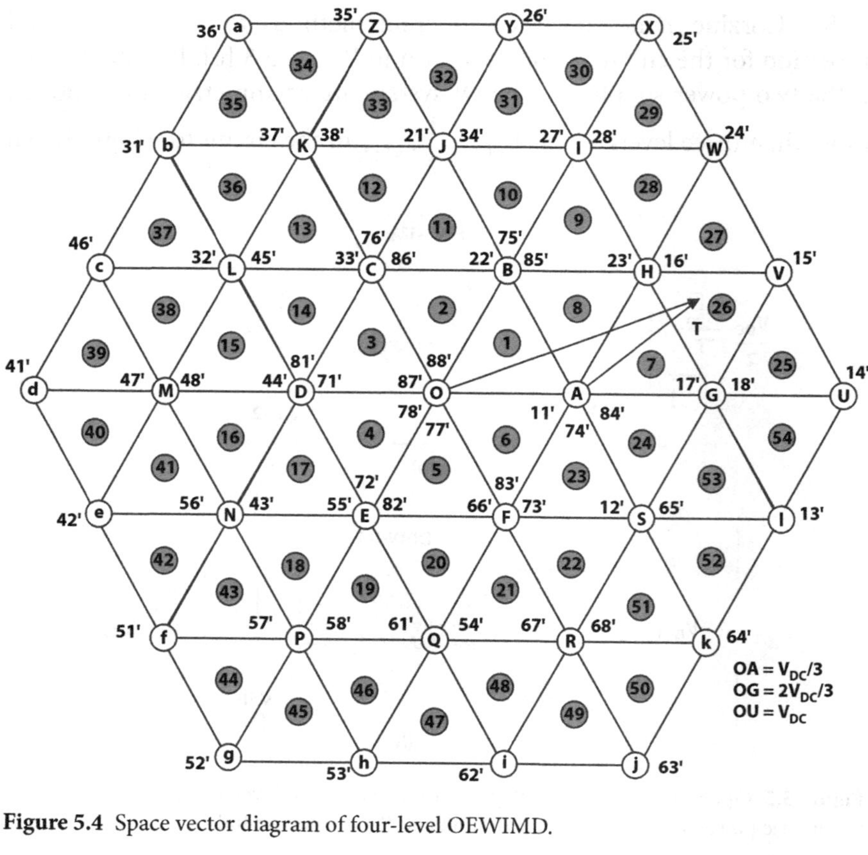

Figure 5.4 Space vector diagram of four-level OEWIMD.

Figure 5.4, which exhibits 64 space-vector combinations and is spread over 37 locations.

An elegant and straightforward SVPWM strategy was proposed for the circuit topology in Figure 5.4 with asymmetric voltage sources by VT Somasekhar *et al.* [7]. The proposed SVPWM technique requires only the instantaneous phase references in all sectors, and the modulation waveform was controlled based on the speed of the OEWIM drive. However, it uses the hysteresis comparators to identify the sector information. A controlled dc-link voltage is required. In the proposed SVPWM, exploiting the redundancy of vector combinations, those problematic space vectors are avoided to get four-level inversion.

V. Oleschuk *et al.* suggested novel space-vector-based continuous and discontinuous PWM strategies, named synchronized PWMs, to control dual-inverter-fed OEWIM drives [8]. The proposed PWM technique provides a continuously controlled power sharing between the two VSIs, based on their voltage magnitudes. It also avoids even harmonics and sub-harmonics in the output voltage of the OEWIM drive. However, in the proposed method with continuous PWM, both inverters are switched, resulting in an increased switching power loss. On the other hand, the discontinuous PWM technique results in a reduced switching loss and better THD than the continuous PWM.

B. Venugopal Reddy *et al.* suggested two decoupled SVPWM techniques to avoid the overcharging phenomenon of the lower dc-link voltage capacitor [9]. The two PWM techniques are (i) equal switching duty SVPWM (ESDSVPWM) and (ii) proportional switching duty PWM (PSDSVPWM). It is shown that the later PWM scheme reduces the switching losses by 10% compared to the former. However, the proposed techniques can avoid the overcharging phenomenon, causing higher switching power losses because of switching two inverters. To reduce the switching power losses of the DI system, different SVPWM techniques were suggested in [10–12].

To eliminate the overcharging phenomenon in the four-level OEWIMD, a new topology was suggested by BV Reddy *et al.*, named the *nested rectifier-inverter* topology [13]. In this topology, the dc rails are common to both inverters; hence, there is a provision for circling zero-sequence currents. The DSAZE and SAZE PWM techniques initially proposed for the conventional OEW-IMD shown in Figure 5.2(a) are also extended to this power circuit configuration [13, 14]. The advantage of the nested rectifier–inverter topology is that an overall DI system dc-link voltage of 0.77 PU is sufficient to apply 1.0 PU voltage to the motor phases for the four-level OEW-IMD. However, the nested rectifier–inverter topology uses the three-isolated dc power supplies, which makes the drive expensive and bulky. A new nested topology only with two isolation transformers was proposed in [15].

A new power circuit configuration was proposed to avoid overcharging the lower dc-link voltage capacitor and achieve four-level inversion for the OEW-IMD [16]. The proposed topology segments the motor winding into two groups, which are in the ratio of 2:1 and forming an isolated neutral to deny a path for the circulation of ZSCs with each group. Also, each group of motor phase windings are fed with two two-level VSIs with common dc-rails. The topology in [16] increases the dc-bus utilization by 33% and uses the decoupled SVPWM technique. However, the proposed topology is suitable only for IMs having a number of poles = 6-*n* (n = 1, 2, ...). Because of the employment of the decoupled strategy, both VSIs are to be switched, which increases the switching power loss of the DI system.

S. Chowdhury *et al.* proposed a new configuration for the three-phase OEW-IMD [17] to realize four-level inversion. The proposed topology uses a single dc-voltage source for the main inverter and a floating capacitor for the second inverter. Thus, the proposed topology avoids one bulky isolation transformer and results in the cost reduction of the drive system. However, the modulation index is limited to 0.66 only. As a result, the usable phase-voltage levels are restricted to nine, compared to 14 for the circuit topology shown in Figure 5.3. Thus, 33% of the dc-link voltage utilization is reduced compared to the topology shown in Figure 5.3.

From the literature, it may observe that the four-level OEWIMD (Figure 5.3) is mainly plagued with two problems. They are (i) the higher dc-link voltage capacitor overcharges the lower dc-link voltage capacitor and (ii) the presence of zero-sequence current (ZSC) in the motor phase windings. However, as mentioned earlier, the ZSC can be avoided by eliminating a path for the ZSCs using isolated dc supplies.

Owing to the above-mentioned observations, the performance of the four-level OEW-IMD with isolated power supplies can be improved by reducing losses in the DI system. This objective should be achieved along with (a) avoidance of overcharging of VSI-2 dc-link capacitor, (b) suppression of ZSCs. This chapter investigates the possibility of switching only one inverter of the DI system, as switching both could lead to higher switching power loss.

The circuit illustrated in Figure 5.3 has a significant flaw. To use the minimum-ripple SVPWM approach [9], inverter-1 must be clamped to a fixed state for each sample time while inverter-2 is switched. However, the switching combinations 11′, 12′, 23′, 22′, 34′, 33′, 44′, 45′, 55′, 16′, 66′, and 56′ (Figure 5.4), which are critical for implementing this method (i.e., switching the vertices nearest to the sample), cannot be used [13, 16].

When these switching vectors are active, the capacitor which feeds VSI-2 gets overcharged by the higher voltage (which provides inverter-1). As a result, the essential dc-link voltage ratio of 2:1 is disrupted, distorting the OEW-IM phase voltages and currents and leaving the circuit inoperable.

The present chapter suggests an easily implemented SVPWM technique to enrich the performance of the OEW-IMD. It has been demonstrated that switching VSI-1, which operates at a higher dc-link voltage, and clamping VSI-2, which operates at a lower dc-link voltage, will produce superior results and prevent the overcharging of the VSI-2 capacitor in a DI system.

5.2 Proposed Biasing SVPWM Scheme

The SVPWM scheme clamps VSI-2, producing the vector offset around which VSI-1 is switched. This means that VSI-2 is the biasing inverter to provide the required vector offset. It may be noted that this simple alteration avoids deploying the troublesome vector combinations and results in a considerably better performance of the dual-inverter-fed OEWIMD.

Figure 5.4 depicts a core hexagon ABCDEF, centered on O, which may be used to represent the suggested PWM system. The sub-hexagonal designations for the six extra hexagons in Figure 5.4 are OASRQE, OCJIHA, OBHGSF, OENMLC, ODLKJB, and OFQPND. The A....F are the centers of these hexagons. The sub-hexagonal centers provide the biasing inverter, sometimes referred to as inverter-2, with its vector offset.

The DI system in Figure 5.3 is intended to synthesize the reference space vector ($\boldsymbol{v}_{sr}$) $\boldsymbol{OT}$ using modulation index m_a and is defined as:

$$m_a = \frac{\boldsymbol{OT}}{\boldsymbol{OU}} = \frac{|\boldsymbol{v}_{sr}|}{v_{DC}} \tag{5.1}$$

where OU is the overall dc-link voltage (V_{DC}).

The SVPWM approach scales the DI system to provide the rated line-line voltage (400 V) and frequency (50 Hz) at the verge of $\boldsymbol{m}_a$, i.e. ,0.866. The fundamental frequency of the voltage component of the OEW-IM becomes:

$$f_1 = \frac{m_a}{\sqrt{3/2}} \times 50 \tag{5.2}$$

Furthermore, regardless of frequency, the reference OEW-IM phase voltages comprising the DI system's ***OT*** are 42 times sampled every cycle. This yields seven samples per sector. Samples at sector borders are avoided to acquire waveform symmetries [9]. The difference in angle between subsequent samples is 8.57° (i.e., 360°/42). The following is the sample period:

$$T_s = \frac{1}{f_1 \times 42} \tag{5.3}$$

The **OT** may be divided into AT and OA in Figure 5.4, representing the switching and biasing vectors. $[v_a\ v_b\ v_c]^T$ and $[v_a^*\ v_b^*\ v_c^*]^T$ are the column voltage vectors of AT and OT, respectively. It may be demonstrated using a suitable transformation of coordinates that:

$$[v_a\ v_b\ v_c]^T = I_3[v_a^*\ v_b^*\ v_c^*]^T + \frac{2V_{dc}}{9}\begin{bmatrix} -\cos\left((n-1)\times\pi/3\right) \\ \cos(n\pi/3) \\ \cos\left((n-2)\times\pi/3\right) \end{bmatrix} \tag{5.4}$$

where

$$I_3 = \text{unity diagonal matrix } (3\times 3)$$

n = 1 to 6

The OA vector is output by VSI-2 within a particular sampling period. AT vector synthesized by VSI-1 by switching between the vertices W, U, and A (see Figure 5.4). During the sample time, VSI-2 is clamped at state 4 (- + +), while VSI-1 rotates between states 8 (---), 1 (+ - -), 2 (++-), and 7 (+++). With this switching approach, the space vector combinations 74, 24, 14, and 84 are consecutively deployed. The effective period $'T_{eff}'$ is defined by the total periods during which states 14 and 24 are exchanged. The total number of periods during which the vector groups 74 and 84 are swapped is represented by the symbol $'T_z'$. Based on imaginary switching periods, the switching algorithm described in [18] generates gate pulses for VSI-1. Figure 5.5 depicts the flow chart of the proposed SVPWM scheme.

Three alternatives emerge from the $'T_z'$ distribution: (i) the $'T_z'$ is allocated evenly among 74 and 84, (ii) the 84 combinations are switched for the $'T_z'$ duration, and (iii) the 74 combinations are switched for the $'T_z'$ duration [18]. The first of these three types is named the center-spaced SVPWM

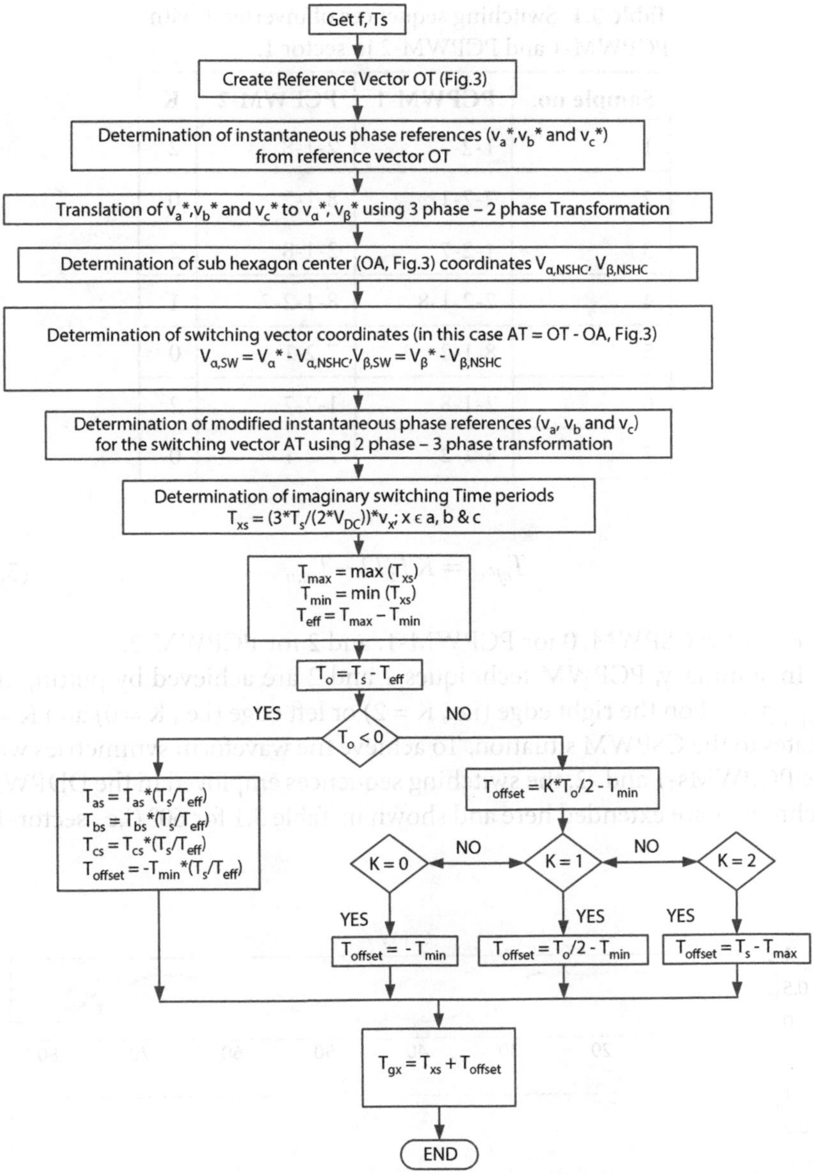

Figure 5.5 A simplified representation of the proposed SVPWM strategy.

(CSPWM) strategy, while the second and third are collectively referred to as the phase-clamped PWM-1 (PCPWM-1) and phase-clamped PWM-2 (PCPWM-2) methods, respectively. The universal T_{offset} in Equation 5.5 is used to actualize the expression and yields the following three variants:

Table 5.1 Switching sequences of inverter-1 with PCPWM-1 and PCPWM-2 in sector 1.

Sample no.	PCPWM-1	PCPWM-2	K
1	1-2-7	2-1-8	2
2	7-2-1	8-1-2	0
3	1-2-7	2-1-8	2
4	7-2-1-8	8-1-2-7	1
5	8-1-2	7-2-1	0
6	2-1-8	1-2-7	2
7	8-1-2	7-2-1	0

$$T_{offset} = KT_z/2 - T_{min} \tag{5.5}$$

K = 1 for CSPWM, 0 for PCPWM-1, and 2 for PCPWM-2.

In summary, PCPWM techniques 1 and 2 are achieved by putting the T_{offset} period on the right edge (i.e., K = 2) or left edge (i.e., K = 0) and K = 1 relates to the CSPWM situation. To achieve the waveform symmetries with the PCPWMs-1 and -2, the switching sequences employed in the DDPWM techniques are extended here and shown in Table 5.1 for 60° (i.e., sector-1).

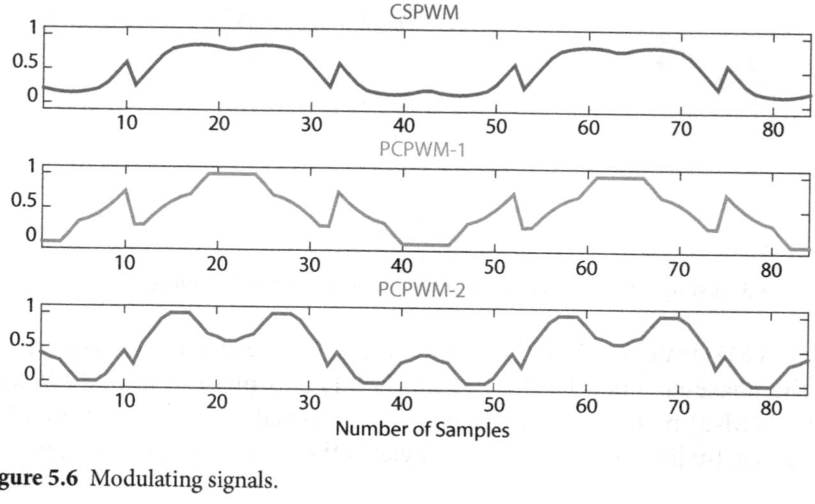

Figure 5.6 Modulating signals.

When $|OT| > |OU|$, the **OT** vector lies outside the space vector diagram (Figure 5.4). In such a case, the period $'T_z'$ would become negative, which calls for the necessity to implement over modulation. The negative value of $'T_z'$ is re-adjusted to zero by multiplying the imaginary switching periods and the offset period with a factor of T_S/T_{eff}.

At a modulation index of 0.7, the modulating waveforms for CSPWM, PCPWM-1, and PCPWM-2 are shown in Figure 5.6.

The traces in Figure 5.6 show that the specific phase is clamped to both the negative or positive states, and switching power losses are reduced with PCPWM-1 and PCPWM-2.

5.3 Experimental and Simulation Results

The suggested SVPWM technique mimics the four-level OEW-IMD and is realized by experimentation with the dSPACE-1104 controller. The OEW-IM drive is operated in an open loop with volts/Hz control. The DI system's total dc-link voltage is 300 V and is distributed asymmetrically between inverter-1 and inverter-2 in the ratio of 2.1

The four-level OEWIMD applies throughout the speed spectrum and at various modulation indices. The results from simulations and experiments for PCPWM-1 and PCPWM-2 at $m_a = 0.7$ are shown in Figures 5.7–5.12. Figures 5.7 and 5.8 display the pole voltages of VSI-1 and VSI-2. Figure 5.9 depicts the zero-sequence voltages between O and O'. Figures 5.10 and 5.11 show the OEW-IM phase voltage and its harmonic spectrum. The no-load phase current of the OEW-IMD is depicted in Figure 5.12. Figures 5.13–5.18 present the simulated and experimental findings using PCPWM-2.

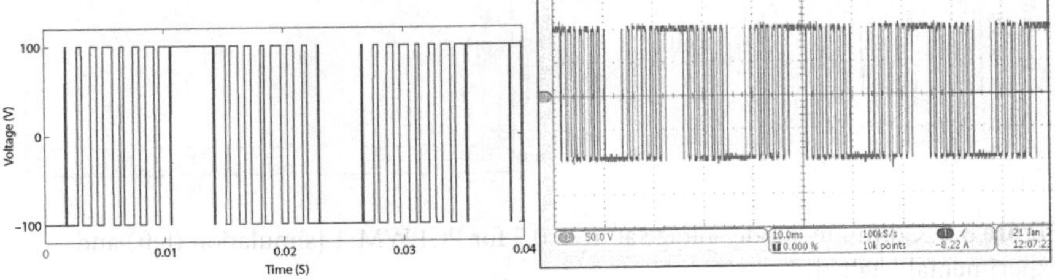

Figure 5.7 VSI-1 pole voltage at $m_a = 0.7$ for PCPWM-1 [simulation (left) and experimental (right)].

The simulation and experimental results are consistent with each other. According to Figures 5.7–5.18, the SVPWM approach suggested in this chapter avoids overcharging the capacitor connected to VSI-2. Figure 5.19 depicts the simulated OEW-IM voltage waveforms, zooming in on the center portion. The PWM approaches suggested in this chapter result in significantly lower dv/dt throughout the motor windings and accomplish all the waves from symmetries.

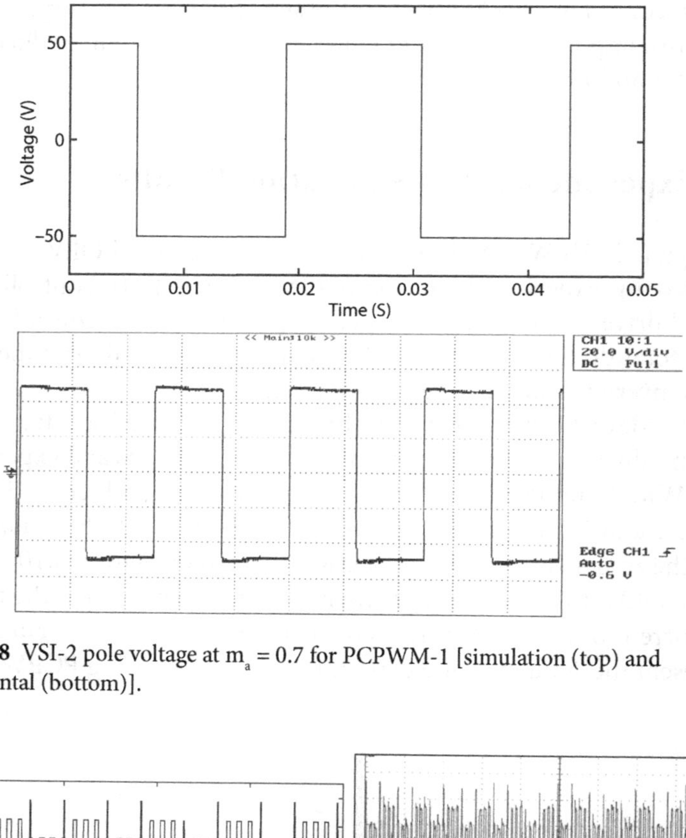

Figure 5.8 VSI-2 pole voltage at $m_a = 0.7$ for PCPWM-1 [simulation (top) and experimental (bottom)].

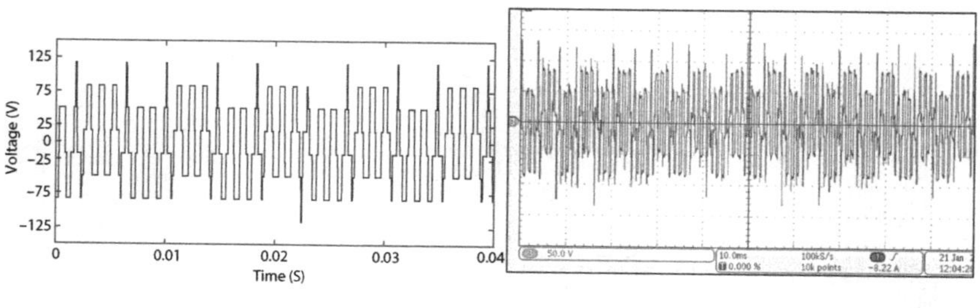

Figure 5.9 Common-mode voltage at $m_a = 0.7$ for PCPWM-1 [simulation (left) and experimental (right)].

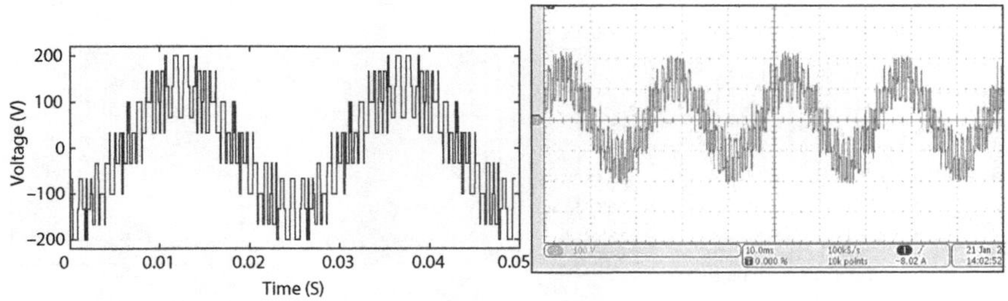

Figure 5.10 OEW-IM phase voltage at $m_a = 0.7$ for PCPWM-1 [simulation (left) and experimental (right)].

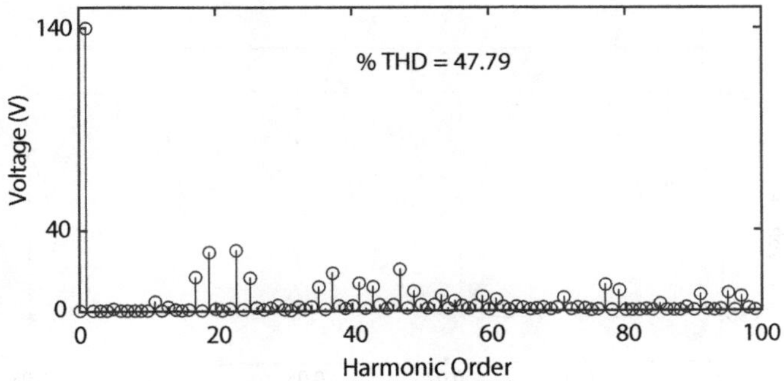

Figure 5.11 OEW-IM phase voltage harmonic spectrum at $m_a = 0.7$ for PCPWM-1.

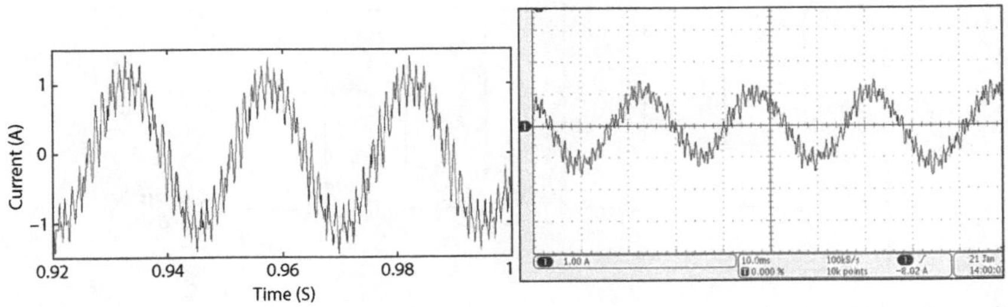

Figure 5.12 OEW-IM no-load phase current at $m_a = 0.7$ for PCPWM-1 [Simulation (left) and experimental (right)].

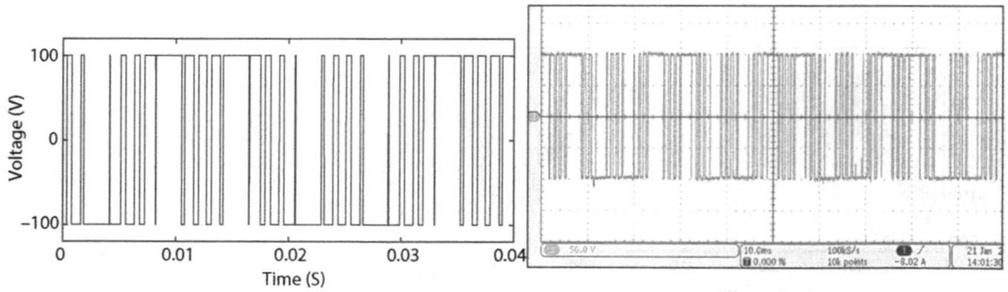

Figure 5.13 Pole-voltage of VSI-1 at $m_a = 0.7$ for PCPWM-2 [simulation (left) and experimental (right)].

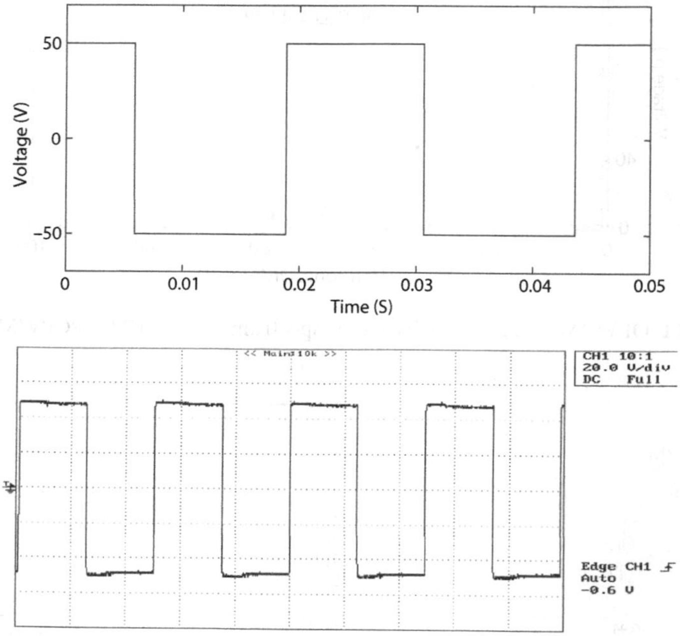

Figure 5.14 VSI-2 pole voltage at $m_a = 0.7$ for PCPWM-2 [simulation (top) and experimental (bottom)].

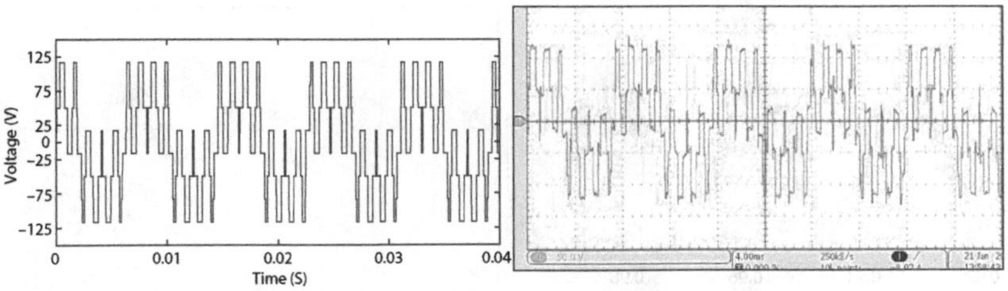

Figure 5.15 Voltage between O and O′ at $m_a = 0.7$ for PCPWM-2 [simulation (left) and experimental (right)].

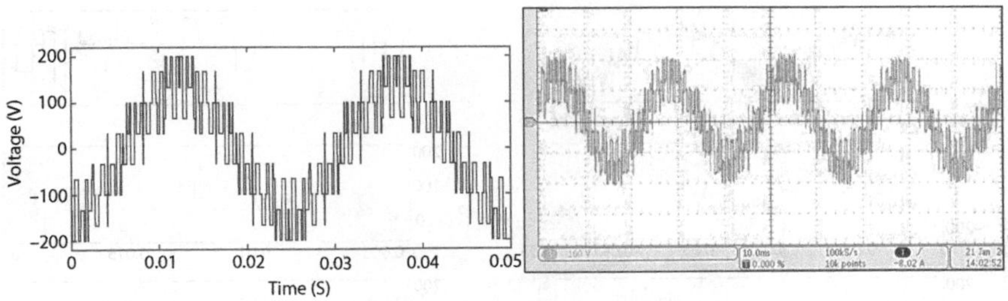

Figure 5.16 OEW-IM phase voltage at $m_a = 0.7$ for PCPWM-2 [simulation (left) and experimental (right)].

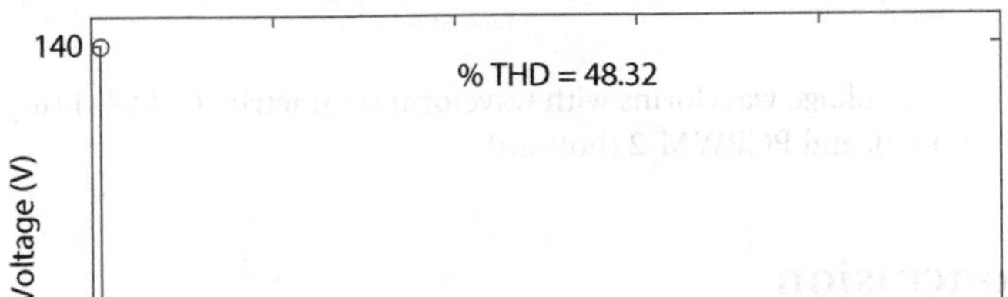

Figure 5.17 OEW-IM phase voltage harmonic spectrum at $m_a = 0.7$ for PCPWM-2.

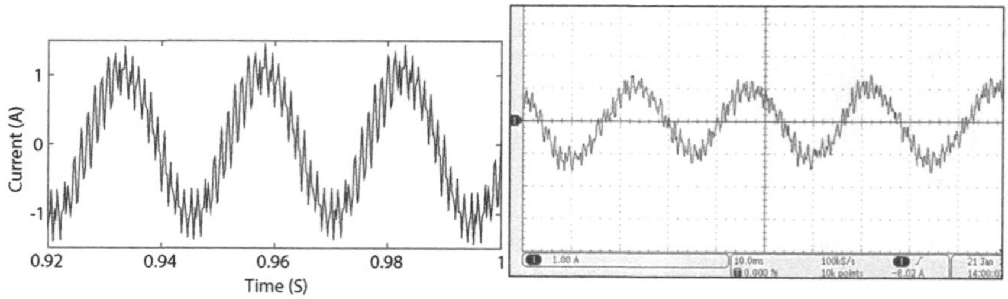

Figure 5.18 OEW-IM no-load phase current at $m_a = 0.7$ for PCPWM-2 [simulation (left) and experimental (right)].

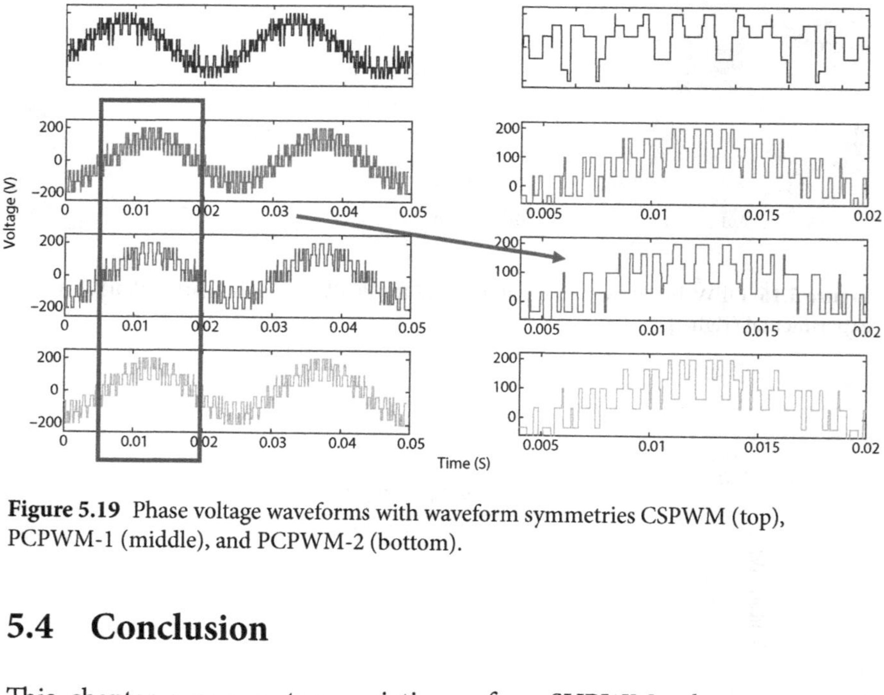

Figure 5.19 Phase voltage waveforms with waveform symmetries CSPWM (top), PCPWM-1 (middle), and PCPWM-2 (bottom).

5.4 Conclusion

This chapter proposes two variations of an SVPWM scheme to avoid overcharging an inverter-2 capacitor of a four-level OEW-IMD. Variable dc-link voltages power the component inverters in a 2:1 ratio. These PWM techniques also aim to reduce the switching power loss of the DI system of OEW-IM.

As one of the inverters is clamped and the switching inverter is one phase clamped, the suggested PWM approaches result in reduced switching device losses and a lower overall loss of the DI system from both modelling and experimental data. Furthermore, the suggested PWM approaches significantly reduce dv/dt across the OEW-IM phase windings. This extends the life of the insulation and bearings and reduces bearing currents.

References

1. S. Kouro, J. Rodriguez, B. Wu, S. Bernet and M. Perez, "Powering the Future of Industry: High-Power Adjustable Speed Drive Topologies," in *IEEE Ind. Appl. Mag.*, vol. 18, no. 4, pp. 26-39, July-Aug. 2012, doi: 10.1109/MIAS.2012.2192231.
2. J. Rodriguez, Jih-Sheng Lai and Fang Zheng Peng, "Multilevel inverters: a survey of topologies, controls, and applications," in *IEEE Trans. Ind. Electron.*, vol. 49, no. 4, pp. 724-738, Aug. 2002, doi: 10.1109/TIE.2002.801052.
3. Jih-Sheng Lai and Fang Zheng Peng, "Multilevel converters-a new breed of power converters," in *IEEE Trans. Ind. Appl.*, vol. 32, no. 3, pp. 509-517, May-June 1996, doi: 10.1109/28.502161.
4. J. Rodriguez, S. Bernet, B. Wu, J. O. Pontt and S. Kouro, "Multilevel Voltage-Source-Converter Topologies for Industrial Medium-Voltage Drives," in *IEEE Trans. Ind. Electron.*, vol. 54, no. 6, pp. 2930-2945, Dec. 2007, doi: 10.1109/TIE.2007.907044.
5. H. Stemmler and P. Guggenbach, "Configurations of high-power voltage source inverter drives," *1993 Fifth European Conference on Power Electronics and Applications*, Brighton, UK, 1993, pp. 7-14 vol.5.
6. K. A. Corzine, S. D. Sudhoff and C. A. Whitcomb, "Performance characteristics of a cascaded two-level converter," in *IEEE Trans. Energy Convers.*, vol. 14, no. 3, pp. 433-439, Sept. 1999, doi: 10.1109/60.790893.
7. V T Somasekhar, E G Shivakumar, K Gopakumar & Andre Pittet (2002) Multi Level Voltage Space Phasor Generation for an Open-End Winding Induction Motor Drive Using a Dual Inverter Scheme with Asymmetrical DC-Link Voltages, *EPE J.*, 12:3, 59-77, DOI: 10.1080/09398368.2002.11463508.
8. V. Oleschuk, R. Bojoi, G. Griva and F. Profumo, "Dual Inverter-Fed Traction Drive with DC Sources Power Balancing Based on Synchronized PWM," *2007 IEEE International Electric Machines & Drives Conference*, Antalya, Turkey, 2007, pp. 260-265, doi: 10.1109/IEMDC.2007.382676.
9. B. Venugopal Reddy, V. T. Somasekhar and Y. Kalyan, "Decoupled Space-Vector PWM Strategies for a Four-Level Asymmetrical Open-End Winding Induction Motor Drive With Waveform Symmetries," in *IEEE Trans. Ind. Electron.*, vol. 58, no. 11, pp. 5130-5141, Nov. 2011, doi: 10.1109/TIE.2011.2116759.
10. S. Lakhimsetty, N. Surulivel and V. T. Somasekhar, "Improvised SVPWM Strategies for an Enhanced Performance for a Four-Level Open-End Winding

Induction Motor Drive," in *IEEE Trans. Ind. Electron.*, vol. 64, no. 4, pp. 2750-2759, April 2017, doi: 10.1109/TIE.2016.2632059.
11. Lakhimsetty, S. and Somasekhar, V.T. (2018), Discontinuous decoupled SVPWM schemes for a four-level open-end winding induction motor drive with waveform symmetries. *IET Power Electron.*, 11: 280-292. https://doi.org/10.1049/iet-pel.2017.0096.
12. Lakhimsetty, S, Hema Kumar, P, Somasekhar, VT. Hybrid space-vector pulse width modulation strategies for a four-level open-end winding induction motor drive with an improvised harmonic performance and balanced DC-link capacitors. *Int. Trans. Electr. Energ. Syst.*, 2021; 31:e12814. https://doi.org/10.1002/2050-7038.12814.
13. B. V. Reddy and V. T. Somasekhar, "A Dual Inverter Fed Four-Level Open-End Winding Induction Motor Drive With a Nested Rectifier-Inverter," in *IEEE Trans. Ind. Inf.*, vol. 9, no. 2, pp. 938-946, May 2013, doi: 10.1109/TII.2012.2223224.
14. B. Venugopal Reddy and V. T. Somasekhar, "An SVPWM Scheme for the Suppression of Zero-Sequence Current in a Four-Level Open-End Winding Induction Motor Drive With Nested Rectifier–Inverter," in *IEEE Trans. Ind. Electron.*, vol. 63, no. 5, pp. 2803-2812, May 2016, doi: 10.1109/TIE.2015.2511083.
15. S. Lakhimsetty and V. T. Somasekhar, "A Four-Level Open-End Winding Induction Motor Drive With a Nested Rectifier–Inverter Combination With Two DC Power Supplies," in *IEEE Trans. Power Electron.*, vol. 34, no. 9, pp. 8894-8904, Sept. 2019, doi: 10.1109/TPEL.2018.2884023.
16. V. T. Somasekhar, B. Venugopal Reddy and K. Sivakumar, "A Four-Level Inversion Scheme for a 6n-Pole Open-End Winding Induction Motor Drive for an Improved DC-Link Utilization," in *IEEE Trans. Ind. Electron.*, vol. 61, no. 9, pp. 4565-4572, Sept. 2014, doi: 10.1109/TIE.2013.2288212.
17. S. Chowdhury, P. W. Wheeler, C. Patel and C. Gerada, "A Multilevel Converter With a Floating Bridge for Open-End Winding Motor Drive Applications," in *IEEE Trans. Ind. Electron.*, vol. 63, no. 9, pp. 5366-5375, Sept. 2016, doi: 10.1109/TIE.2016.2561265.
18. Dae-Woong Chung, Joohn-Sheok Kim and S. -K. Sul, "Unified voltage modulation technique for real-time three-phase power conversion," in *IEEE Trans. Ind. Appl.*, vol. 34, no. 2, pp. 374-380, March-April 1998, doi: 10.1109/28.663482.

6

Field-Oriented Control (FOC) of Motor Drives With Multilevel Converter

Arif Iqbal[1]* and S. P. Singh[2]

[1]Department of Electrical Engineering, Rajkiya Engineering College, Ambedkar Nagar (U.P.), India
[2]Indian Institute of Technology Roorkee, Roorkee, Haridwar, India

Abstract
Field-oriented control (FOC) used in induction motor drives is well adopted in many applications, where higher performance is of prime importance. The improved performance of a motor drive using FOC is accounted by decoupled control of the current component of torque and field flux. Hence, the operational control of the motor becomes similar to a DC motor (separately excited). With the ongoing research for a better system, the converter used in the motor drive is extensively investigated, and their different topologies are used. In view of this, multilevel converters are preferably used in the last two decades. Among the various available topologies, a diode-clamped multi-level inverter is proposed in this chapter for the closed-loop operation of an induction motor drive using FOC. This chapter deals with the induction motor operation in motoring and reversed regenerating modes, fed through a diode-clamped three-phase three-level inverter circuit. Complete drive systems are developed and analyzed by using MATLAB/Simulink.

Keywords: Motor drives, field-oriented control, multilevel converter

6.1 Introduction

A variety of advanced control schemes have been adopted in motor drives for the high performance of different applications. Among the various available

**Corresponding author*: arif.iqbal.in@gmail.com

Salman Ahmad, Farhad Ilahi Bakhsh, and P. Sanjeevikumar (eds.) Multilevel Converters, (117–126)

schemes, field-oriented control (FOC) is well adopted in induction motor drives when compared with other types of available machines. In FOC, improved performance is achieved due to the decoupled control of torque and field flux analogous to that of a separately excited DC motor [1].

With the recent ongoing research and development, the popularity of multilevel inverters is increased in motor drive applications like traction, conveyors, pump, fans, propulsion drives, etc. This is due to an improved output voltage in a stepped waveform with reduced harmonic content and suitability for high-voltage applications [2, 3]. The basic concept of such inverters is to use power electronic switches connected in series with DC sources (low voltage) during power conversion with a staircase waveform of voltage. Multiple DC sources can be in the form of batteries, capacitors, and renewable energy sources [4].

This chapter deals with the analysis of induction motor drive by using indirect, field-oriented control in which a three-phase diode-clamped three-level inverter is used. The topology of the employed inverter is shown in Figure 6.1, whose switching state of one out of three phases (phase *a*) is given in Table 6.1. The remaining chapter is structured as follows: Section 6.2 explains the mathematical modeling and closed-loop control scheme applicable to an induction motor. In Section 6.3, the complete motor drive system is investigated, having the inverter (three-phase diode-clamped three-level) connected to an induction machine. The key concluding remarks are given in Section 6.4.

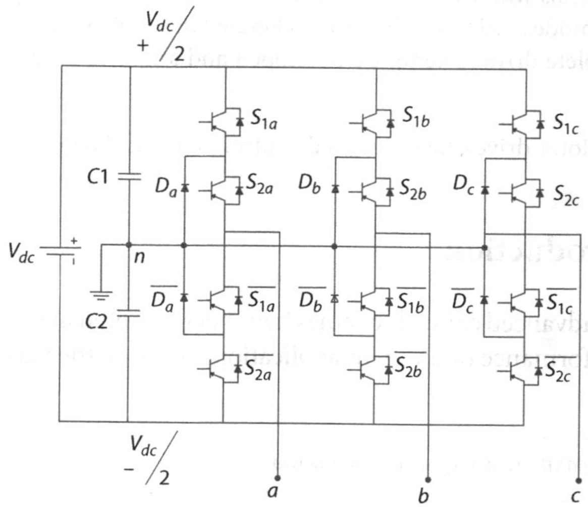

Figure 6.1 Topology of a three-phase diode-clamped three-level inverter.

Table 6.1 Switching states of a diode-clamped three-level inverter.

S_{a1}	S_{a2}	$\overline{S_{a1}}$	$\overline{S_{a2}}$	Pole voltage (V_{ao})	Phase voltage (V_{an})
1	1	0	0	$+V_{dc}$	$+V_{dc}/2$
0	1	1	0	$+V_{dc}/2$	0
0	0	1	1	0	$-V_{dc}/2$

6.2 Mathematical Modeling

6.2.1 Induction Motor

The stator and rotor circuit voltages in an arbitrary reference frame at speed ω_k are given as [5, 6]:

$$v_{qs} = R_s i_{qs} + \omega_k F_{ds} + pF_{qs} \tag{6.1}$$

$$v_{ds} = R_s i_{ds} - \omega_k F_{qs} + pF_{ds} \tag{6.2}$$

$$v_{qr} = R_r i_{qr} + (\omega_k - \omega_r)F_{dr} + pF_{qr} \tag{6.3}$$

$$v_{dr} = R_r i_{dr} - (\omega_k - \omega_r)F_{qr} + pF_{dr} \tag{6.4}$$

The quantity with subscripts "*s*" and "*r*" is associated with stator and rotor, respectively. In the expression above, flux linkage (*F*) is determined as

$$F_{qs} = L_{ls} i_{qs} + L_m \left(i_{qs} + i_{qr} \right) \tag{6.5}$$

$$F_{ds} = L_{ls} i_{ds} + L_m (i_{ds} + i_{dr}) \tag{6.6}$$

$$F_{qr} = L_{lr} i_{qr} + L_m \left(i_{qs} + i_{qr} \right) \tag{6.7}$$

$$F_{dr} = L_{lr}i_{dr} + L_m(i_{ds} + i_{dr}) \tag{6.8}$$

Motor torque is developed and rotor speed is given by

$$T_e = \left(\frac{3}{2}\right)\left(\frac{P}{2}\right)\left(F_{ds}i_{qs} - F_{qs}i_{ds}\right) \tag{6.9}$$

$$\omega_r = \frac{P}{2J}\int(T_e - T_l)dt \tag{6.10}$$

where the developed torque and rotor speed are shown by symbols T_e and ω_r, respectively.

6.2.2 Indirect Field-Oriented Control

This control is conceptually based on the alignment of rotor flux on one axis (*d*-axis) and the torque component of current on the other axis (*q*-axis) of the rotating *d-q* reference frame at synchronous speed $(\omega_k = \omega_e)$ [1, 7–12]. Hence, the motor expressions are simplified as shown below:

$$0 = R_r i_{qr} + (\omega_e - \omega_r)F_{dr} + pF_{qr} \tag{6.11}$$

$$0 = R_r i_{dr} - (\omega_e - \omega_r)F_{qr} + pF_{dr} \tag{6.12}$$

where

$$\omega_{sl} = \omega_e - \omega_r \tag{6.13}$$

Assuming the alignment of rotor flux only along *d*-axis, therefore

$$F_r = F_{dr} \tag{6.14}$$

$$F_{qr} = 0 \tag{6.15}$$

$$p\omega_r = 0 \tag{6.16}$$

The substitution of Equations (6.14)–(6.16) into (6.11) and (6.12) yields,

$$R_{qr}i_{qr} + \omega_{sl}F_r = 0 \tag{6.17}$$

$$R_{dr}i_{dr} + pF_{dr} = 0 \tag{6.18}$$

The values of i_{qr} and i_{dr} obtained from Equation (6.7) and (6.8), substituted in Equations (6.17) and (18), yield

$$\omega_{sl} = K_{sl}i_T \tag{6.19}$$

$$i_F = (1 + T_r p)F_r \tag{6.20}$$

where

$$K_{sl} = \frac{L_m}{T_r F_r} \tag{6.21}$$

$$i_T = i_{q1} + i_{q2} \tag{6.22}$$

$$i_F = i_{d1} + i_{d2} \tag{6.23}$$

$$T_r = \frac{L_r}{R_r} \text{ is rotor time constant.} \tag{6.24}$$

The substitution of the rotor circuit current in the torque equation yields

$$T_e = K_T F_r i_T = K_1 i_T \tag{6.25}$$

$$K_1 = K_T F_r \tag{6.26}$$

$$K_T = c\frac{L_m}{L_r} \tag{6.27}$$

6.3 Simulation Results

A model of an induction motor drive using vector control technique (i.e., field-oriented control) was developed in a synchronously rotating reference frame ($\omega = \omega_e$), which was analytically implemented, in which a three-level diode-clamped inverter was used. The motor parameters used are given in Table 6.2. The developed Simulink model to implement the induction motor drive using indirect FOC is shown in Figure 6.2. During the implementation with simplicity, variation in DC link voltage was ignored, in consideration of the constant flux operation of the motor. Hence, a constant value of reference current along d-axis, i_{ds}^*, was used. The reference q-axis current, i_{qs}^*, was synthesized by using speed error through a PI controller. Hence, the generated reference current along the d–q axes in a synchronous frame is transformed to a stationary frame using reference

Table 6.2 Parameters of four poles, two HP, three-phase induction machine.

$r_s = 10\,\Omega$	$r_r = 6.3\,\Omega$
$L_{ls} = L_{lr} = 0.04$ H	$L_m = 0.42$ H
$J = 0.03\ \text{kg m}^2$	

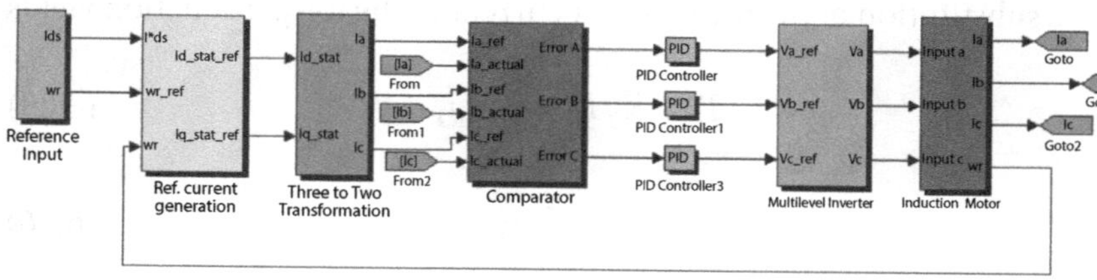

Figure 6.2 Simulink model of an induction motor drive system.

frame transformation concept [1, 6]. The reference phase current i_a^*, i_b^*, i_c^* is compared with the actual motor current i_a, i_b, i_c for the current error to generate the reference voltage v_a^*, v_b^*, v_c^* in a stationary frame through three voltage controllers (PI controllers). The reference voltages are then fed to the block of a "multilevel inverter" which constitutes the generation of required switching signal and three-phase inverter circuit [2, 4]. In the simulation, the speed command was initialized at time t = 0.21 s, which increases proportionally and settles to a steady state at 314 rad/s. At time t = 2.0 s, a mechanical load torque of 8 Nm, due to which a small rotor speed dip was noted, is shown in the zoomed portion of Figure 6.3(a). The complete generated response of torque is shown in Figure 6.3(b). The step increase in load torque (i.e., active output power) results in the increase in q-axis, i.e., active component of the stator current as shown in Figure 6.4. Since constant flux operation of the machine is assumed, a variation in the d-axis component (i.e., reactive component) of the stator current

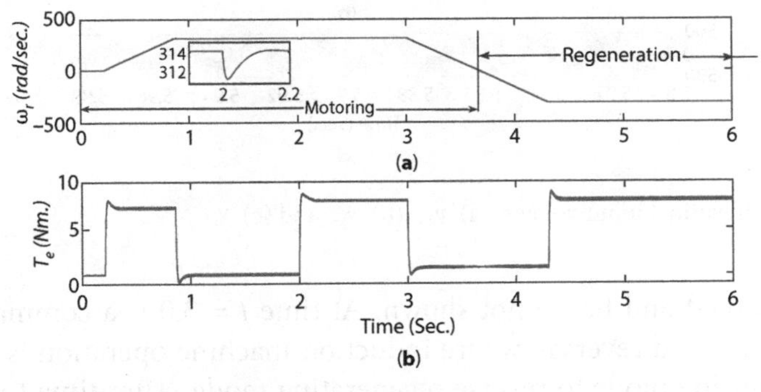

Figure 6.3 Rotor dynamics: (a) speed ω_r and (b) T_e.

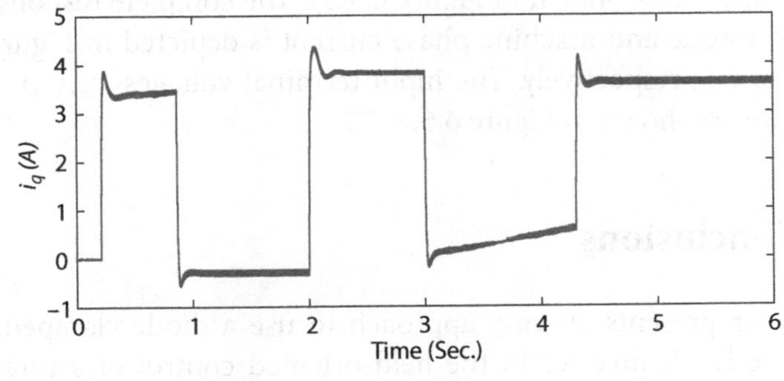

Figure 6.4 q-component of the stator current of an induction machine.

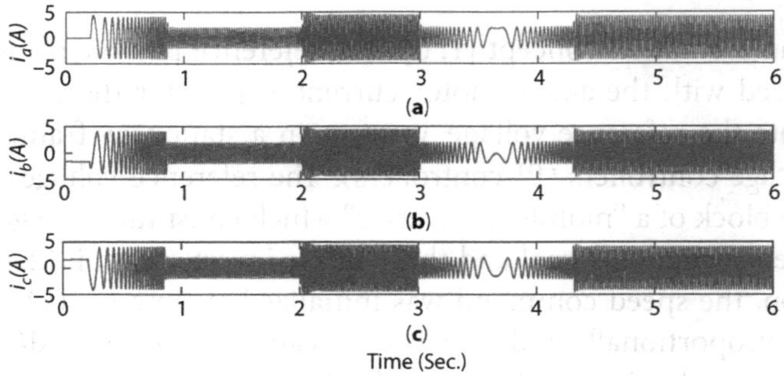

Figure 6.5 Input motor current: (a) i_a, (b) i_b, and (c) i_c.

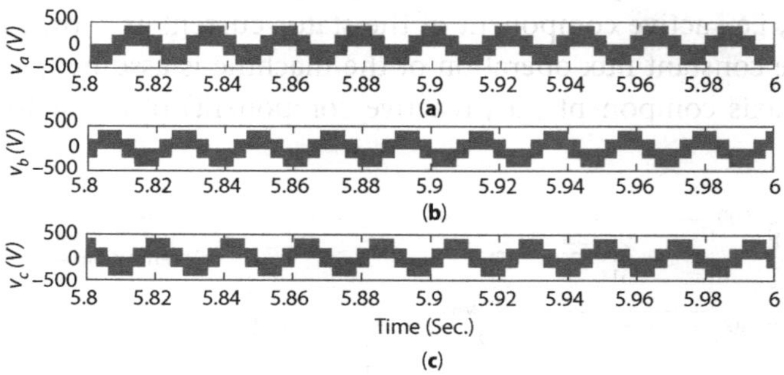

Figure 6.6 Terminal input voltage: (a) v_a, (b) v_b, and (c) v_c.

was not noted and hence not shown. At time t = 3.0 s, a command was started for speed reversal, where induction machine operation is shifting from motoring mode to reverse regenerating mode. After time t = 3.65 s, the direction of the rotor speed is reversing and the machine operates as a grid-connected induction generator. The operational mode of the induction machine is also shown in Figure 6.3(a). The complete response of the generated torque and machine phase current is depicted in Figure 6.3(b) and Figure 6.5, respectively. The input terminal voltages v_a, v_b, v_c during steady state are shown in Figure 6.6.

6.4 Conclusions

This chapter presents an easy approach to use a diode-clamped, multilevel (three-level) inverter in the field-oriented control of a three-phase induction motor drive. For this purpose, a MATLAB Simulink model of a complete drive system was developed and operated in two modes: motoring and reverse regeneration. During the implementation and operation process, the developed system was found to be precise, with good tracking characteristics and a smooth transition from one mode to another. The present study may be extended by using various other types of multilevel inverter circuit on different machines.

References

1. Bose, B. K., *Modern power electronics and AC drives*, N J, Upper Saddle River, Prentice Hall, Inc., 2002.
2. Chaturvedi, P. K., Jain, S. Agarwal, P., Reduced switching loss pulse width modulation technique for three-level diode clamped inverter. *IET Power Electron*, 4, pp. 393-399, 2010. https://doi.org/10.1049/iet-pel.2010.0311
3. Khomfoi S., Tolbert, L. M., Chapter 17: Multilevel Power Converters, in: *Power Electronics Handbook*, 2nd Edition Elsevier, 2007, ISBN 978-0-12-088479-7, pp. 451-482, 2007. https://doi.org/10.1016/B978-012088479-7/50035-3
4. Nzongo, D. L. M., Leugoue, E., Zhang, J. H., Ekemb, G., Improved Field oriented control of PWM Multi-level Inverter fed Induction Motor Drives. *Indones. J. Electr. Eng. Comput. Sci.*, 9, pp. 481-492, 2018. https://doi.org/10.11591/ijeecs.v9.i2.pp481-492
5. Krause, P. C., Thomas, C. H., Simulation of Symmetrical Induction Machinery. *IEEE Trans. Power Appar. Syst.*, 84, pp. 1038-1053, 1965. https://doi.org/10.1109/TPAS.1965.4766135
6. Krause, P. C., Wasynczuk O., Sudhoff S. D., *Analysis of electrical machinery and drive system*. 2nd edition, IEEE Press, Piscataway, NJ, 2004.
7. Krishnan, R., *Electric motor drives: modeling, simulation and control*. N J, Upper Saddle River, Prentice Hall, Inc., 2001.
8. Imtiyaz, T., Prakash A., Bakhsh, F. I., Jain, A., Modelling and Analysis of Indirect Field-Oriented Vector Control of Induction Motor (IM), in: *Power Electronics and High Voltage in Smart Grid. Lecture Notes in Electrical Engineering*, Gupta, A.R., Roy, N.K., Parida, S.K. (eds), Vol 817. Springer, Singapore, 2022. https://doi.org/10.1007/978-981-16-7393-1_18
9. Imtiyaz, T., Prakash A., Bakhsh, F. I., Jain, A., Modelling and Analysis of Vector Controlled Doubly Fed Induction Generator (DFIG), in: *Power Electronics and High Voltage in Smart Grid. Lecture Notes in Electrical Engineering*, Gupta, A.R., Roy, N.K., Parida, S.K. (eds), Vol 817. Springer, Singapore, 2022. https://doi.org/10.1007/978-981-16-7393-1_21
10. Iqbal, A., Bakhsh, F. I., Singh, G. K., Operation and Testing of Indirect Field Oriented Control of Asymmetrical Six-Phase Open-Ended Winding Induction Machine Using Hardware-in-Loop (HIL) Emulator. *Int. Trans.*

Electr. Energy Syst., 2023, Article ID 4623140, 2023. https://doi.org/10.1155/2023/4623140
11. Iqbal, A., Singh, G. K., Integrated battery charger of EVs using six-phase propulsion system. *Electr. Eng.*, 2023. https://doi.org/10.1007/s00202-023-01807-5
12. Iqbal, A., Singh, G. K., Pant, V., Vector control of asymmetrical six-phase synchronous motor. *Cogent Eng.*, 3, Article: 1134040. https://doi.org/10.1080/23311916.2015.1134040

7

A Review on Self-Balanced Switched-Capacitor Multilevel Converter

Dhananjay Kumar[1]*, Kasinath Jena[2], Jitendra Kumar Tandekar[3], Niraj Kumar Dewangan[4] and Vishal Rathore[5]

[1]*Department of Electrical Engineering, Government Engineering College, Siwan, Bihar, India*

[2]*Department of Electrical and Electronic Engineering, Arka Jain University, Jamshedpur, India*

[3]*Department of Electrical Engineering, Samrat Ashok Technological Institute, Vidisha, India*

[4]*Department of Mechatronics Manipal Institute of Technology, Manipal Academy of Higher Education, Manipal, Karnataka, India*

[5]*Department of Electrical and Electronics Engineering, Bhopal Institute of Technology and Science, Bhopal, India*

Abstract

The multilevel converter technology presents a promising solution across a spectrum of power applications, ranging from low to high power. This technology offers several advantages, including a relatively low total harmonic distortion (THD) and the generation of high-quality output waveforms. Multilevel inverter (MLI) topologies have garnered significant attention due to their distinct benefits, such as reduced voltage stress on switches, smaller filter requirements, and enhanced electromagnetic compatibility (EMC). Recent developments have focused on MLI configurations with reduced device counts (RDC), which promise improved reliability and performance over traditional MLIs. MLIs have found extensive applications in various industries, including electric vehicles, ship propulsion, rolling mills, the paper industry, and metal forming processes. Despite their utility, conventional MLI topologies like cascaded H-bridge (CHB), neutral point-clamped (NPC), flying capacitor (FC) inverters, T-type inverters, and modular multilevel converters (MMC) face inherent limitations. These limitations include the lack of inherent voltage boosting capabilities and challenges related to voltage balancing

*Corresponding author: dhananjayk00@gmail.com

Salman Ahmad, Farhad Ilahi Bakhsh, and P. Sanjeevikumar (eds.) Multilevel Converters, (127–146)

among capacitors, resulting in complex circuitry, control systems, and increased costs. To address these challenges, researchers and engineers are exploring novel MLI topologies with reduced component counts, aiming for optimal efficiency and cost-effectiveness. Switched capacitor (SC)-based MLIs have emerged as a promising alternative, offering inherent voltage boosting and capacitor voltage self-balancing capabilities, alongside high-resolution voltage waveforms. This work focuses on the development of innovative MLI topologies with minimal component counts, high voltage gain, cost-effectiveness, and reduced voltage stress on power switches. Specifically, five-level SCMLI topologies are proposed for medium- and high-power applications, featuring bipolar voltage generation without a backend HB, and capacitor voltage self-balancing through a series/parallel approach. PWM control techniques are employed to regulate power switches effectively in the proposed single-phase five-level topology.

Keywords: Multilevel inverter, self-balanced, switched capacitor, pulse width modulation, total harmonic distortion

7.1 Introduction

The multilevel converter is a promising technology suitable for low, medium, and high-power applications, delivering a relatively low total harmonic distortion (THD) and producing high-quality output waveforms. MLI topologies have gained widespread popularity because of their obvious benefits, which include lower voltage stress across the switches, smaller filter size, and improved electro-magnetic compatibility (EMC). In recent times, MLI topologies with reduced device count (RDC) have become a subject of interest among researchers. Traditional MLIs provide the same number of output levels, but these topologies overcome certain limits and display better reliability. Thus, recent studies have centred on developing topologies that achieve maximum output levels with a minimal number of components. MLIs have found widespread applications over the years, particularly in electric vehicles (EVs), ship propulsion drives, rolling mills, the paper industry, and metal forming processes. However, conventional MLI topologies, such as cascaded H-bridge (CHB), neutral point-clamped (NPC), flying capacitor (FC) inverters, as well as other popular structures like T-type inverters and modular multilevel converters (MMC), suffer from two main limitations. Firstly, they lack inherent voltage boosting capabilities. Secondly, achieving voltage balancing among the capacitors in these inverters is challenging. Moreover, the conventional MLIs require a high number of devices, which adds complexity to the circuitry and control systems, leading to higher expenses and reduced reliability. In addition to the previously mentioned limitations, these topologies also pose challenges related to the necessity of isolated power supplies, intricate design requirements, and the implementation of switching control circuits to effectively eliminate lower-order harmonics. Consequently, researchers and engineers have been actively exploring the development of newer MLI topologies that can significantly reduce the number of required devices compared to conventional designs. The objective is to achieve an optimum component count, which would not only lead to lower costs but also ensure high efficiency in operation. In view of these challenges, switched capacitors (SC)-based MLIs are emerging as a new class of MLIs with the ability to boost the voltage with inherent self-balancing of capacitors' voltages, and high-resolution voltage waveform. This work primarily concentrates on creating innovative topological designs that possess a low number of components, high voltage gain, cost-effectiveness, and minimal voltage stress on power switches. The proposed five-level SCMLI topologies for generating multilevel voltage waveforms with high efficiency for medium- and high-power applications have been proposed. In the topology, a bipolar voltage is generated without the usage of a back-end HB and capacitor voltage self-balancing with a series/parallel approach. Furthermore, the proposed single-phase five-level topology utilizes a PWM control technique to regulate the power switches effectively.

The Multilevel Converter is a promising technology suitable for low, medium, and high-power applications, delivering a relatively low Total Harmonic Distortion (THD) and producing high-quality output waveforms. MLI topologies have gained widespread popularity because to their obvious benefits, which include lower voltage stress across the switches, smaller filter size, and improved Electro-Magnetic Compatibility (EMC). In recent times, MLI topologies with Reduced Device Count (RDC) have become a subject of interest among researchers. Traditional MLIs provide the same number of output levels, but these topologies overcome certain limits and display better reliability. Thus, recent studies have centered on developing topologies that achieve maximum output levels with a minimal amount of components.

MLIs have found widespread applications over the years, particularly in electric vehicles (EVs), ship propulsion drives, rolling mills, the paper industry, and metal forming processes. However, conventional MLI topologies, such as cascaded H-bridge (CHB), neutral point clamped (NPC), flying capacitor (FC) inverters, as well as other popular structures like T-type inverters and modular multilevel converters (MMC), suffer from two main limitations. Firstly, they lack inherent voltage boosting capabilities. Secondly, achieving voltage balancing among the capacitors in these inverters is challenging. Moreover, the conventional MLIs require a high number of devices, which adds complexity to the circuitry and control systems, leading to higher expenses and reduced reliability. In addition to the previously mentioned limitations, these topologies also pose challenges related to the necessity of isolated power supplies, intricate design requirements, and the implementation of switching control circuits to effectively eliminate lower-order harmonics. Consequently, researchers and engineers have been actively exploring the development of newer MLI topologies that can significantly reduce the number of required devices compared to conventional designs. The objective is to achieve an optimum component count, which would not only lead to lower costs but also ensure high efficiency in operation. In view of these challenges, switched capacitors (SC) based MLIs are emerging as a new class of MLIs with the ability to boost the voltage with inherent self-balancing of capacitors' voltages, and high-resolution voltage waveform.

This work primarily concentrates on creating innovative topological designs that possess a low number of components, high voltage gain, cost-effectiveness, and minimal voltage stress on power switches. The proposed 5-level SCMLI topologies for generating multilevel voltage waveforms with high efficiency for medium and high-power applications have been proposed. In the topology, a bipolar voltage is generated without the usage of a back-end HB, and capacitor voltage self-balancing with a series/parallel approach. Furthermore, the proposed single-phase 5-level topology utilizes a PWM control technique to regulate the power switches effectively. Through experimental testing, the viability of each of the suggested topologies is confirmed and validated. The significant benefits of the proposed MLIs are reduced voltage stress, a high gain, the ability to operate at any power factor, the reduction of switching components, and high efficiency. Each of the suggested topologies has undergone simulation using the MATLAB/Simulink environment and practical experimentation utilizing the dSPACE-1104 controller.

7.2 Literature Review

Within the domain of SCMLIs, numerous review papers and literature are available, each focusing on advancements in topological modifications that aim to augment the voltage levels while minimizing the quantity of switching components in the system. A thorough examination of the existing literature concerning SCMLIs with reduced device counts has been carried out, and the significant findings are detailed in the subsequent section.

Growing demands of RES not only reduce reliance on fossil fuels but also increase the efficient conversion and satisfy increasing energy demand. This is mostly due to widespread worry about environmental concerns and market price volatility of fuels. Recently, significant attention has been given to the development of SCMLIs. The widespread popularity of SCMLIs is largely due to the boosting ability, natural balancing of capacitors voltage, the reduced voltage stress on the switching components, and compact size. SC topologies are commonly divided into two categories: single-stage and two stages.

In the case of two-stage topologies, the first stage generates unipolar voltage, and the second stage generates bipolar voltage *via* an HB. Nevertheless, single-stage topologies eliminate the need for any HB circuitry to generate bipolar voltage. Two-stage SC topologies have the potential to enhance supply voltage and operate under inductive loading conditions with inherent self-balancing capabilities [5–7]. The non-isolated DC supply and unidirectional switches also have other advantages. The primary limitation of these designs is that the voltage stress on the HB switch is equal to the peak load voltage. Due to this characteristic, these are unsuitable for medium- and high-voltage applications. Moreover, the utilization of high-rated HB switches results in an escalation of the inverter's overall cost. Multiple SC topologies that can be used with renewable power and contemporary EVs are described in [8–10]. These topologies require the implementation of unidirectional switches and are designed in a modular fashion. The presence of HB increases the peak inverse voltage equal to the load voltage. Additional DC supply and more uncontrolled switches are required.

A new modular SC multilevel converter is proposed in [11, 12], which uses fewer power electronic components to significantly boost the number of levels. This configuration requires a combination of unidirectional and uncontrolled switches. HBs are also utilized to generate bipolar voltage levels on the rear end. [13, 14] suggest setups of five-level SC topologies that are useful for medium voltage applications. Both topologies can be cascaded to increase their voltage level. The augmented quantity of part components and voltage stress in the back-end half-bridge, equivalent to the maximum values of the load terminal voltage, renders them costly and prone to unreliability. In [15], a dual T-type SC structure is introduced as a solution to reduce the impulse charging current and enhance the input voltage. The design incorporates unidirectional and bidirectional switches, and the voltage stress across the switches remains within the supply voltage range. Five levels of SC boost inverter for electric vehicles, grid-tied systems, and variable-speed drives have been presented in [16]. The reduced voltage stress is one of the key benefits of this inverter. However, the switching components stand high. Five levels of ANPC topology with unity gain are proposed in [17]. The inverter is suitable for photovoltaic applications.

The proposed topology's shortcomings include a lack of boost capability and many switching components.

A T-type five-level topology with self-balanced capacitors is proposed in [18]. This structure incorporates both unidirectional and bidirectional switches as well as uncontrolled switches. Additionally, the stress associated with switching components remains high. [19] proposed a generalized SC structure. However, when more voltage gain is required, these inverters suffer from a higher component count and experience high PIV across the switching components. A transformerless inverter for solar applications has been introduced to reduce leakage current [20]. The inverter offers several advantages, including full utilization of the DC bus voltage, self-balancing capacitors, and zero leakage current. However, it lacks voltage boosting capability and has a drawback of requiring a few switches with high voltage equal to or greater than the peak load voltage. Furthermore, these facts contribute to the increase in power losses and expenses associated with the inverter.

The principal research of anxiety for transformerless PV systems is the problem of safety. These safety issues are mainly due to the leakage current. The leakage current in these inverters leads to several issues such as grid current ripples, system losses, and electromagnetic interference. In [21], a solution called the HB zero-voltage state rectifier (HBZVR-D) is proposed, which combines the advantages of the low-loss alternating current decoupling approach with the total leakage current reduction achieved through the CMV clamping method.

To address the problem of leakage current, a widely employed strategy [22–24] involves isolating the photovoltaic (PV) system from the grid during null or freewheeling states. These topologies incorporate additional switches in the full-bridge configuration to achieve source isolation. Consequently, the common-mode path is disconnected during zero modes, effectively reducing the leakage current.

A novel three-phase transformerless H7 topology has been developed with the aim of diminishing leakage current and enhancing performance in various aspects, including total harmonic distortion (THD), voltage linearity, ripples, and voltage stability [25]. There is a risk of leakage current going directly from the photovoltaic system to the grid because there is no galvanic isolation between the two. When the PV system is grounded, stray capacitance forms, which is charged and discharged in response to variations in potential and causes a large amount of leakage current. This issue is addressed in [26–29]. The topology outlined in [30] employs a half-bridge configuration to produce a multilevel waveform, resulting in a reduced output voltage peak compared to the full-bridge topology. Additionally,

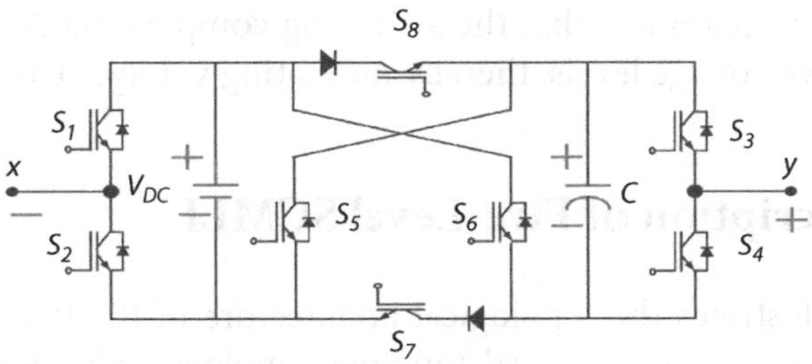

Figure 7.1 Proposed five-level SC topology.

the maximum modulation index is limited. A unity gains seven-level SC topology presented in [31]. The suggested topology is associated with HB for bipolar voltage generation. Voltage stress corresponds to the maximum load voltage experienced across the HB switches. Hence, such an inverter is suitable for low-voltage applications. Moreover, it requires multiple series capacitors and uncontrolled switches. In [32], an eight-switch, seven-level ANPC inverter with lower voltage stresses has been described. The inverter can handle high and medium voltages, making it useful for a variety of uses. The requirement for 50% of the dc-link voltage is a critical characteristic of the anticipated topology.

The applicability of a seven-level (7L) inverter design has been showcased for grid-connected renewable energy applications [33]. The notable features of the suggested topology are its ability to minimize voltage stress, enhance input voltage capability, and lower leakage currents. By contrast, this inverter achieves the desired voltage level with the employment of a huge number of switches. In, [34–39] propose seven-level ANPC topologies with boosting capacities of 1.5 times the supply voltage. These

Table 7.1 Valid switching combination of the proposed five-level topology.

State	S_1	S_2	S_3	S_4	S_5	S_6	S_7	S_8	Effect of "C"	$v_{xy}(t)=v_0$
1	0	1	0	1	0	0	1	1	C	0
	1	0	1	0	0	0	1	1	C	0
2	0	1	1	0	0	0	1	1	C	$+1V_{DC}$
3	0	1	1	0	0	1	0	0	D	$+2V_{DC}$
4	0	0	0	1	0	0	1	1	C	$-1V_{DC}$
5	1	0	0	1	1	0	0	0	D	$-2V_{DC}$

configurations guarantee that the switching components do not experience excessive voltage levels, thereby mitigating voltage stress on them.

7.3 Description of Five-Level SCMLI

Figure 7.1 illustrates the topological architecture of the five-level boost-type SC inverter. The suggested topology employs eight switches and a capacitor to produce a voltage waveform with five levels. It utilizes two distinct types of switches: IGBTs with a series-connected diode and IGBTs with a parallel-connected diode.

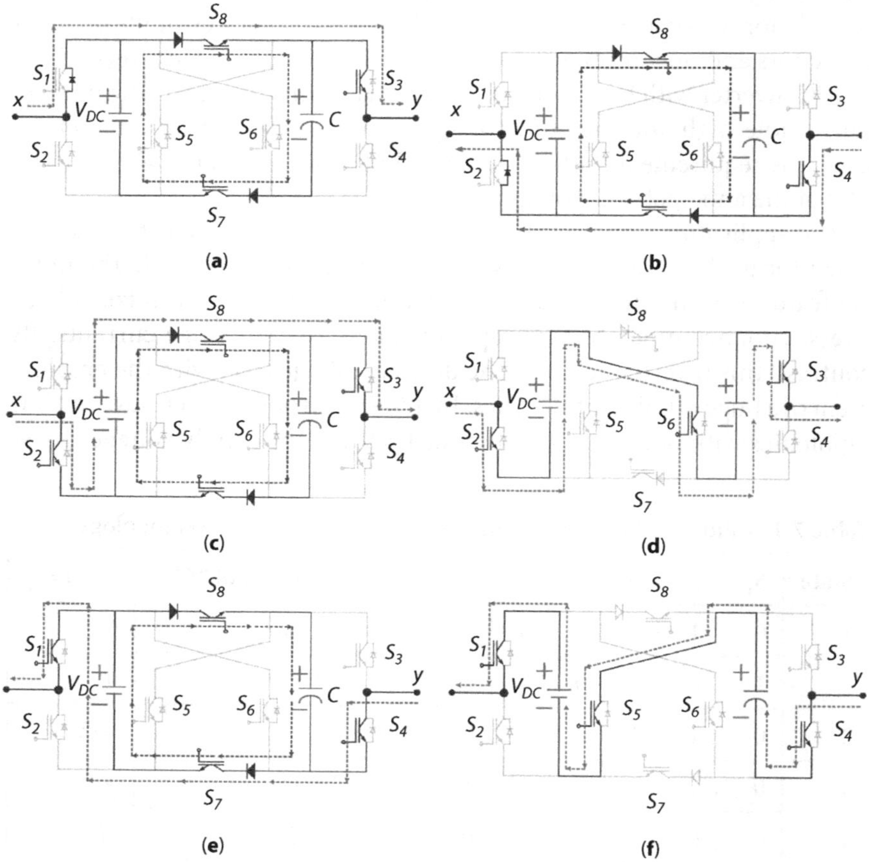

Figure 7.2 Various operational conditions of the suggested SCMLI (a, b) v_o=0, (c) v_o = − V_{DC}, (d) v_o = − V_{DC}, (e) v_o = − V_{DC}, and (f) v_o = − V_{DC}.

The output voltage (Vo) between terminals x and y is expressed in five levels according to the valid switching combinations mentioned in Table 7.1. These levels are $\pm 1V_{DC}$, $\pm 2V_{DC}$, and 0. The charging and discharging procedures of the capacitors are represented by the letters "C" and "D", correspondingly.

The key features of the proposed topology are as follows:

a) Two times the source voltage makes up the voltage gain.
b) To attain the same voltage level, a reduced quantity of switching components is necessary.
c) Capacitor voltage is not required to be balanced.
d) 50% switches are operated for any voltage level.
e) HBs are not needed to generate the bipolar voltage.
f) The voltage stress on the switching components is decreased.

7.3.1 Operational Analysis

The suggested five-level multilevel inverter is analyzed in Figure 7.2 to demonstrate the various output voltage states. The following are brief studies of various output voltage levels.

State-1: Output voltage (v_0=0)

The load terminals are shortened in this voltage state, as shown in Figure 7.2(a, b). Additionally, the capacitor C is charged *via* a parallel connection to the dc voltage source V_{DC}. As a result of this state, v_0 equals to 0, and V_C equals to V_{DC}.

State-2: Output voltage v_0=+1V_{DC}

In this state of operation, the switches S_2, S_3, S_7, S_8 are turned on and S_1, S_4, S_5, S_6 are turned off simultaneously to achieve this voltage level (+2V_{DC}). In this state, the capacitor is charged up to a voltage V_{DC} through the turning of the switches S_7, S_8. The equivalent circuit diagram of this state is shown in Figure 7.2(c).

State-3: Output voltage v_0=+2V_{DC}

In this working state, the switches S_2, S_3, S_6 are turned on and switches S_1, S_4, S_5, S_7, S_8 are turned off simultaneously to achieve this desired voltage level (+2V_{DC}). In this state of operation, the capacitor is brought in series with the supply source; hence, the capacitor releases its stored energy along with the supply to the load. The equivalent circuit diagram for the load current is shown in Figure 7.2(d).

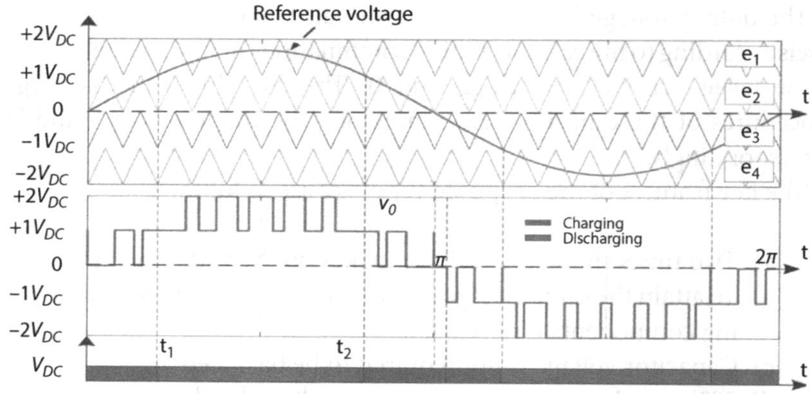

Figure 7.3 Discharging time period of switched capacitors.

State-4: output voltage v_0=-V_{DC}

In this state, the switches S_1, S_4, S_7, S_8 are turned on and switches S_2, S_3, S_5, S_6 are turned off simultaneously to achieve this desired level (-1V_{DC}). During this state of operation, the capacitor is brought in parallel with the DC source through the power switches S_7, S_8 and charged up to a voltage equivalent to the supply source (V_{DC}). The equivalent circuit diagram for this state is shown in Figure 7.2(e).

State-5: output voltage v_0=-2V_{DC}

In this working state, the switches S_1, S_4, S_5 are turned on and S_2, S_3, S_7, S_8 is turned off simultaneously to obtain this desired level v_0=-2V_{DC}. In this operating state, the capacitor is in series with the supply voltage, and hence it discharges its stored energy to the load along with the supply voltage. The equivalent circuit diagram for this operating state is shown in Figure 7.2(f).

7.3.2 Self-Balancing Mechanisms of Capacitor and Capacitance Calculation

In Figure 7.3, the capacitor "C" is depicted to be undergoing charging and discharging processes at voltage levels of 0, ±V_{DC}, and ±2V_{DC}, respectively. To charge the capacitor in parallel with the input supply voltage (V_{DC}), switches S_7 and S_8 are utilized in the proposed design. Subsequently, when the capacitor is connected in series with the load, it discharges its stored energy. Throughout one cycle of the output voltage, multiple charging durations of the capacitor take place, resulting in the dynamic maintenance of the source voltage V_{DC} by the capacitor voltages, albeit with some voltage ripples. This characteristic facilitates the automatic balancing of the capacitor voltage.

The capacitance discharge values are dependent on the load, the longest discharge time, and the load's power factor, where t_1, and t_2 denote the start and end of the ith capacitor's longest discharging period, respectively. In this context, $t_{x,I}$ and $t_{y,i}$ represent the beginning and end points of the longest discharge period for the i_{th} capacitor, respectively. Consequently, the total discharge amount of capacitor C within the time interval $[t_1, t_2]$ can be approximated in the following manner:

$$\Delta Q_o = \int_{t_{x,i}}^{t_{y,i}} I_m sin(2\pi f_o t)dt \tag{7.1}$$

where ϕ denotes power factor angle. Also, f_0 and I_m denotes the frequency of the output voltage and current, respectively. For the suggested topology, the capacitor's longest discharge durations ($t_{y,i}$) are (t_1,t_2). The values of "t1" and "t2" can be determined using the equations provided in [21].

$$t_1 = \frac{\sin^{-1}\left(\frac{1}{2}\right)}{2\pi f_o} \tag{7.2}$$

$$t_2 = \pi - \frac{\sin^{-1}\left(\frac{1}{2}\right)}{2\pi f_o} \tag{7.3}$$

The capacitance value of the capacitor is critical in the inverter because as the capacitance value increases, the inverter becomes larger and bulkier. As a result, the optimal capacitance value is as follows:

$$C \geq \frac{\Delta Q_C}{\Delta V_C} \tag{7.4}$$

Additionally, when a pure resistive (R) load is used, the capacitor ripple voltage is calculated using (7.5)

Table 7.2 List of the parameters used for the simulation and experimental studies.

Parameter	Specification
Direct current source	50 V
Output frequency (f_0)	50 Hz
Switching frequency	2 kHz and 100 Hz
Capacitors (C)	1,100 μF
Modulation index (M)	0.95
RL-load	R=40 Ω, L=100 mH, and R=20 Ω, L=50 mH

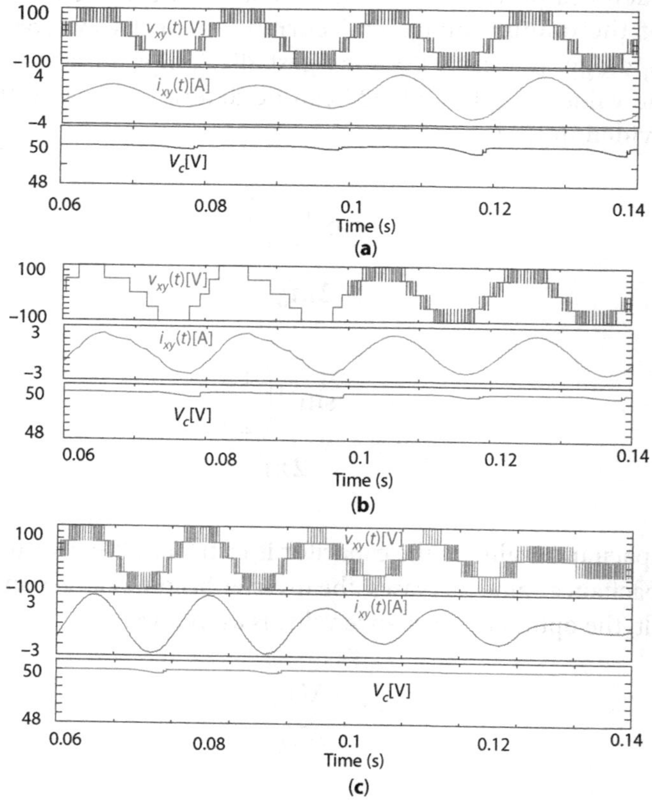

Figure 7.4 showcases the simulation results for (a) a step change in load, (b) switching frequency, and (c) a change in modulation index.

$$\Delta V_C = \frac{2V_{DC}}{2\pi f_o RC}[\pi - 2t] \tag{7.5}$$

7.4 Results

The suggested five-level structure is verified in MATLAB/Simulink to assess the performance of the proposed SC topology.

7.4.1 Simulation Results

Table 7.2 displays the simulation and experimental parameters of the suggested work. The simulation result obtained under dynamic circumstances is shown in Figure 7.4.

Figure 7.4(a) illustrates the load variation transitioning from (R = 40 Ω, L = 100 mH) to (R = 20 Ω, L = 50 mH). As a result of this load parameter change, the peak load current rises from 1.7 to 3.3 A. Notably, the capacitor voltages, which are maintained at 50 V, remain unaffected by the magnitude of the output current. As a result, the preceding results prove that the anticipated topology is self-balancing.

As explained in Figure 7.4(b), the five-level structure performs satisfactorily under dynamic conditions, i.e., change in switching frequency. The capacitor voltage retained its self-balancing property under dynamic conditions, and the magnitude of output voltage waveform demonstrates its boosting capabilities.

Additionally, as indicated in Figure 7.4(c), a change in the modulation index is considered while validating the suggested topology. Modulation index is modified from 0.95 to 0.75 and then from

0.75 to .02. With a modulation index of 0.75, five levels are obtained and a peak voltage of 100 V is reached. With a modulation index of 0.75 and 0.2, there are three levels. All of these modifications in modulation indexes have resulted in a good result. Regardless of the modulation index change, the capacitor voltage is maintained at 50 V.

7.4.2 Findings From the Experiments

The experimental results presented in Figure 7.5 illustrate the performance of the proposed SC inverter topology under dynamic conditions. The tests reveal that the SC inverter exhibits rapid and satisfactory responses

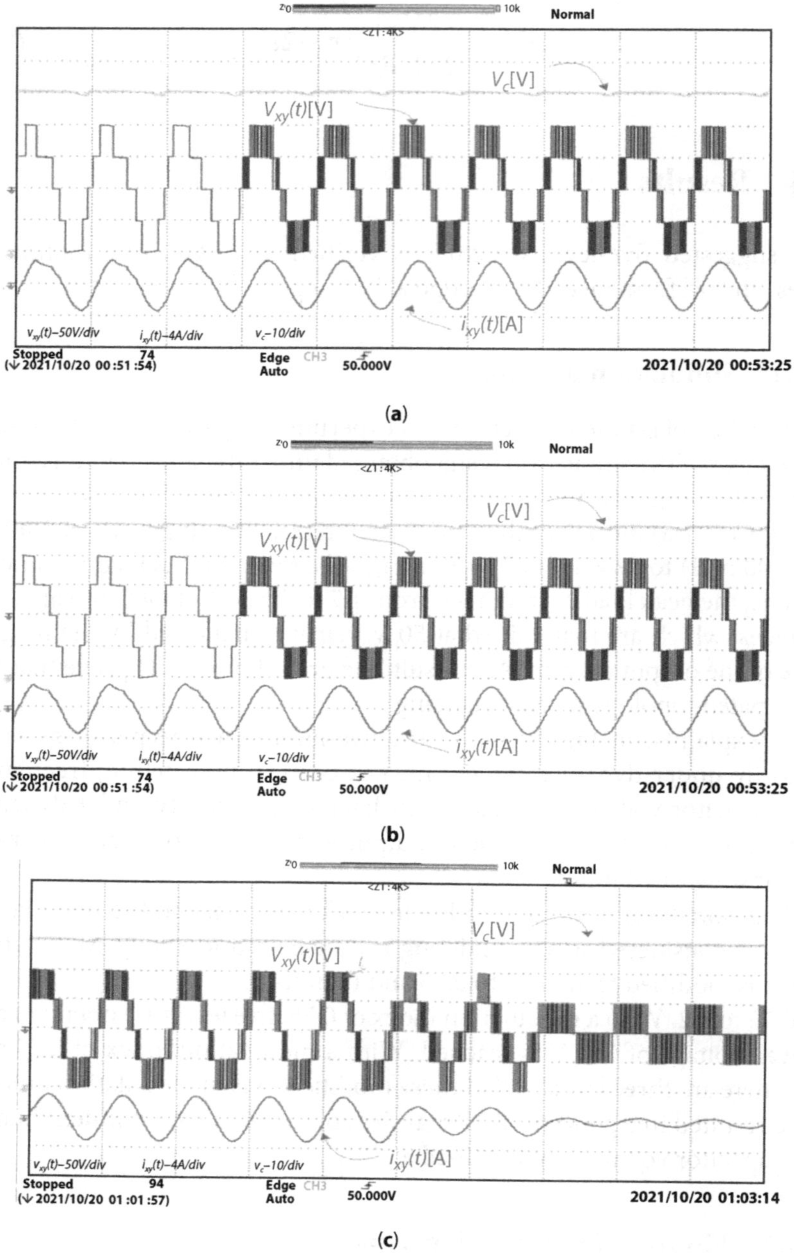

Figure 7.5 The experimental outcomes include (a) load change with a step change, (b) switching frequency variations, and (c) changes in modulation index.

to variations in load, frequency, and modulation index when subjected to dynamic conditions.

The experimental waveforms for a step change in load are shown in Figure 7.5(a). It is achievable to observe a five-level output, which is identical to the simulation result. With a 50-V input, the output peak voltage approaches 100 V. Clearly, the capacitor voltage balances at 50 V without the use of sensors or other additional ways. As a result, the five output levels, natural balance, and ability to increase the output voltage can all be demonstrated.

Indicated in Figure 7.5(b) is the output voltage due to the change in switching frequency. Additionally, capacitors are self-balancing and exhibit very low voltage ripples. A quick transient response from the inverter can be observed in both cases (100 to 2 kHz).

The prototype is designed to function across various modulation index (M) values, specifically 0.95, 0.75, and 0.2. This capability allows for the modification of the prototype's output level. Furthermore, the output waveform and amplitude exhibit variations with different values of Ma. In Figure 7.5(c), the five output levels are condensed into three.

As can be observed, the voltages across capacitor remain constant at various modulation index or output levels. Consequently, modulation does not impact the feasibility of preserving the natural balance of the capacitor. Furthermore, the simulation outcomes align closely with the experimental results of the prototype, particularly in dynamic scenarios.

7.5 Conclusion

MLIs have been widely investigated and developed for a wide range of industrial and home applications. Many studies and descriptions of the topological and control challenging scheme of MLIs have been published over the years. Significant efforts are being made to reduce the complex issues associated with the topological design, operation, and control scheme of MLIs. It is necessary to explore, understand, and enhance current topologies to enhance their suitability for specific applications. Moreover, these topologies come with different limitations, including high voltage stress, an increased number of switching components, restricted gain capabilities, and the necessity for multiple isolated sources. These factors contribute to their complexity and higher cost. This necessitates the development of SCMLIs with reduced part count and voltage stresses. The inverter operated both low switching frequency (f_{sw}<100 Hz) and high switching frequency. Capacitors are self-balanced. The proposed work has reduced part count. As a result, this thesis made significant contributions in the development and validation of three distinct SCMLIs tailored for low- and medium-power applications, along with the implementation of versatile control techniques.

References

1. J. Rodriguez, Jih-Sheng Lai, and Fang ZhengPeng, "Multilevel inverters: a survey of topologies, controls, and applications," *IEEE Trans. Ind. Electron.*, vol. 49, no. 4, pp. 724-738, Aug. 2002, DOI: 10.1109/TIE.2002.801052
2. J. I. Leon, S. Vazquez, and L. G. Franquelo, "Multilevel Converters: Control and Modulation Techniques for Their Operation and Industrial Applications," in *Proc. IEEE*, vol. 105, no. 11, pp. 2066-2081, Nov. 2017, doi: 10.1109/JPROC.2017.2726583.
3. K. K.Gupta, and S.Jain, "A novel multilevel inverter based on switched DC sources." *IEEE Trans. Ind. Electron.*, vol.61, no.7 , pp.3269-3278,2013.
4. K. K. Gupta, A. Ranjan, P. Bhatnagar, L. K. Sahu, and S. Jain, "Multilevel Inverter Topologies With Reduced Device Count: A Review," in *IEEE Trans. Power Electron.*, vol. 31, no. 1, pp. 135-151, Jan. 2016, doi: 10.1109/TPEL.2015.2405012.
5. Y. Hinago and H. Koizumi, "A Switched-Capacitor Inverter Using Series/Parallel Conversion With Inductive Load," *IEEE Trans. Ind. Electron.*, vol. 59, no. 2, pp. 878-887, Feb. 2012, doi: 10.1109/TIE.2011.2158768
6. E. Babaei and S. S. Gowgani, "Hybrid Multilevel Inverter Using Switched Capacitor Units," in *IEEE Trans. Ind. Electron.*, vol. 61, no. 9, pp. 4614-4621, Sept. 2014, doi: 10.1109/TIE.2013.2290769
7. Y. Ye, K. W. E. Cheng, J. Liu, and K. Ding, "A Step-Up Switched-Capacitor Multilevel Inverter With Self- Voltage Balancing," *IEEE Trans. Ind. Electron.*, vol. 61, no. 12, pp. 6672-6680, Dec. 2014, doi: 10.1109/TIE.2014.2314052.
8. S. R. Raman, Y. C. Fong, Y. Ye, and K. W. Eric Cheng, "Family of Multiport Switched-Capacitor Multilevel Inverters for High-Frequency AC Power Distribution," in *IEEE Trans. Power Electron.*, vol. 34, no. 5, pp. 4407-4422, May 2019, DOI: 10.1109/TPEL.2018.2859030.
9. S. R. Raman, K. W. E. Cheng, and Y. Ye, "Multi-Input Switched-Capacitor Multilevel Inverter for High- Frequency AC Power Distribution," in *IEEE Trans. Power Electron.*, vol. 33, no. 7, pp. 5937-5948, July 2018, doi: 10.1109/TPEL.2017.2742525.
10. S. Ramaiah, N. Lakshminarasamma, and M. K. Mishra, "Multisource Switched-Capacitor Based Boost Multilevel Inverter for Photovoltaic-Based Systems," *IEEE Trans. Power Electron.*, vol. 35, no. 3, pp. 2558-2570, March 2020, DOI: 10.1109/TPEL.2019.2924649

11. J. Ebrahimi, E. Babaei and G. B. Gharehpetian, "A New Multilevel Converter Topology With Reduced Number of Power Electronic Components," *IEEE Trans. Ind. Electron.*, vol. 59, no. 2, pp. 655-667, Feb. 2012, doi: 10.1109/TIE.2011.2151813
12. M. N. H. Khan, M. Forouzesh, Y. P. Siwakoti, L. Li and F. Blaabjerg, "Switched-Capacitor Integrated (2n + 1)-Level Step-Up Single-Phase Inverter," *IEEE Trans. Power Electron.*, vol. 35, no. 8, pp. 8248-8260, Aug. 2020, doi: 10.1109/TPEL.2019.2963344.
13. M.Saeedian, S. M.Hosseini, and J.Adabi, "Step-up switched-capacitor module for cascaded MLI topologies". *IET Power Electron.*, vol.11, no.7, pp.1286-1296 ,2018..
14. M. Saeedian, S. M. Hosseini and J. Adabi, "A Five-Level Step-Up Module for Multilevel Inverters: Topology, Modulation Strategy, and Implementation," *IEEE J. Emerging Sel. Top. Power Electron.*, vol. 6, no. 4, pp. 2215-2226, Dec. 2018, doi: 10.1109/JESTPE.2018.2819498.
15. S. S. Lee, C. S. Lim, Y. P. Siwakoti and K. Lee, "Dual-T-Type Five-Level Cascaded Multilevel Inverter With Double Voltage Boosting Gain," *IEEE Trans. Power Electron.*, vol. 35, no. 9, pp. 9522-9529, Sept. 2020, doi: 10.1109/TPEL.2020.29736
16. N. Sandeep, J. S. M. Ali, U. R. Yaragatti and K. Vijayakumar, "A Self-Balancing Five-Level Boosting Inverter With Reduced Components," *IEEE Trans. Power Electron.*, vol. 34, no. 7, pp. 6020-6024, July 2019, doi: 10.1109/TPEL.2018.2889785.
17. M. J. Sathik, Z. Tang, Y. Yang, K. Vijayakumar and F. Blaabjerg, "A New 5-Level ANPC Switched Capacitor Inverter Topology for Photovoltaic Applications," *IECON 2019 - 45th Annual Conference of the IEEE Industrial Electronics Society*, 2019, pp. 3487-3492, doi: 10.1109/IECON.2019.8926832
18. D. Niu *et al.*, "A Novel Switched-Capacitor Five-Level T-Type Inverter," *2019 2nd International Conference on Smart Grid and Renewable Energy (SGRE)*, 2019, pp. 1-6, doi: 10.1109/SGRE46976.2019.9021066.
19. M. D.Siddique, B. P.Reddy, A.Iqbal,andS.Mekhilef, "Reduced switch count-based N-level boost inverter topology for higher voltage gain". *IET Power Electron.*, vol.13, no.15, pp. 3505-3509,2020.
20. M. J. Sathik, A. Hota, N. Sandeep and D. Almakhles, "A Single-Stage Common Ground Type Transformerless Five-Level Inverter Topology," *IEEE J. Emerging Sel. Top. Ind. Electron.*, doi: 10.1109/JESTPE.2021.3095125.
21. T. K. S. Freddy, N. A. Rahim, W. Hew and H. S. Che, "Comparison and Analysis of Single-Phase Transformerless Grid-Connected PV Inverters," in *IEEE Trans. Power Electron.*, vol. 29, no. 10, pp. 5358-5369, Oct. 2014, doi: 10.1109/TPEL.2013.2294953.
22. M. Victor, F. Greizer, S. Bremicker, and U. Hübler, "Method of converting a direct current voltage from a source of direct current voltage, more specifically from a photovoltaic source of direct current voltage, into a alternating current voltage," Aug. 12 2008, uS Patent 7,411,802.

23. L. Zhang, K. Sun, Y. Xing, and M. Xing, "H6 transformerless full-bridge pv grid-tied inverters," *IEEE Trans. Power Electron.*, vol. 29, no. 3, pp. 1229–1238, 2013.
24. J. Ketterer, H. Schmidt, and C. Siedle, "Inverter for transforming a dc voltage into an ac current or an ac voltage," *Europe Patent*, vol. 1, no. 369, p. 985, 2003.
25. T. K. S. Freddy, N. A. Rahim, W. Hew and H. S. Che, "Modulation Techniques to Reduce Leakage Current in Three-Phase Transformerless H7 Photovoltaic Inverter," in *IEEE Trans. Ind. Electron.*, vol. 62, no. 1, pp. 322-331, Jan. 2015, doi: 10.1109/TIE.2014.2327585.
26. M. C. Cavalcanti, A. M. Farias, K. C. de Oliveira, F. A. S. Neves, and J.L. Afonso, "Eliminating leakage currents in neutral point clamped inverters for photovoltaic systems," *IEEE Trans. Ind. Electron.*, vol. 59, no. 1, pp. 435-443, Jan. 2012.
27. O. Lopez, F. D. Freijedo, A. G. Yepes, P. Fernandez-Comesaa, J. Malvar, R. Teodorescu, and J. Doval- Gandoy, "Eliminating ground current in a transformerless photovoltaic application," *IEEE Trans. Energy Convers.*, vol. 25, no. 1, pp. 140-147, Mar. 2010.
28. Y. Wang, and R. Li, "Novel high-efficiency three-level stacked-neutral-point- clamped grid-tied inverter," *IEEE Trans. Ind. Electron.*, vol. 60, no. 9, pp. 3766-3774, Sept. 2013.
29. F. Bradaschia, M. C. Cavalcanti, P. E. P. Ferraz, F. A. S. Neves, E. C.Santos, and J. H. G. M. Silva, "Modulation for three-phase transformerless z-source inverter to reduce leakage currents in photovoltaic systems," *IEEE Trans. Ind. Electron.*, vol. 58, no. 12, pp. 5385-5395, Dec. 2011.
30. A. Kadam and A. Shukla, "A multilevel transformerless inverter employing ground connection between pv negative terminal and grid neutral point," *IEEE Trans. Ind. Electron.*, vol. 64, no. 11, pp. 8897–8907, 2017.
31. J.S Choi and F. Kang, "Seven-Level PWM Inverter Employing Series-Connected Capacitors Paralleled to a Single DC Voltage Source," *IEEE Trans. Ind. Electron.*, vol. 62, no. 6, pp. 3448-3459, June 2015, doi: 10.1109/TIE.2014.2370948.
32. Y. P. Siwakoti, A. Mahajan, D. J. Rogers and F. Blaabjerg, "A Novel Seven-Level Active Neutral-Point- Clamped Converter With Reduced Active Switching Devices and DC-Link Voltage," *IEEE Trans. Power Electron.*, vol. 34, no. 11, pp. 10492-10508, Nov. 2019, doi: 10.1109/TPEL.2019.2897061.
33. M. J. Sathik, N. Sandeep and F. Blaabjerg, "High Gain Active Neutral Point Clamped Seven-Level Self- Voltage Balancing Inverter," *IEEE Trans. Circuits Syst. II Express Briefs*, vol. 67, no. 11, pp. 2567-2571, Nov. 2020, doi: 10.1109/TCSII.2019.2955318.
34. A. Iqbal, M. D. Siddique, J. S. M. Ali, S. Mekhilef and J. Lam, "A New Eight Switch Seven Level Boost Active Neutral Point Clamped (8S-7L-BANPC) Inverter," in *IEEE Access*, vol. 8, pp. 203972-203981, 2020, doi: 10.1109/ACCESS.2020.3036483.

35. S. S. Lee, Y. Bak, S. -M. Kim, A. Joseph and K. -B. Lee, "New Family of Boost Switched-Capacitor Seven-Level Inverters (BSC7LI)," *IEEE Trans. Power Electron.*, vol. 34, no. 11, pp. 10471-10479, Nov. 2019, doi: 10.1109/TPEL.2019.2896606
36. S. S. Lee, C. S. Lim, Y. P. Siwakoti and K. -B. Lee, "Hybrid 7-Level Boost Active-Neutral-Point- Clamped (H-7L-BANPC) Inverter," *IEEE Trans. Circuits Syst. II Express Briefs*, vol. 67, no. 10, pp. 2044-2048, Oct. 2020, doi: 10.1109/TCSII.2019.2946860
37. S. S. Lee and K. Lee, "Dual-T-Type Seven-Level Boost Active-Neutral-Point-Clamped Inverter," *IEEE Trans. Power Electron.*, vol. 34, no. 7, pp. 6031-6035, July 2019, doi: 10.1109/TPEL.2019.2891248.
38. J. Zeng, W. Lin and J. Liu, "Switched-Capacitor-Based Active-Neutral-Point-Clamped Seven-Level Inverter With Natural Balance and Boost Ability," *IEEE Access*, vol. 7, pp. 126889-126896, 2019, doi: 10.1109/ACCESS.2019.2927351.
39. Yu, B. Chen, W. Yao and Z. Lu, "Hybrid Seven-Level Converter Based on T-Type Converter and H- Bridge Cascaded Under SPWM and SVM," *IEEE Trans. Power Electron.*, vol. 33, no. 1, pp. 689-702, Jan. 2018, doi: 10.1109/TPEL.2017.2664068.

8

13 Level Switched-Capacitor Multilevel Converter with High Gain for Grid Connected Solar Photovoltaic Applications

Hasan Iqbal[1,2*], Mohammad Tayyab[1,3†], Haroon Rehman[1], Adil Sarwar[1] and Md Reyaz Hussan[1]

[1]*Department of Electrical Engineering, Aligarh Muslim University, Aligarh, Uttar Pradesh, India*

[2]*Department of Electrical and Computer Engineering, Florida International University, Miami, FL, USA*

[3]*Electrical and Computer Engineering Department, Texas A & M University at Qatar, Doha, Qatar*

Abstract

Solar photovoltaic (SPV) systems have traditionally been used as standalone autonomous systems, but recently there has been a shift in focus to their integration with the grid. PV string can be formed by connecting multiple PV modules in series, which is used in an SPV application to generate a high voltage and then that voltage is converted to an ac voltage through a conventional two-level inverter, thus producing voltage with large harmonic content. To address this issue, multilevel converters (MLCs) have been studied due to their ability to increase power quality, minimize harmonic distortion, and shrink the size of the filtering unit. As the output voltage level increases, more switching components are required but with better power quality. As a result, switched-capacitor multilevel converters (SCMLCs) are now becoming increasingly popular for grid integration of solar photovoltaic systems. The focus now is towards the minimal devices for SCMLC. Consequently, less money is spent, and smaller converters are made. Its greater voltage boosting makes it appropriate for grid integration of low-voltage sources like SPV and fuel cells. A 13-level SCMLC topology is considered for the grid-connected operation of SPV in our case. It produces thrice voltage boosting by utilizing reduced component count.

**Corresponding author*: hiqba003@fiu.edu

†*Corresponding author*: mtayyab1@myamu.ac.in

Salman Ahmad, Farhad Ilahi Bakhsh, and P. Sanjeevikumar (eds.) Multilevel Converters, (147–164)

Keywords**:** SCMLI, solar PV system, grid, THD, power quality, NLC

Nomenclature

CHB	Cascaded H-bridge
DSP	Digital signal processor
FC	Flying capacitor
HIL	Hardware in loop
MLC	Multilevel converter
MLI	Multilevel inverters
MPC	Model predictive control
NLC	Nearest level control
NPC	Neutral point-clamped
PR	Proportional resonant
RDC	Reduced device count
SCMLC	Switched capacitor multilevel inverter
SHE	Selective harmonic elimination
SMC	Sliding mode control
SPV	Solar photovoltaic
VSI	Voltage source inverter

8.1 Introduction

Traditional energy sources are being quickly depleted due to the skyrocketing demand for electrical energy [1]. Due to issues including the depletion of conventional resources, environmental risks, high fuel prices, and growing energy demand, there has been a push toward alternative energy sources. A few examples of renewable energy sources that have gained popularity recently and are now making a significant contribution to the world's energy production include solar, wind, fuel cells, ocean energy, and biogas. The current emphasis is on increasing the usage of such "non-conventional" grid sources to meet anticipated energy requirements [2]. This has led to a lot of study into the potential of power plants that run on renewable energy sources (RES). Advances in power electronics have increased the demand for energy from renewable alternatives like wind and solar power [3, 4]. The development of power electronics equipment and technology has made it possible for RES to contribute significantly to the power grid. Power electronic inverters are needed because renewable energy sources cannot directly power loads or the grid. DC–AC and DC–DC converters are interfaces responsible for integrating energy from renewable sources into the grids or loads [5, 6].

A power inverter is an essential part of renewable energy power conversion systems, which convert DC electricity into AC power to be used by the grid and loads [7]. Inverters are being studied meticulously to determine which are most suitable for integrating renewable energy sources into the grid or meeting local load requirements [3]. Voltage source inverters (VSIs) with two levels of voltage are typically used for medium- and low-power applications since power semiconductors have voltage and power rating limitations [2]. Considering that these inverters generate a great deal of harmonics, it is necessary to install costly and bulky passive lowpass filters before they can be used with the utility grid [1]. In addition, the high-frequency switching function of these inverters makes them vulnerable to significant switching losses. It is also important to mention that these inverters are not designed for use in high-power applications because of the high voltage stress and significant switching losses they incur. Due to these constraints, multilevel inverters (MLIs) were developed, which allow for the generation of more than two distinct output voltages [2].

For power conversion systems with medium and high loads, multilevel inverters (MLIs) have quickly become the standard. Its ability to provide a clean, stepped voltage output from a DC supply with manageable harmonics has led to their widespread adoption. An MLI's primary purpose is to create a nearly sinusoidal AC voltage from a different DC voltage source [3, 8].

Baker and Bannister initially proposed the idea of MLIs in 1975. As the topology they came up with in [9], they arranged the power cells in a full-bridge configuration, each of them having its own isolated DC source and four switches. The output of this circuit design, known as a cascaded H-bridge (CHB) inverter, was an ac voltage waveform with discrete steps. It is a three-level voltage source inverter (VSI) design that employs diodes to clamp the switch voltage to the halfway DC link voltage or the voltage across a DC link capacitor. The diode-clamped or neutral-point-clamped (NPC) inverter was invented by Baker in 1980 [10]. The first MLI based on flying capacitors (FC) was introduced in 1992 by Meynard and Foch [11]. The FC MLI is a modest variant of the NPC MLI in which clamping capacitors are used instead of clamping diodes. Traditional MLI topologies include CHB, NPC, and FC configurations. A classical MLI is gaining increasing attention because it operates at high voltage, has low switching losses, and is highly efficient as well as has low electromagnetic interference. This allows MLIs to not only keep up with rising power needs but also boost power quality by lowering harmonic distortion. There are advantages and disadvantages associated with each MLI topology that are different. As a result of the CHB MLI design's requirement for a wide range of voltages, it is required that a great deal of DC voltage sources be incorporated, such as batteries, PV cells, or rectifiers. NPC MLI's drawbacks include the fact that it can only support a fixed number of voltage levels, the need for many clamping diodes, voltage imbalances, inefficient switch use, and unequal power loss. High clamping capacitor counts and capacitor charge imbalance at low switching frequencies are the primary problems with FC MLI. The need for many power semiconductor switches is a frequent drawback of conventional MLIs. As gate driver circuits are required for each switch, the system's complexity and expense rise. It was due to the constraints imposed by classical topologies that novel and hybrid MLI topologies were developed for use in a wide range of contexts to overcome these constraints. The development of MLIs with low component counts and high output voltages is a major area of study [1, 12].

It has been common practice in recent years to develop several different inverter topologies to address the drawbacks of traditional MLIs. Switched-capacitor multilevel inverters (SCMLIs) are one kind that has seen significant growth in recent years. By using fewer isolated DC sources, capacitors, diodes, and power switches, we aim to create compact, low-loss topologies for reducing overall standing voltage as well as costs by using fewer isolated DC sources, power switches, diodes, and capacitors [3, 13–15].

When developing converters for renewable energy systems, voltage boosting capabilities are essential, but conventional MLIs lack these features. For PV panels and fuel cells to work with grid-connected systems, the voltage they produce must be raised. Voltage enhancement is a feature of several inverter topologies, including SCMLIs. Due to their potential as small, DC-source-efficient voltage booster modules, SCMLIs are now the topic of intensive studies [3].

A recent focus on reducing switches, DC supplies, and gate driver circuits has made MLIs more cost-effective for integrating with renewable energy grids while simultaneously improving power quality and fault tolerance [2, 16].

Low-voltage systems like solar photovoltaics, batteries, and fuel cells must have a voltage boost to function properly. There has been a lot of interest in SCMLIs for these kinds of systems among researchers. SC-MLIs are advantageous since they can increase voltage and self-balance capacitors [17].

The remaining chapters are organized as follows: Section 8.2 discusses the SCMLI topology and its comparison with other topologies. Section 8.3 comprises the SCMLI topology operation along with the modulation scheme applied to generate inverter output voltage. A grid-connected SCMLI is shown in Section 8.4. Furthermore, in Section 8.5, the simulation results of the presented inverter are discussed along with the real-time hardware in loop emulator results, which are also presented and discussed. Finally, in Section 8.6, the conclusion and future work are depicted.

8.2 Switched-Capacitor Multilevel Inverters

Literature offers several configurations based on reduced device count (RDC) [18]. Two significant limitations of RDC MLIs are the necessity for several independent DC power sources and the inability to amplify the input voltage. When using low voltage, as in PV systems, increasing the output voltage is crucial to powering loads. By integrating switched-capacitor (SC) units with DC–AC converters, as opposed to traditional two-stage conversion systems, it may be possible to increase efficiency, optimize switching devices, and reduce the size of passive components like inductors and capacitors [19, 20].

The high voltage gain and self-balancing properties of SCMLI topologies made up of series and parallel capacitor combinations make them a popular choice for use in high-voltage applications. SCMLIs use a DC power supply or another charged capacitor to directly charge the capacitors. A discrete output voltage can be produced by discharging the capacitors in series or parallel with or without a direct current (DC) source [21]. SCMLIs enable the use of low-voltage supplies like solar PV sources in high-voltage environments by eliminating the need for a dc–dc converter at an intermediary step. The system becomes more effective and saves money consequently. Self-voltage balancing capacitors in SCMLIs further reduce control complexity [19].

Numerous alternative MLIs based on switching capacitors have been developed because of the growing focus on the use of renewable energy sources. Ingenious SC-based topologies with the intention of reducing the overall number of components have been revealed by a recent study [22].

While capacitors are already used in MLI topologies based on NPC and FC, capacitors in MLI topologies are not an innovative concept. Despite this, the capacitors that are used in these topologies do not possess any boosting capabilities [19].

One of the most researched areas in power electronics research has been switched capacitor (SC) units in multilayer inductors (MLIs) since the beginning of this decade. Compared to other converter options, SCMLI is an appealing converter option for a range of applications due to its ability to enhance voltage and its self-balancing capacitors. SCMLI structures, which employ a broad range of fundamental switching processes to produce a number of distinct output voltage levels, have made significant developments in recent years [21].

Both a high demand for components to allow a higher number of levels and limited power ratings owing to high current in some switches are obstacles that the SCMLI topologies must overcome. The utilization of numerous DC voltage sources in conjunction with SC modules provides a solution to the problem. The problem with the power rating may be handled by distributing the current over all the different sources and switches. According to the research that has been done, many SC-MLI topologies that consider more than one source are available. Among the research that have been done so far, several topologies that use numerous sources and are based on SC-MLI have been documented [19].

8.3 Switched Capacitor MLI Operation

A 13-level SCMLI inverter's operation, as presented in Figure 8.1, is explained to understand the working of SC MLIs. One DC source, one bidirectional and 12 unidirectional power switches comprise the SC MLI. The topology also consists of three capacitors—C_1, C_2, and C_3. To create different voltage levels, the voltages are split into ($V_{DC}/2$) by capacitors C_2 and C_3. Another capacitor, C_1, plays a role in producing these voltage levels by charging up to V_{DC} during operation and boosting the voltages. This topology has three times boosting capability. SCMLI topology switching states are shown in Table 8.1.

The switching states of the presented SCMLI are tabulated along with charging and discharging of individual for each state in Table 8.1. The output voltage obtained in every state is also presented for each state. In positive half cycle, C_1 is charged at L_3, L_2, and L_1 levels, while C_2 is charged at L_4, L_2, and L_1 levels. The capacitor C_3 is also charged at L_4, L_2, and L_1 levels along with no change at L_5 and L_3 levels.

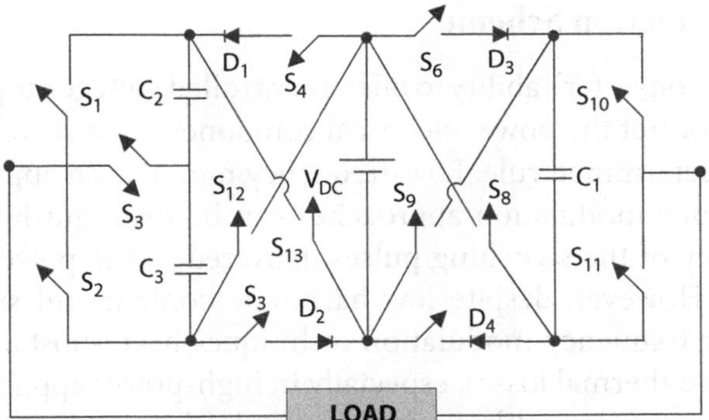

Figure 8.1 13-level switched-capacitor multilevel.

Table 8.1 13-level SCMLI topology's conductivity states.

S_1	S_2	S_3	S_4	S_5	S_6	S_7	S_8	S_9	S_{10}	S_{11}	S_{12}	S_{13}	C_1	C_2	C_3	V_O
0	1	0	0	0	0	0	1	0	1	0	0	1	↓	↓	↓	$6V_{DC}/2$
0	0	1	0	0	0	0	1	0	1	0	0	1	↓	↓	-	$5V_{DC}/2$
0	1	0	1	1	0	0	1	0	1	0	0	0	↓	↑	↑	$4V_{DC}/2$
0	0	1	0	0	1	1	0	0	1	0	0	1	↑	↓	-	$3V_{DC}/2$
0	1	0	1	1	1	1	0	0	1	0	0	0	↑	↑	↑	$2V_{DC}/2$
0	0	1	1	1	1	1	0	0	1	0	0	0	↑	↑	↑	$V_{DC}/2$
1	0	0	1	1	1	1	0	0	1	0	0	0	↑	↑	↑	0
0	1	0	1	1	1	1	0	0	0	1	0	0	↑	↑	↑	0
0	0	1	1	1	1	1	0	0	0	1	0	0	↑	↑	↑	$-V_{DC}/2$
1	0	0	1	1	1	1	0	0	0	1	0	0	↑	↑	↑	$-2V_{DC}/2$
0	0	1	0	0	1	1	0	0	0	1	1	0	↑	-	↓	$-3V_{DC}/2$
1	0	0	1	1	0	0	0	1	0	1	0	0	↓	↑	↑	$-4V_{DC}/2$
0	0	1	0	0	0	0	0	1	0	1	1	0	↓	-	↓	$-5V_{DC}/2$
1	0	0	0	0	0	0	0	1	0	1	1	0	↓	↓	↓	$-6V_{DC}/2$

"0" = off, "1" = on, "↑" = charging, "↓" = discharging, "-" = no change,

8.3.1 Modulation Scheme

A multilevel converter's ability to offer a controlled voltage output depends on whether or not the power electrical components are switched according to a predetermined rule. Low-frequency modulation approaches and high-frequency modulation approaches can be distinguished based on the frequency of the switching pulses delivered to the power electronic equipment. However, despite low harmonic content and smaller filter sizes, higher frequency modulation techniques have substantial voltage transients and thermal losses, especially in high-power applications. This is why we use pulse width modulation and other low-frequency techniques like selective harmonic elimination (SHE) [23–25]. It takes a lot of processing power (since the switching angle is calculated by the solution of a mathematical equation), and SHE-PWM has other drawbacks as well. The controllers in closed-loop systems also face higher computational complexity. A low-frequency modulation scheme called NLC can be used to overcome these limitations. The output voltage was controlled using NLC-PWM.

This design utilizes a low-frequency switching transition to greatly reduce switching transition loss, similar to that experienced by NLC modulation. Compared to SHEPWM, total harmonic distortion (THD) is reduced, even though it may be hard to completely remove all harmonic material with conventional NLC techniques. This is despite the fact that THD is minimized.

Equation (8.1) can be used to easily calculate the output voltage for various modulation indices.

$$V_{\text{out}} = m \times \frac{(N_{\text{level}} - 1)}{2} \times V_{dc} \times \cos \omega t \tag{8.1}$$

where N_{level}—output voltage level and m—modulating index.

According to Equation (8.2), the switching angles are computed as per the following:

$$\theta = \sin^{-1}\left(\frac{i - 0.5}{x}\right) \tag{8.2}$$

where $i = 1, 2, 3, \ldots, x$ and $x = (N_{\text{level}} - 1)/2$.

8.3.2 Comparative Assessment

The comparative assessment based on output voltage level count (N_L), dc sources utilized (V_{dc}), number of power switches (N_{sw}), and capacitors count (N_{cap}) of different SCMLI topologies is presented in Table 8.2. All the topologies presented here are of 13-level output voltage and one dc source. The topology presented in [26] has 12 switches, seven diodes along with two capacitors, also attaining the voltage gain of 1.5, while for the same voltage gain, the topology of [27] needs 14 switches, two more capacitors but only five capacitors.

The topology of [28] also utilizes 12 switches, four capacitors but 12 diodes, while [29] needs the same number of switches and diodes but one less capacitor. In [30] and [31], the number of devices are the same, but the voltage gain are 3 and 6, respectively. [33] and [34] present the topologies with the same number of switches and capacitor to have a gain of six but 12 and 13. In contrast, to achieve the same level at output voltage, the presented single-sourced topology needs only 13 switches, four diodes, along with three capacitors. The voltage gain obtained from the presented architecture is also thrice. Thus, this inverter is suitable for the solar photovoltaic system with high gain along with low device count.

Table 8.2 Comparison of SCMLI topologies.

Ref.	N_L	N_{dc}	N_{sw}	N_d	N_{cap}	Gain
[26]	13	1	12	7	2	1.5
[27]	13	1	14	5	4	1.5
[28]	13	1	12	12	4	3
[29]	13	1	12	12	3	3
[30]	13	1	14	14	3	3
[31]	13	1	14	14	3	6
[32]	13	1	13	13	3	6
[33]	13	1	13	12	3	6
[34]	13	1	13	13	3	6
[P]	13	1	13	4	3	3

8.4 Grid-Connected Operation of SCMLIs

Figure 8.2 displays the block diagram for grid-connected SCMLIs. The current controller's main function is to maximize the output of the PV system while also ensuring optimal power balance, in addition to its use in voltage management. By comparing the standard grid current with the exact grid current, the current error is determined. A PI-controller then responds to the current error that is generated as a result of this comparison. Adjustments are made to the values of the PI-tuning controller (K_p, K_i) to reduce the amount of inaccuracy [12, 24]. In addition to enhancing the effectiveness and dependability of the system, dual feedback control can also be known as current control in the inner loop and voltage control in the outer loop.

When designing an active and reactive power control system, the proportional resonant (PR) controller reduces the computational load on the digital signal processor (DSP). It is possible to use sliding mode control (SMC) and its many refined and flexible forms.

If there are more constraints that need to be controlled, then multi-objective controllers like model predictive control (MPC) and its modified variants can be utilized depending on the multi-level inverter being used to convert the solar PV generated DC power in AC power that is going to be injected into the grid. The active and reactive power control also become more controllable in the case of multi-objective controller. In a single multi-objective controller, the grid current along the voltage maintained across the capacitor and active–reactive power transfer can be controlled efficiently.

The main problem with the MPC is the tuning of the weighing factor which can be mitigated using AI/ML-based adaptive weighing factors.

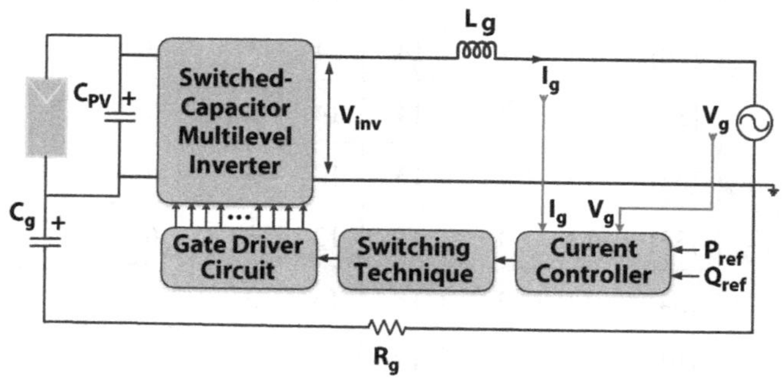

Figure 8.2 Block diagram for grid-connected SCMLI.

The advantage of this over the simplified version is that, according to the necessity of the current scenario, it can adapt and control the variables which need to be controlled and regulated.

8.5 Results and Discussion

The suggested architecture is initially modeled using asymmetrical dc power supply in this part in MATLAB/Simulink. The result values differ according to loading and modulation parameters, also the outcomes with a range of loads and modulation indices. Typhoon-HIL emulators are utilized to demonstrate the given topology in action in real time.

8.5.1 Simulation Results

Tests have been conducted to evaluate and assess the effectiveness of the SCMLI topology. This section presents the simulations that were carried out. The software used to produce the results of the SCMLI topology was MATLAB® 2022b, with the simulation parameters tabulated in Table 8.3. The topology was also simulated while subjected to varying load conditions along with change in modulation index.

In Figure 8.3(a), the voltage and current for a 13-level converter with a constant R-load of 50 Ω are shown. Figure 8.3(b) illustrates the output voltage and current for a load of 50 Ω + 120 mH. Figure 8.3(c) shows the output voltage and current waveform for a load that varies from R-load of 50 Ω to RL-load of 50 Ω + 120 mH. Figure 8.3(d) displays the voltage and current waveform for R-load variation from 50 to 100 Ω, while Figure 8.3(e) shows the same for RL-load variation. Figure 8.3(f) plots the voltage and current waveform for changes in modulation indices for an R-load of 50 Ω

Table 8.3 Simulation parameters.

Parameters	**Values**
V_{DC}	100 V
Capacitors	4,700 µF, 200 V
Resistance at load	50 Ω, 100 Ω
Inductance at load	120 mH
Output frequency	50 Hz

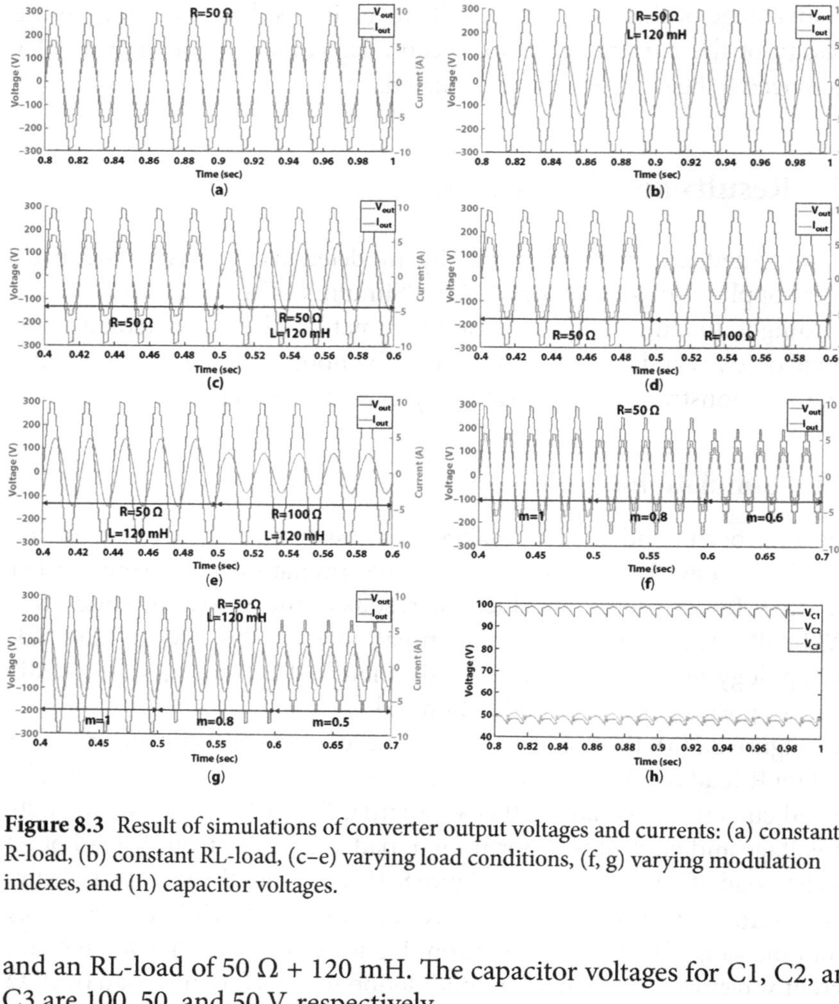

Figure 8.3 Result of simulations of converter output voltages and currents: (a) constant R-load, (b) constant RL-load, (c–e) varying load conditions, (f, g) varying modulation indexes, and (h) capacitor voltages.

and an RL-load of 50 Ω + 120 mH. The capacitor voltages for C1, C2, and C3 are 100, 50, and 50 V, respectively.

8.5.2 Real-Time Results

Real-time emulators for the typhoon-HIL platform are used to demonstrate the provided topology's performance in real time. Different results are shown for each loading scenario. Results are also shown for a range of loads and modulation indices.

A 13-level SC-MLC was validated using Typhoon HIL 402 hardware emulators under hardware-in-the-loop (HIL) conditions. For the

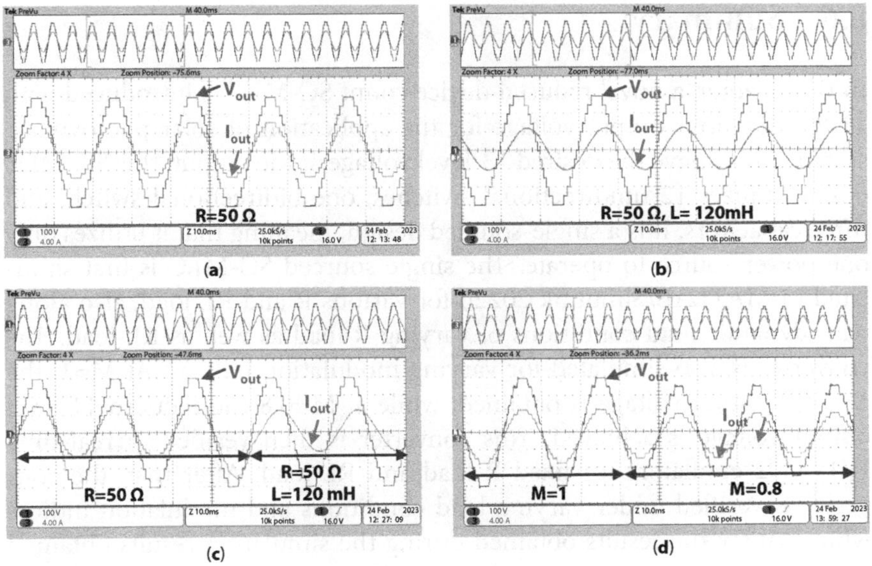

Figure 8.4 Hardware in loop emulators' results of converter output voltage and current for (a) constant R load, (b) constant RL load, (c) varying load, and (d) varying modulation indices.

development of the NLC control system, a HIL system was used in conjunction with Typhoon, an external real-time physical controller, which acted as an external physical controller within a real-time simulation. Based on the results of the MATLAB/Simulink simulation conducted in the previous section, the proposed inverter was found to be effective in the HIL testing.

Figures 8.4(a) and 4(b) depict the converter output voltage and current for a constant R-load of 50 Ω and a constant RL-load of 50 Ω + 120 mH, respectively, which are like the results simulated in Figures 8.3(a) and 8.3(b). In Figure 8.4(c), the output voltage and current are presented for a load condition that varies from 50 to 50 Ω + 120 mH, as obtained using the HIL-402 real-time emulator, which verifies the results obtained in Figure 8.3(c). Figure 8.4(d) shows the voltage and current waveform for varying modulation indices from M=1 to M=0.8, which agree with the results simulated in Figure 8.3(f). It should be noted that even at a reduced modulation index of M=0.8, the proposed inverter can operate effectively, albeit with a reduced voltage level of 11 compared to the 13 levels achieved at a modulation index of M=1.

8.6 Summary

In this chapter, a novel reduced device count SC-MLC is introduced with 13-level inverter output voltage for the application of solar photovoltaic system. To achieve the desired 13-level voltage at the output, this SC-MLC design utilizes 12 unidirectional switches, one bidirectional switch, and three capacitors. It is a single-sourced design, meaning that it utilizes only one power source to operate. The single-sourced SC-MLC is first simulated in MATLAB/Simulink 2022b for various R and RL load, also along with dynamic load conditions of varying R-load as well as RL-load. The converter is also simulated for varying modulation indices. At M=1, the 13-level output voltage is obtained, while at M=0.8 the reduced 11-level output voltage is achieved. This converter is then verified in real-time HIL—402 emulators for fixed R load and RL load. After that, the converter is verified under varying load conditions and modulation indices which satisfy the results obtained during the simulation results obtained in MATLAB/simulation environments. Upon conducting FFT analysis, it was found that the 13-level inverter voltage and grid current exhibit a significantly reduced total harmonic distortion (THD) of 6.34% and 0.55%, respectively. These results are in line with the IEEE-519 standard, which mandates that the THD should be ≤8% and each order should be ≤5%. Therefore, the proposed inverter satisfies this standard, indicating its effectiveness in reducing harmonic distortion. As far as future work is concerned, the 13-level topology can be implemented for grid integration and the reliability analysis can be done, which will be useful. The power loss analysis and efficiency calculations can also be done in future work.

References

1. P. R. Bana, K. P. Panda, R. T. Naayagi, P. Siano, and G. Panda, "Recently Developed Reduced Switch Multilevel Inverter for Renewable Energy Integration and Drives Application: Topologies, Comprehensive Analysis and Comparative Evaluation," *IEEE Access*, vol. 7, pp. 54888–54909, 2019, doi: 10.1109/ACCESS.2019.2913447.
2. P. Kala and S. Arora, "A comprehensive study of classical and hybrid multilevel inverter topologies for renewable energy applications," *Renew. Sustain. Energy Rev.*, vol. 76, pp. 905–931, Sep. 2017, doi: 10.1016/j.rser.2017.02.008.
3. M. Kumari, M. D. Siddique, A. Sarwar, M. Tariq, S. Mekhilef, and A. Iqbal, "Recent trends and review on switched-capacitor-based single-stage boost

multilevel inverter," *Int. Trans. Electr. Energy Syst.*, vol. 31, no. 3, p. e12730, Mar. 2021, doi: 10.1002/2050-7038.12730.

4. M. Tayyab, A. Sarwar, F. I. Bakhsh, A. Al-Durra, and K. M. Siddiqui, "A single-source nine-level inverter with quadratic boost ability for renewable energy applications," *IET Renew. Power Gener.*, 2022, doi: 10.1049/RPG2.12549.
5. R. Mahalakshmi and K. C. S. Thampatty, "Grid Connected Multilevel Inverter for Renewable Energy Applications," *Procedia Technol.*, vol. 21, pp. 636–642, Jan. 2015, doi: 10.1016/j.protcy.2015.10.076.
6. M. Tayyab, A. Sarwar, S. Murshid, M. Tariq, A. Al-Durra, and F. I. Bakhsh, "Active and reactive power control of grid-connected single-phase asymmetrical eleven-level inverter," *IET Gener. Transm. Distrib.*, 2022, doi: 10.1049/GTD2.12442.
7. S. Islam, A. Iqbal, S. De, and F. I. Bakhsh, "Distributed secondary controller to ensure proportional sharing of reactive power in AC microgrid," *Energy Rep.*, vol. 8, pp. 6779–6793, Nov. 2022, doi: 10.1016/J.EGYR.2022.05.014.
8. S. Ahmad, A. Iqbal, I. Ashraf, and M. Meraj, "Improved power quality operation of symmetrical and asymmetrical multilevel inverter using invasive weed optimization technique," *Energy Rep.*, vol. 8, pp. 3323–3336, Nov. 2022, doi: 10.1016/j.egyr.2022.01.122.
9. R. H. Baker and L. H. Bannister, "Electric power converter," 3867643A, 1975
10. R. H. Baker, "Switching circuit," 4210826A, Jul. 1980
11. T. A. Meynard and H. Foch, "Multi-level conversion: high voltage choppers and voltage-source inverters," in *PESC '92 Record. 23rd Annual IEEE Power Electronics Specialists Conference*, IEEE, 1992, pp. 397–403. doi: 10.1109/PESC.1992.254717.
12. H. Iqbal, M. Tariq, M. Sarfraz, M. A. Anees, W. Alhosaini, and A. Sarwar, "Model predictive control of Packed U-Cell inverter for microgrid applications," *Energy Rep.*, vol. 8, 2022, doi: 10.1016/j.egyr.2022.05.188.
13. A. K. Yarlagadda and V. Verma, "A novel multilevel inverter topology with reduced number of component count and total standing voltage for renewable energy conversion system," *Renew. Energy Focus*, vol. 43, pp. 84–92, Dec. 2022, doi: 10.1016/J.REF.2022.09.004.
14. Y. Ye, K. W. E. Cheng, J. Liu, and K. Ding, "A step-up switched-capacitor multilevel inverter with self-voltage balancing," *IEEE Trans. Ind. Electron.*, vol. 61, no. 12, pp. 6672–6680, Dec. 2014, doi: 10.1109/TIE.2014.2314052.
15. Y. Ye, G. Zhang, X. Wang, Y. Yi, and K. W. E. Cheng, "Self-Balanced Switched-Capacitor Thirteen-Level Inverters with Reduced Capacitors Count," *IEEE Trans. Ind. Electron.*, vol. 69, no. 1, pp. 1070–1076, Jan. 2022, doi: 10.1109/TIE.2021.3050378.
16. H. Rehman, M. Tariq, A. Sarwar, W. Alhosaini, M. A. Hossain, and S. M. Batiyah, "Single-Phase Fault Tolerant Multilevel Inverter Topologies—Comprehensive Review and Novel Comparative Factors," *Energies (Basel)*, vol. 15, no. 24, p. 9319, Dec. 2022, doi: 10.3390/en15249319.

17. T. Roy and P. K. Sadhu, "A Step-Up Multilevel Inverter Topology Using Novel Switched Capacitor Converters with Reduced Components," *IEEE Trans. Ind. Electron.*, vol. 68, no. 1, pp. 236–247, Jan. 2021, doi: 10.1109/TIE.2020.2965458.
18. H. Rehman *et al.*, "Fault Tolerant Operation in Multilevel Inverter for More Electric Aircraft," pp. 1–6, Aug. 2022, doi: 10.1109/ICEFEET51821.2022.9848055.
19. A. Iqbal, M. D. Siddique, B. Prathap Reddy, P. K. Maroti, and R. Alammari, "A New Family of Step-up Hybrid Switched-Capacitor Integrated Multilevel Inverter Topologies with Dual Input Voltage Sources," *IEEE Access*, 2020, doi: 10.1109/ACCESS.2020.3046192.
20. K. Wang, Y. Li, Z. Zheng, and L. Xu, "Voltage balancing and fluctuation-suppression methods of floating capacitors in a new modular multilevel converter," *IEEE Trans. Ind. Electron.*, vol. 60, no. 5, pp. 1943–1954, 2013, doi: 10.1109/TIE.2012.2201433.
21. R. Barzegarkhoo, M. Forouzesh, S. S. Lee, F. Blaabjerg, and Y. P. Siwakoti, "Switched-Capacitor Multilevel Inverters: A Comprehensive Review," *IEEE Trans. Power Electron.*, vol. 37, no. 9, pp. 11209–11243, Sep. 2022, doi: 10.1109/TPEL.2022.3164508.
22. M. Tayyab, A. Sarwar, F. I. Bakhsh, A. Al-Durra, and K. M. Siddiqui, "A single-source nine-level inverter with quadratic boost ability for renewable energy applications," *IET Renew. Power Gener.*, 2022, doi: 10.1049/RPG2.12549.
23. H. Iqbal *et al.*, "Selective harmonic mitigation based two-scale frequency control of cascaded modified packed U-Cell inverters," *Energy Rep.*, vol. 8, pp. 1009–1020, Nov. 2022, doi: 10.1016/J.EGYR.2022.10.180.
24. J. Napoles, J. I. Leon, R. Portillo, L. G. Franquelo, and M. A. Aguirre, "Selective harmonic mitigation technique for high-power converters," *IEEE Trans. Ind. Electron.*, vol. 57, no. 7, pp. 2315–2323, Jul. 2010, doi: 10.1109/TIE.2009.2026759.
25. B. Adineh, R. Keypour, P. Davari, and F. Blaabjerg, "Review of Harmonic Mitigation Methods in Microgrid: From a Hierarchical Control Perspective," *IEEE J. Emerg. Sel. Top. Power Electron.*, vol. 9, no. 3, pp. 3044–3060, Jun. 2021, doi: 10.1109/JESTPE.2020.3001971.
26. M. Ali, M. Tariq, A. Sarwar, and B. Alamri, "A 13-, 11-, and 9-Level Boosted Operation of a Single-Source Asymmetrical Inverter With Hybrid PWM Scheme," *IEEE Trans. Ind. Electron.*, vol. 69, no. 12, pp. 12817–12828, Dec. 2022, doi: 10.1109/TIE.2022.3144576.
27. J. Zeng, W. Lin, D. Cen, and J. Liu, "Novel k-Type multilevel inverter with reduced components and self-balance," *IEEE J. Emerg. Sel. Top. Power Electron.*, vol. 8, no. 4, pp. 4343–4354, Dec. 2020, doi: 10.1109/JESTPE.2019.2939562.
28. K. P. Panda, P. R. Bana, and G. Panda, "A Reduced Device Count Single DC Hybrid Switched-Capacitor Self-Balanced Inverter," *IEEE Trans. Circuits*

Syst. II Express Briefs, vol. 68, no. 3, pp. 978–982, Mar. 2021, doi: 10.1109/TCSII.2020.3018333.

29. M. Tayyab *et al.*, "A Single Source Switched-Capacitor 13-Level Inverter with Triple Voltage Boosting and Reduced Component Count," *Electron. 2021, Vol. 10, Page 2321*, vol. 10, no. 19, p. 2321, Sep. 2021, doi: 10.3390/ELECTRONICS10192321.
30. M. P. Nayak, D. Kumar, B. Patnaik, and R. K. Nema, "Thirteen Level Boost Inverter with Low Switch Count," *2022 2nd International Conference on Emerging Frontiers in Electrical and Electronic Technologies, ICEFEET 2022*, 2022, doi: 10.1109/ICEFEET51821.2022.9847727.
31. K. M. Kim, J. K. Han, and G. W. Moon, "A High Step-Up Switched-Capacitor 13-Level Inverter with Reduced Number of Switches," *IEEE Trans. Power Electron.*, vol. 36, no. 3, pp. 2505–2509, Mar. 2021, doi: 10.1109/TPEL.2020.3012282.
32. Y. Ye, G. Zhang, X. Wang, Y. Yi, and K. W. E. Cheng, "Self-Balanced Switched-Capacitor Thirteen-Level Inverters with Reduced Capacitors Count," *IEEE Trans. Ind. Electron.*, vol. 69, no. 1, pp. 1070–1076, Jan. 2022, doi: 10.1109/TIE.2021.3050378.
33. P. Bhatnagar, A. K. Singh, K. K. Gupta, and Y. P. Siwakoti, "A switched-capacitors-based 13-level inverter," *IEEE Trans. Power. Electron.*, vol. 37, no. 1, pp. 644–658, Jan. 2022, doi: 10.1109/TPEL.2021.3098827.
34. K. P. Panda, P. R. Bana, and G. Panda, "A Switched-Capacitor Self-Balanced High-Gain Multilevel Inverter Employing a Single DC Source," *IEEE Trans. Circuits Syst. II Express Briefs*, vol. 67, no. 12, pp. 3192–3196, Dec. 2020, doi: 10.1109/TCSII.2020.2975299.

9

Multilevel Inverter for Renewable Energy Source-Based Grid Integration

Akhlaque Ahmad Khan[1]*, Ahmad Faiz Minai[1], Mohammed Aslam Husain[2] and Mohammad Naseem[1]

[1]*Department of Electrical Engineering, Integral University, Lucknow, Uttar Pradesh, India*

[2]*Department of Electrical Engineering, REC, Ambedkar Nagar, Uttar Pradesh, India*

Abstract

Due to the increasing cost of energy, it becomes ever more prevalent to generate electricity using renewable energy sources like solar, wind, hydro, and other natural resources. For an entire system to become more energy efficient, it needs to utilize renewable energy sources (RES). Utilizing power produced by RES for grid feed is complicated. In order to connect RES to the grid or system demand, multilevel converters are required. Grid-connected photovoltaic systems (GCPVS) are the most efficient way to incorporate solar energy into the grid. The inverter is a critical part of the GCPVS. The inverter has control over performance, total harmonic distortion (THD), and maximum power point (MPP) tracking. One of the best viable converters for attaining minimal THD is indeed the multilevel inverter (MLI). This chapter explores various MLI configurations with the objective of improving quality attributes, the evolution of their benefits and drawbacks, an outline of the MLIs used in different RES, PV structures and their attributes, followed by the MLIs used for SPV systems, and a brief description of appropriate MPPT strategies for a grid-connected SPV/hybrid system.

Keywords: SPV system, total harmonic distortion, MPPT, multilevel inverter, renewable energy sources, grid integration

**Corresponding author*: akhlaque@iul.ac.in; https://orcid.org/0000-0001-8728-8243

Salman Ahmad, Farhad Ilahi Bakhsh, and P. Sanjeevikumar (eds.) Multilevel Converters, (165–184)

9.1 Introduction

The dwindling supply of fossil fuels has made conventional sources of energy untenable, which has led to a sharp rise in the use of sustainable green technology and the production of electricity. The International Energy Agency (IEA) projects that by 2023 renewable energy sources will meet over 30% of the world's energy needs [1]. However, even though the majority of renewable energy sources (RES) offer DC power, the majority of loads and the power transmission network almost exclusively use AC energy, making it impossible for them to directly supply the power. RES produce DC power, which inverters convert into AC power before sending it to the electrical grid. The signal's frequency that is sent into the electrical grid must match that of the grid [2]. The signal injected into the grid ought to contain a minimum amount of harmonic components; however, conventional two-level inverters yield rectangular or -rectangular signals, and a rectangular signal with 50% duty cycle is close to a sine wave with 48% THD, which is completely inadequate for grid integration. For grid integration, a signal's THD must be less than 5%, according to the IEEE standard. The IEEE standard 519 states that, for grid integration, a signal's THD needs to be below 5% [3]. By utilizing a multilevel inverter (MLI), the drawbacks of traditional two-level inverters may be solved. With the proper THD rating for grid integration, MLIs are able to generate AC power from a variety of DC voltage sources in their input. In order to generate an electrical output that alternates in a staircase pattern, these inverters employ strategically positioned solid-state switches and DC voltage inputs [4]. MLIs are emerging as an effective and viable method for converting HVDC energy to AC energy. They are becoming more and more popular due to their ability to operate at medium and high voltages as well as their greater power quality, lower-order harmonics, increased resistance to EMI, and decreased switching losses [5]. In a MLI, the overall operating voltage level is greater than the voltage stress caused by the switching components [6]. In low- and medium-voltage application areas, MLIs might be fed to the grid without the use of transformers and filters, while high-voltage applications necessitate the use of high-frequency transformers in order to integrate two-level inverters with the power system using step-up transformers and filters [7]. Being substantially lighter and smaller than a fundamental frequency transformer, MLIs with a high frequency transformer can be regarded as transformer-free systems. The most prominent and widely used MLI configurations are CHBMLI, FCMLI, and DCMLI/NPCMLI [7, 8]. Since MLIs have a multilevel output waveform, they also have the capacity to run at high voltages, generate minimal EMI, are very efficient, and have excellent power quality characteristics (minimal THD formation) [9].

The MLI, on the other hand, is built to generate massive amounts of power by processing a stepped voltage waveform with an extensive range of low-power-rated switching components and plenty of DC sources. As there are numerous DC sources, you might need batteries, SPV panels, capacitors, and fuel cells, among other power sources. An algorithm controls the power switches after combining numerous DC sources to produce a single, high-voltage result [9, 10]. MLI topologies have several industrial uses, including as a cutting-edge power converter architecture in grid-connected RES. Current advances in MLIs aim to enhance resilience to faults and power quality while simultaneously lowering the overall price of the structure for on-grid RES [11], with the primary focus on reducing the number of switches, the number of gate controller circuits, and the DC power source.

In terms of accuracy, dependability, effectiveness, ease, and tracking speed, MPPT controllers have made a number of advancements. The best MPPT approaches often exhibit rapid speed and little PPP oscillation, which would go in tandem with the output power changes that are made quickly. Even though there are many uses for MPPT regulation approaches right now [12, 13], their primary application is for monitoring either a SPV system with MMPP or the MPP under uniform irradiation. The MPP for each SPV system must be distinct and independent of any complicated algorithm [12].

The remaining sections are as follows, excluding the "Introduction": Section 9.2 provides an overview of the chapter, detailing the structure and functioning of three typical MLI arrangements and discussing the pros and cons of each. Brief descriptions of the MLIs employed by various RES can be found in Section 9.3. Section 9.4 provides an introduction to SPV systems and their features, followed by a discussion of the MLIs used in PV systems and a brief description of appropriate MPPT methods. Challenges and future research avenues are discussed in Section 9.5, and the findings are summarized in Section 9.6.

9.2 Multilevel Inverters (MLI)

MLIs are now prevalent in electrical networks owing to their ability to meet the requirements for voltage stability and rated power, in addition to their decreased levels of harmonic distortion (HD) and EMI. There are several

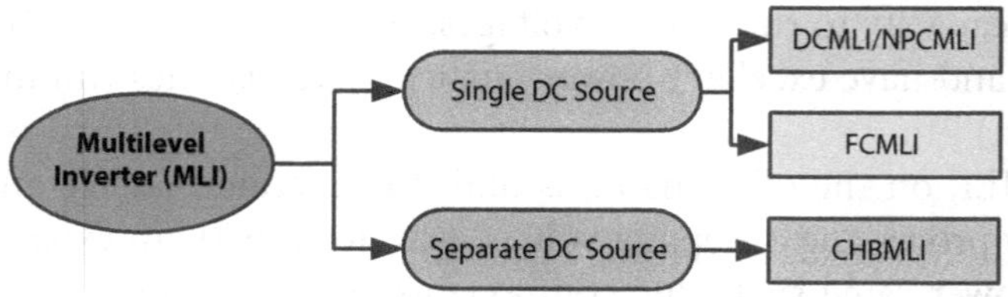

Figure 9.1 Types of MLIs.

benefits to using an MLI instead of a traditional two-level inverter that uses high-frequency switching PWM [14]. MLIs are being evaluated as a possible industrial solution to structures demanding remarkable voltage consistency and dynamic flexibility, and their power ratings range from 1 to 30 MW [36]. Due to their ability to generate low THD output voltage patterns and greater voltages while retaining an inadequate device specification [15], MLIs are particularly appropriate for use in high-voltage situations. Various interactions between an MLI system and RES, such as wind generators, fuel cells, and SPV cells, are possible [2, 16]. The sort of management method employed to establish MLIs' functionality, efficacy, power ratings, and usage is what essentially regulates their PWM [17]. Over the past couple of decades, studies have put forth a number of MLI configurations [18]. Figure 9.1 splits the MLIs into two distinct categories according to the number of DC sources employed in their architecture: those that are currently the most popular configurations in the industry [19].

9.2.1 Cascaded H-Bridges MLI (CHBMLI)

Sequentially coupling a number of single-phase HBIs with various DC sources results in CHBMLIs. One DC source and four power switches make up an H-bridge, as depicted in Figure 9.2 [14]. Each inverter level is built to produce three output voltages ($+V_{dc}$, 0, and $-V_{dc}$) by the connection of the battery to the AC output; the required output is obtained by linking the four switches (S_{W1}–S_{W4}) in various ways. When the S_{W1} and S_{W4} switches are in the ON state, +Vdc output is obtained; however, when the S_{W2} and S_{W3} switches are in the ON state, -Vdc output is obtained. The zero voltage output can only be produced if S_{W1} and S_{W2} or S_{W3} and S_{W4} are both turned on. The voltage indication that results from connecting the full-bridge inverter's AC outputs in series will be a precise representation of the inverters' combined AC power output. The equation b = 2a + 1, where "a" is the total number of DC power sources, indicates that cascade inverters have a variety of output voltage levels [20]. In contrast to DCMLI

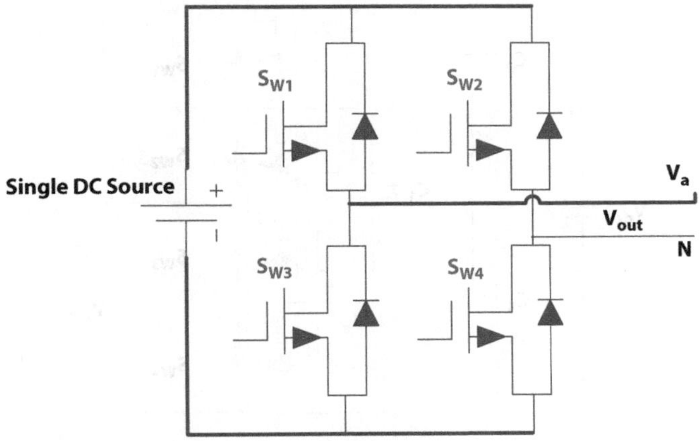

Figure 9.2 Configuration of a three-level CHB-MLI.

and FCMLI, this architecture uses fewer components because it does not employ clamping capacitors and diodes.

It is also possible to use a cascade inverter for static VAR mitigation by directly connecting it in series with the power system. They are excellent for integrating RES into the utility grid because they are used with FCs and SPVs, both of which require separate DC sources [21]. It has also been suggested for usage as the primary traction drive in electric cars since many batteries or ultra-capacitors work as SDCSs in these applications [22]. This topology's structure is adaptable and may be employed with various numbers of inverter levels. By using varying ratios of DC sources and minimizing switching redundancy caused by inner voltage levels, multiple output voltages may be generated.

9.2.2 Flying Capacitor MLI (FCMLI)

Although the FCMLI relies on floating capacitors instead of clamping diodes, their architectures are highly comparable. The FCMLI's output waveform has a direct relationship between the size of the voltage steps and the voltage fluctuation in the nearby capacitors [23]. DC link capacitors with a count of "b-1" make up the FCMLI architecture for the "b" level inverter. The three-level FCMLI topology is depicted in Figure 9.3. The design additionally incorporates two unidirectional power switches and a flying capacitor in conjunction with the voltage levels ($+V_{dc}/2$, 0, and $-V_{dc}/2$) and the two DC feed capacitors. Switches S_{W1} and S_{W2} turn on to create a +ve polarity voltage output, whereas S_{W3} and S_{W4} turn on a -ve

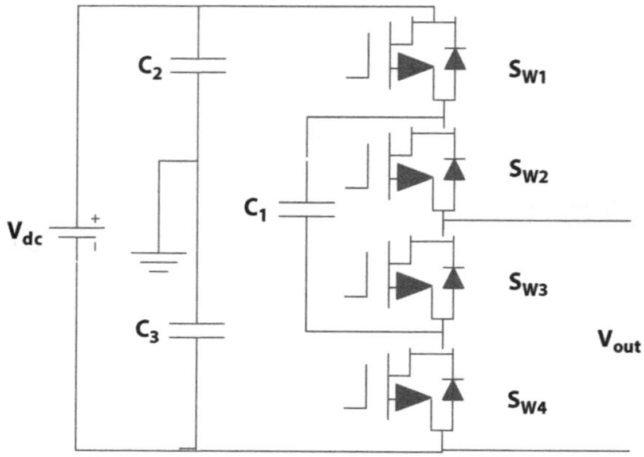

Figure 9.3 Configuration of a three-level FC-MLI.

polarity output voltage. To produce the output voltage at the zero level, either switches S_{W1} and S_{W3} or S_{W2} and S_{W4} are triggered. FCMLI's voltage composition is more malleable than that of DC-MLI. Whenever one has over five levels, the challenge of voltage management can be resolved by meticulously choosing the switching arrangement [15]. The advantages of this design include control over reactive and active power [23], but drawbacks include the need for numerous capacitors, which push up the system's expenses and complexity. Furthermore, active power transmission in such configurations suffers from large switching frequency losses [24].

9.2.3 Diode-Clamped MLI (DC-MLI/NPC-MLI)

The DC-MLI arrangement shown in Figure 9.4 is designed to deliver a three-stage output voltage. Two diodes, two capacitors, and four one-directional power switches comprise this topology's design. To split the obstructing voltage, the clamping diodes are placed in series. Three output voltage levels are feasible in this configuration: $V_{dc}/2$, 0, and $-V_{dc}/2$. Turning on the switches S_{W3} and S_{W4} results in $-V_{dc}/2$, while keeping S_{W1} and S_{W2} on produces $+V_{dc}/2$. Switches S_{W2} and S_{W3} are turned on to provide the zero-voltage state. Every operational switching device will experience voltage stress when the corresponding voltage passes through the DC link capacitors, which diode clamping will then clamp to the voltage of each capacitor [24]. In real-life situations, the clamping diodes are sequentially connected to share the blocking voltage. To put it simply, all active devices must block the voltage V/(b-1). Adjusting these clamping diodes' voltage specifications achieves reversed blocking. When combining the DC-MLI

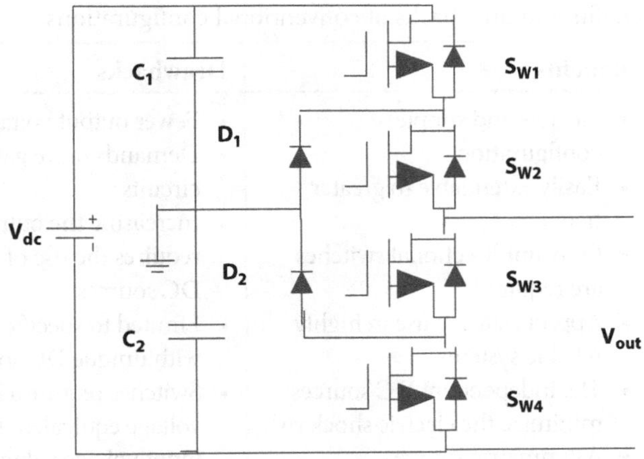

Figure 9.4 Configuration of a three-level DC-MLI.

with PWM, the most significant configuration problem in high-voltage settings is the diode reverse recovery associated with these clamping diodes [15].

Owing to its large delivery capabilities, versatility, and effectiveness, the DC-MLI has growing manufacturing utility compared to the other MLI schemes. Static VAR compensators (SVC) and high-voltage system interfaces are examples of fields where it has been deployed [25].

9.2.4 Evaluation of the DC-MLI, FC-MLI, and CHB-MLI Configuration

The three major MLI configurations, also known as the conventional configurations, were compared in this section based on their benefits and drawbacks, as illustrated in Table 9.1 [15, 20–25].

9.3 Solar Photovoltaic Systems (SPVs)

Due to its unique properties, such as its affordability, longevity, lack of noise and pollution, speed of execution, high flexibility and element adaptability, and the need to create adequate power to satisfy peak loads, PV-based power generation has attracted a lot of interest. As a result, PV arrays have now been employed in a number of industrial operations, such as water pumping machinery, solar-powered hybrid automobiles, battery charging systems, and spacecraft power systems. Due to this, several nations are

Table 9.1 Benefits and drawbacks of conventional configurations.

Types	Benefits	Drawbacks
CHB-MLI	• Flexible and simple configuration • Easily extendable to greater tiers • Only unidirectional switches are required • Appropriate for use in highly reliable systems • The independent DC sources minimize the electric shock risk • Asymmetric source configurations are possible • Building a single DC source arrangement is feasible	• Fewer output voltage levels • Demands more gate driver circuits • Increasing the output voltage requires the use of several DC sources • Limited to specific scenarios with unique DC sources • Switches require a blocking voltage equivalent to the input voltage value • Metaprogramming failure (asymmetric source architecture) • Implementation costs are expensive (asymmetric source architecture) • The voltage ratings of switches vary (asymmetric source architecture)
DC-MLI	• A smart choice for commercial applications. • Demonstrates great switching efficiency at fundamental frequencies • Able to pre-charge the capacitors collectively • Effortless techniques of control • Less DC sources are required • Suitable for applications requiring fault tolerance • Solve the problem of unbalanced voltage and unequal loss distribution across switching devices	• The circuitry used in voltage balancing is sophisticated • The uneven distribution of losses among inner and exterior switches • An increase in clamping diodes as the level rose

(Continued)

Table 9.1 Benefits and drawbacks of conventional configurations. (*Continued*)

Types	Benefits	Drawbacks
FC-MLI	• Requires fewer DC sources overall • Phase redundancy is generated to equalize the voltage levels of the capacitors • Perfect for applications that need to be fault-tolerant • Modulable real and reactive power consumption • Reducing harmonics without the use of filters is possible	• The circuitry used in voltage balancing is sophisticated • Needs a lot of capacitors at high levels • High switching frequencies and losses while transmitting actual power • Exorbitant installation costs

making significant investments in projects that use renewable energy sources. However, there are specific issues with PV-generating systems, including poor conversion efficiency and interference from environmental factors. Solar radiation/intensity directly affects the power of PV cells, but temperature has a less significant impact [32–35]. Batteries, DC–DC voltage converters, regulators, solar cells, and other components make up a comprehensive PV system.

Figure 9.5 demonstrates the nonlinearity of the solar power attributes. V_{oc} and I_{sc} are two variables that are used to examine the properties of a PV cell. The greatest amount of current that a cell may produce is known as Isc, and irradiance has an impact on this parameter. However, Voc is the highest possible voltage that a solar cell can generate while utilizing

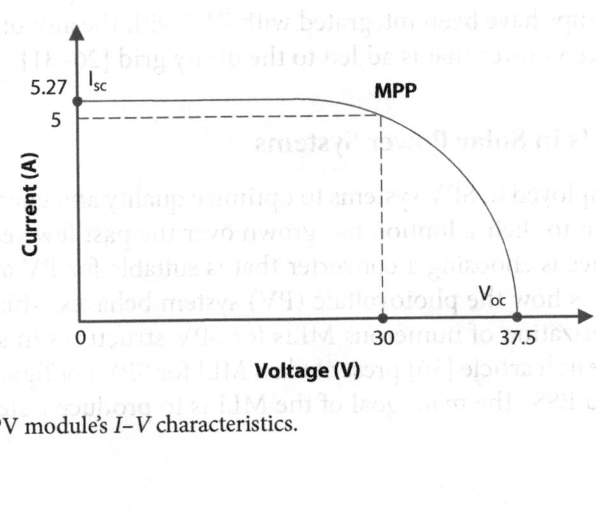

Figure 9.5 PV module's *I–V* characteristics.

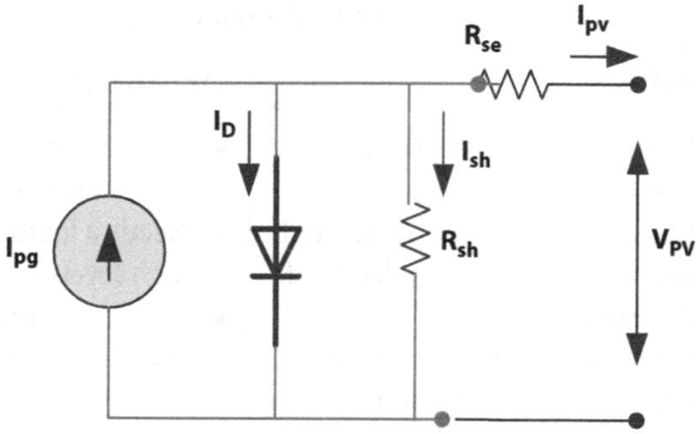

Figure 9.6 A PV cell's equivalent circuit.

no current [35]. There is a broad, similar network concept that may be used to describe PV cells. As illustrated in Figure 9.6, it consists of just one diode with a series resistance and a shunt resistance. The current source is a depiction of the photon-generated current (I_{pg}), which would be inversely proportional to solar energy. The parallel resistance of the diode is used to depict the leakage currents (R_{sh}), while the resistance (R_{se}) depicts the losses associated with the contacts and connections.

9.4 Applications of MLIs in RES

For medium- and high-power application scenarios, the MLI is suggested because it can produce high-quality output voltages while functioning at a lower switching frequency [18, 26]. MLIs are extensively used in many diverse sectors, including power supply (PSs), RES, such as SPV systems, WECSs, and FCs, EVs, IHs, and numerous others. In on-grid systems, various MLI setups have been integrated with RES with the aim of generating additional RES power that is added to the utility grid [26–31].

9.4.1 MLIs in Solar Power Systems

MLIs are employed in SPV systems to optimize quality and energy efficacy, and attention to their adoption has grown over the past few years. One of the difficulties is choosing a converter that is suitable for PV applications since it affects how the photovoltaic (PV) system behaves. This has led to the characterization of numerous MLIs for SPV structures in studies [35, 36]. The research article [36] presented an MLI for SPV configurations that incorporated ESS. The main goal of the MLI is to produce a steady power output in all kinds of climates. The suggested MLI revealed a seamless changeover between the charging and discharging processes. In light of its flexible characteristics, excellent output patterns, and capacity to interface with medium-voltage grids, the CHB inverter may indeed be used for grid interfaces. It has been postulated [37] that large-scale SPV systems benefit from a CHB-MLI, including DC–DC circuit isolation. The CHB-MLI that was suggested offered better current flow with minimal distortion. Its small size makes it suitable for use in high-voltage environments where power quality is an issue. By illustrating an improved CMLI with a robust architecture, [37] showed how current leakages in transformerless SPV structures can be diminished. The drawbacks of switching and conduction have been mitigated, enabling high switching frequencies.

9.4.2 MLIs for On-Grid SPV Systems

In order to identify the best configurations for SPV applications that may govern at basic and high power frequencies, yield superior waveforms with lesser THD, attain the best electromagnetic compatibility, and demonstrate high efficiency, various MLC topologies have been examined and contrasted in the literature [38]. A redesigned MLI architecture was introduced in [39]. This structure produces a voltage waveform with seven levels and little harmonic distortion. For the purpose of delivering grid electricity from SPV modules with different voltage/current specifications, this inverter has indeed been granted clearance to be used in SPV settings. The studies of [40] identified an innovative MLI architecture that differs from the typical MLI configuration in that it uses fewer switches and does not require any extra components like inductors and capacitors. Due to the asymmetry of its SPV sources, this design is appropriate. Additionally, it demonstrates fault tolerance and modularity. This arrangement yields greater voltage levels and an appropriate sinusoidal voltage waveform because of the larger number of sources. This inverter is significantly smaller and lighter than other inverters since it does not need a filter.

Despite appearing to have fewer components than conventional MLI arrangements, [41] designed a novel basic asymmetric MLI for use in PV systems. The process produces 31 output levels successfully, and it operated without a hitch under diverse load scenarios. Additionally, the AC output from this architecture is very effective and has minimal THD. Some of the recently proposed MLIs for use in grids connected with fewer parts have been closely examined by [42] in aspects of component selection, evaluation criteria, and usability. For deployments that connect to the typical one- or three-phase grid, [43] suggests a modular CHB-MLI design.

This architecture was designed to increase solar energy collection while also boosting the performance and adaptability of PV systems. It also ensures greater PV module use.

Using two MLI topologies, a performance study of the on-grid solar system has been carried out [44]. An ANN approach was employed to construct the MPPT controller, which optimizes the electrical output of the SPV systems with regard to changes in the surrounding environment. When used with cascaded inverters, the examined inverter topologies worked admirably and produced lower current THD than the NPC architecture, but since it needs so many power-switching transistors, it is more costly.

A hybrid modular MLC (H-MMLC) for solar electric grid integration was presented in the study by [45]. The suggested architecture is appropriate

Table 9.2 Evolution of ongoing MLIs with on-grid SPV systems.

Type	MPPT	Voltage level	No. of switches	THD (%)	References
NPC/DC	PCT	9	10	13.99	[49]
NPC/DC	ANN	5	6	2.35	[50]
NPC/DC	IC	4	11	4.61	[51]
CHB	P&O	17	-	3.23	[52]
CHB	DMPPT control	7	12	3.3	[53]
CHB	ANN	5	8	2.75	[50]
CHB	P&O	7	24 (6H)	10.75	[37]
CHB	P&O	5	8	3.2	[54]
Modular CHB	DMPPT control	7	12/phase	4.2	[55]
Modular CHB	P&O	7	6	11.81	[56]
Modified CHB	P&O	7	6	3.1	[39]
VLB	DMPPT control	15	10	5.54	[43]

for high power utilization since the PV panels are connected in each converter sub-module, but it necessitates increasing the quantity of sub-modules in each phase. In comparison to regular MMLC, hybrid MMLC has a number of benefits, including a reduced cap count, greater voltage tiers, improved efficacy, no looping electricity, and a smaller surface area. Such a converter has a GMPPT algorithm that only needs the voltage and current inputs from a single sub-module to function. The flaws of the GMPPT under partial shade were solved by the scattered nature of the MPPT. The category, voltage ratings, MPPT approach, number of switches implemented, and THD are some of the features that Table 9.2 presents for comparison among the most recent MLIs for on-grid SPV systems [46–48].

9.5 Challenges and Future Work

The versatility of MLIs has led to their growing acceptance in on-grid systems as well as manufacturing processes. Hybrid architectures, which are descended from conventional architectures, have only recently been proposed by scholars in order to comply with stringent grid code requirements and successfully tackle power quality challenges [9, 11]. It is necessary to look into how well these novel topologies function in grid-integrated applications because the majority of them have not yet been tested in RES that are linked to the grid.

- To effectively address the challenges of on-grid RES deployments, further research into the performance evaluation of current MLIs is necessary. The role that MLIs play in smart grid technologies is important and should be the focus of future research.
- While reduced switch MLI configurations can be used in the incorporation of RES, especially for SPV programmes, more RS-MLI configurations need to be built to increase their attractiveness when utilized for linking RES such as wind and solar.
- Grid systems have been steadily transitioning to the future smart grid through the integration of RES employing appropriate MLIs. While there are many difficulties involved, there is also some potential for MLI topology creation and control. As a result, this industry has seen a lot of innovation.
- The nonlinearity of the SPV framework, the system's operating parameters, and variations in the surrounding

environment are just a few of the variables that affect the MPPT algorithm's performance [46].

In majority of the times, algorithm developers do not take these modifications into account, and in some working settings, this has been the cause of system failure. Because of this, creating the SPV panel regulator is a complex procedure that requires extra caution to assure stability.

9.6 Conclusion

The purpose of this chapter's brief overview of MLI characteristics is to draw attention to the necessity of developing novel inverters or altered arrangements of inverters for on-grid SPV systems. There has been much discussion of MLIs, with attention paid to their categorization, advantages, downsides, and potential to improve conversion efficiency in contemporary power systems. This chapter also suggests that an improved method that uses conventional MLIs for various levels should use them to decrease the switching number. In regard to size, price, decreased THD, and highly efficient power conversion, modified MLIs are potential alternatives for SPV and other RES. In order to entice and encourage the community to focus on developing an efficient and cost-effective structure, the challenges and helpful solutions for accomplishing this were highlighted. The majority of the deployed and assessed inverters' functionalities are integrated into MLI's on-grid SPV system.

When appropriate MLIs are used to integrate RES into electricity grids, smart grids may be implemented in a phased manner. This poses various difficulties for power grid management and security. In spite of this, there are possibilities for MLI topology design and oversight in RES-enhanced intelligent grids that demand more investigation and evaluation.

References

1. Birol Dr. F., Renewables 2018, *Int. Energy Agency*, Accessed on: Nov. 5, 2022. [Online]. Available: https://www.iea.org/renewables2018/
2. Khan, A.A.; Minai, A.F.; Pachauri, R.K.; Malik, H. Optimal Sizing, Control and Management Strategies for Hybrid Renewable Energy Systems: A Comprehensive Review. *Energies*, 2022, 15(17), 6249. https://doi.org/10.3390/en15176249

3. "Power Inverter," Sep. 24, 2019. Accessed on: Nov. 5, 2022. [Online]. Available: https://en.wikipedia.org/wiki/Power_inverter.
4. Ebrahimi J., Babaei E., and Gharehpetian G. B., "A new topology of cascaded multilevel converters with reduced number of components for high-voltage applications," *IEEE Trans. Power Electron.*, vol. 26, no. 11, pp. 3109–3118, Nov. 2011.
5. Minai A. F., Husain M. A., Naseem M., Khan A. A., (2021) 'Electricity Demand Modeling Techniques for Hybrid Solar PV System', *Int. J. Emerg. Electr. Power Syst. (IJEEPS)*, ISSN: 1553-779X, Published by De Gruyter July 7, 2021 https://doi.org/10.1515/ijeeps-2021-0085.
6. Khan A. A., Minai A. F., Devi L., Alam Q. and Pachauri R. K., "Energy Demand Modelling and ANN Based Forecasting using MATLAB/Simulink," *2021 International Conference on Control, Automation, Power and Signal Processing (CAPS)*, 2021, pp. 1-6, doi: 10.1109/CAPS52117.2021.9730746.
7. Blaabjerg F., Chen Z., Kjaer S.B., "Power electronics as efficient interface in dispersed power generation systems," *IEEE Trans. Power Electron.*, vol. 19(5), 2004, pp. 1184–1194.
8. Naseem M., Husain M. A., Kumar J. D., Ahmad M. N., Minai A. F., Khan A. A., "Particle Swarm Optimization based Maximum Power Tracking Techniques for Solar PV System under Partially Shaded Conditions," *2021 International Conference on Control, Automation, Power and Signal Processing (CAPS)*, 2021, pp. 1-6, doi: 10.1109/CAPS52117.2021.9730746.
9. Omer, P.; Kumar, J.; Surjan, B.S. A Review on Reduced Switch Count Multilevel Inverter Topologies. *IEEE Access*, 2020, 8, 22281–22302.
10. Tolbert, L.M.; Shi, X. Multilevel power converters. In *Power Electronics Handbook*; Elsevier: Amsterdam, the Netherlands, 2018; pp. 385–416.
11. Prabaharan, N.; Palanisamy, K. A comprehensive review on reduced switch multilevel inverter topologies, modulation techniques and applications. *Renew. Sustain. Energy Rev.*, 2017, 76, 1248–1282.
12. Danandeh, M. Comparative and comprehensive review of maximum power point tracking methods for PV cells. *Renew. Sustain. Energy Rev.*, 2018, 82, 2743–2767.
13. Kumar M., Minai A. F., Khan A. A. and Kumar S., "IoT based Energy Management System for Smart Grid," *2020 International Conference on Advances in Computing, Communication & Materials (ICACCM)*, 2020, pp. 121-125, doi: 10.1109/ICACCM50413.2020.9213061.
14. Khan A. A., Minai A. F., Alam Q. & M.A. Mallick, "Performance Analysis of Various Switching Scheme in Multilevel Inverters using MATLAB/SIMULINK" in *International Journal of Current Engineering and Technology*, Vol.4, No.2 (April 2014) Page No. 718-724 (E- ISSN 2277 – 4106, P-ISSN 2347-5161)
15. Rodriguez, J.; Franquelo, L.G.; Kouro, S.; Leon, J.I.; Portillo, R.C.; Prats, M.A.M.; Perez, M.A. Multilevel converters: An enabling technology for high-power applications. *Proc. IEEE*, 2009, 97, 1786–1817.

16. Minai, A. F., Usmani, T., & Mallick, M. A. (2016). Performance analysis of multilevel inverter with SPWM strategy using MATLAB/SIMULINK. *J. Electr. Eng. (JEE)*, 16(4), 428-433.
17. Minai A. F., Usmani T. and Mallick M. A., "Optimum Sizing and Estimation of a 30kWp Hybrid Solar Photovoltaic System with Multilevel Inverter" in *International Journal of Research and Scientific Innovation (IJRSI), Research and Scientific Innovation Society*, vol. 2, no. 7, pp. 31-36, July 2015, ISSN 2321-2705.
18. Hasan, N.S.; Rosmin, N.; Osman, D.A.A.; Musta'amal, A.H. Reviews on multilevel converter and modulation techniques. *Renew. Sustain. Energy Rev.*, 2017, 80, 163–174.
19. Minai A. F., Tariq A. and Alam Q., "Theoretical and experimental analysis of photovoltaic water pumping system," *India International Conference on Power Electronics 2010 (IICPE2010)*, 2011, pp. 1-8, doi: 10.1109/IICPE.2011.5728130.
20. Sumithira, T.; Nirmal Kumar, A. Elimination of harmonics in multilevel inverters connected to solar photovoltaic systems using ANFIS: An experimental case study. *J. Appl. Res. Technol.*, 2013, 11, 124–132.
21. Minai A. F. and Tariq A., "Analysis of Cascaded multilevel inverter," *India International Conference on Power Electronics 2010 (IICPE2010)*, 2011, pp. 1-6, doi: 10.1109/IICPE.2011.5728129.
22. Tolbert, L.M.; Peng, F.Z.; Habetler, T.G. Multilevel inverters for electric vehicle applications. In *Proceedings of the Power Electronics in Transportation* (Cat. No. 98TH8349), Gdansk, Poland, 27–30 June 2011; pp. 79–84.
23. Naseem, M., Husain, M.A., Minai, A.F. *et al.* Assessment of Meta-Heuristic and Classical Methods for GMPPT of PV System. *Trans. Electr. Electron. Mater.*, 22, 217–234 (2021). https://doi.org/10.1007/s42341-021-00306-3
24. Chen, A.; He, X. Research on hybrid-clamped multilevel-inverter topologies. *IEEE Trans. Ind. Electron.*, 2006, 53, 1898–1907.
25. El-Hosainy, A.; Hamed, H.A.; Azazi, H.Z.; El-Kholy, E. A review of multilevel inverter topologies, control techniques, and applications. In *Proceedings of the 2017 Nineteenth International Middle East Power Systems Conference (MEPCON)*, Cairo, Egypt, 19–21 December 2017; pp. 1265–1275.
26. Faiz Minai, A.; Khan, A. A.; Pachauri, R.K.; Malik, H.; García Márquez, F.P.; Arcos Jiménez, A. Performance Evaluation of Solar PV-Based Z-Source Cascaded Multilevel Inverter with Optimized Switching Scheme. *Electronics*, **2022**, *11*, 3706. https://doi.org/10.3390/electronics11223706.
27. Khan A. A., Minai A. F. (2023), Introduction to Grid-Forming Inverters (GFMIs), In B. (Eds.). (2023). *Grid-Forming Power Inverters: Control and Applications* (1st Ed.). CRC Press. Page 1-14, ISBN: 9781003302520. https://doi.org/10.1201/9781003302520-1.
28. Khan A. A., Minai A. F. (2023) "A Strategic Review: The Role of Commercially Available Tools for Planning, Modeling, Optimization, and Performance

Measurement of Photovoltaic Systems", *Energy Harvesting Syst., (EHS)*, ISSN: 2329-8766. https://doi.org/10.1515/EHS-2022-0157.

29. Gopi, R.R.; Sreejith, S. Converter topologies in photovoltaic applications–A review. *Renew. Sustain. Energy Rev.*, 2018, 94, 1–14.
30. Amir, M., Zaheeruddin, Haque, A., Bakhsh, F.I., Kurukuru, V.S.B., Sedighizadeh, M.: Intelligent energy management scheme-based coordinated control for reducing peak load in grid-connected photovoltaic-powered electric vehicle charging stations. *IET Gener. Transm. Distrib.*, 00, 1–18 (2023). https://doi.org/10.1049/gtd2.12772
31. Tayyab, M., Sarwar, A., Murshid, S., Tariq, M., Al-Durra, A., Bakhsh, F.I.: Active and reactive power control of grid-connected single-phase asymmetrical eleven-level inverter. *IET Gener. Transm. Distrib.*, 17, 632–644 (2023). https://doi.org/10.1049/gtd2.12442
32. Pandey, Y., Hasan, N., Husain, M. A., Khan, A. N., Bakhsh, F. I., Minai, A. F., & Tabrez, M. (2022). An Environment Friendly Energy-Saving Dispatch Using Mixed Integer Linear Programming Relaxation in the Smart Grid with Renewable Energy Sources. *Distrib. Gener. Altern. Energy J.*, *37*(4), 1239–1258. https://doi.org/10.13052/dgaej2156-3306.37414.
33. Islam Wani N. U., Bakhsh F. I., Choudekar P. and Ruchira, "Active Power Control of Grid Connected SPV Plant Based Microgrid Using Active Power Regulating Scheme," *2021 Emerging Trends in Industry 4.0 (ETI 4.0)*, Raigarh, India, 2021, pp. 1-8, doi: 10.1109/ETI4.051663.2021.9619213.
34. Ahmad S., Bakhsh F. I., "Transformer-Based DC-to-AC Grid-Connected Multi-Level Inverter Topology", In: Iqbal A., Malik H., Riyaz A., Abdellah K., Bayhan S. (eds) *Renewable Power for Sustainable Growth*. Lecture Notes in Electrical Engineering, Vol 723, 2021, Springer, Singapore. doi: 10.1007/978-981-33-4080-0_28
35. Lashab, A.; Sera, D.; Martins, J.; Guerrero, J.M. Multilevel DC-link converter-based photovoltaic system with integrated energy storage. In *Proceedings of the 2018 5th International Symposium on Environment-Friendly Energies and Applications (EFEA)*, Rome, Italy, 24–26 September 2018; pp. 1–6.
36. A. F. Minai, A. A. Khan, M. A. Siddiqui, F. I. Bakhsh, M. A. Hussain and R. K. Pachauri, "Genetic Algorithm Based SPV System with Cascaded H-Bridge Multilevel Inverter," *2023 International Conference on Power, Instrumentation, Energy and Control (PIECON)*, Aligarh, India, 2023, pp. 1-6, doi: 10.1109/PIECON56912.2023.10085864.
37. Jain, S.; Sonti, V. A Highly Efficient and Reliable Inverter Configuration Based Cascaded Multilevel Inverter for PV Systems. *IEEE Trans. Ind. Electron.*, 2017, 64, 2865–2875.
38. Jiang, X.; Doumbia, M.L. Comparative Study of Grid-Connected Multilevel Inverters for High Power Photovoltaic Systems. In *Proceedings of the IEEE 7th International Conference on Smart Energy Grid Engineering (SEGE)*, Oshawa, ON, Canada, 12–14 August 2019; pp. 184–190.

39. Vahedi, H.; Sharifzadeh, M.; Al-Haddad, K. Modified seven-level pack U-cell inverter for photovoltaic applications. *IEEE J. Emerg. Sel. Top. Power Electron.*, 2018, 6, 1508–1516.
40. Abdoli, H.; Khorsandi, A.; Eskandari, B.; Moghani, J.S. A new reduced switch multilevel inverter for pv applications. In *Proceedings of the 11th Power Electronics, Drive Systems, and Technologies Conference (PEDSTC)*, Tehran, Iran, 4–6 February 2020; pp. 1–5.
41. Hamidi, M.N.; Ishak, D.; Zainuri, M.A.A.M.; Ooi, C.A. An asymmetrical multilevel inverter with optimum number of components based on new basic structure for photovoltaic renewable energy system. *Sol. Energy*, 2020, 204, 13–25.
42. Agrawal, R.; Jain, S. Comparison of reduced part count multilevel inverters (RPC-MLIs) for integration to the grid. *Int. J. Electr. Power Energy Syst.*, 2017, 84, 214–224.
43. Bana, P.R.; Panda, K.P.; Naayagi, R.; Siano, P.; Panda, G. Recently developed reduced switch multilevel inverter for renewable energy integration and drives application: Topologies, comprehensive analysis and comparative evaluation. *IEEE Access.*, 2019, 7, 54888–54909.
44. Alhamrouni, I.; Rahmat, M.K.; Ismail, F.A.; Salem, M.; Jusoh, A.; Sutikno, T. Design and development of SEPIC DC-DC boost converter for photovoltaic application. *Int. J. Pow. Elec. Dri. Syst.*, 2019, 1, 406–413.
45. Basu, T.S.; Maiti, S. A hybrid modular multilevel converter for solar power integration. *IEEE Trans. Ind. Appl.*, 2019, 55, 5166–5177.
46. Baba, A.O.; Liu, G.; Chen, X. Classification and Evaluation Review of Maximum Power Point Tracking Methods. *Sustain. Futures*, 2020, 2, 100020.
47. Mohammad, S.S., Iqbal, S.J. (2021). Improved Dispatchability of Solar Photovoltaic System with Battery Energy Storage. In: Bose, M., Modi, A. (eds) *Proceedings of the 7th International Conference on Advances in Energy Research.* Springer Proceedings in Energy. Springer, Singapore. https://doi.org/10.1007/978-981-15-5955-6_96.
48. S. S. Mohammad and S. J. Iqbal, "Power Management, Control and Design of Supercapacitor Assisted Fuel Cell-based Micro Power System for Electric Vehicles," *2022 Second International Conference on Power, Control and Computing Technologies (ICPC2T)*, Raipur, India, 2022, pp. 1-6, doi: 10.1109/ICPC2T53885.2022.9776880.
49. Mahfuz-Ur-Rahman, A.; Islam, M.R.; Muttaqi, K.M.; Sutanto, D. Model Predictive Control for a New Magnetic Linked Multilevel Inverter to Integrate Solar Photovoltaic Systems With the Power Grids. *IEEE Trans. Ind. Appl.*, 2020, 56, 7145–7155.
50. Keddar, M.; Doumbia, M.L.; Della, M.; Belmokhtar, K.; Midoun, A. Interconnection Performance Analysis of Single Phase Neural Network Based NPC and CHB Multilevel Inverters for Grid-connected PV Systems. *Int. J. Renew. Energy Res.*, 2019, 9, 1451–1461.

51. Taghvaie, A.; Haque, M.E.; Saha, S.; Mahmud, M.A. A new step-up switched-capacitor voltage balancing converter for NPC multilevel inverter-based solar PV system. *IEEE Access.*, 2020, 8, 83940–83952.
52. Tayari, M.; Guermazi, A.; Ghariani, M. Cascaded Multilevel Inverter for PV-Active Power Filter Combination into the Grid-Tied Solar System. *Int. J. Renew. Energy Res.*, 2020, 10, 1810–1819.
53. Kumar, A.; Verma, V. Performance enhancement of single-phase grid-connected PV system under partial shading using cascaded multilevel converter. *IEEE Trans. Ind. Appl.*, 2018, 54, 2665–2676.
54. Sharma, B.; Nakka, J. Single-phase cascaded multilevel inverter topology addressed with the problem of unequal photovoltaic power distribution in isolated dc links. *IET Power Electron.*, 2018, 12, 284–294.
55. Xiao, B.; Hang, L.; Mei, J.; Riley, C.; Tolbert, L.M.; Ozpineci, B. Modular cascaded H-bridge multilevel PV inverter with distributed MPPT for grid-connected applications. *IEEE Trans. Ind. Appl.*, 2014, 51, 1722–1731.
56. Karthikeyan, D.; Vijayakumar, K.; Sathik, J. Generalized cascaded symmetric and level doubling multilevel converter topology with reduced THD for photovoltaic applications. *Electron.*, 2019, 8, 161.

10

Modeling and Analysis of Bidirectional Electric-Drive-Reconstructed On-Board Converter for Plug-In Electric Vehicles

Faizan Fayaz Bhat[1], Zahid Ahmad Tantry[1], Md Ibrahim[2]* and Farhad Ilahi Bakhsh[1]

[1]Department of Electrical Engineering, National Institute of Technology Srinagar, Hazratbal, Srinagar (J&K), India

[2]Department of Electrical Engineering, Integral University, Lucknow, India

Abstract

A typical part of charging systems is the electric-drive-reconstructed onboard converter (EDROC), which can function in either the grid-to-vehicle (G2V) or vehicle-to-grid (V2G) mode. It is built on a network of DC side switches. They do not require any extra equipment (filter or relay) on the AC side when it connects to the electrical grid through the plug at home or at work. In comparison to traditional EDROC, the novel EDROC that is being described has cost and volume advantages. This technique can achieve unity power factor while operating in both charging and driving modes. It starts by connecting a pulse width modulated voltage source rectifier, which was formerly employed in six inverter legs, directly to an AC grid. The three-phase buck-type converter is then built using the three-phase inverter's leftover machine windings and legs. The electromagnetic torque produced during charging mode is lessened by the decoupling transformation matrix by utilizing the rotor parking position. With the help of the simulation results, the proposed EDROC and its control are validated.

Keywords: Electric drive, onboard converter, PWM, electric vehicle, PMSM

**Corresponding author*: ibrahimalig3893@gmail.com

Salman Ahmad, Farhad Ilahi Bakhsh, and P. Sanjeevikumar (eds.) Multilevel Converters, (185–202)

10.1 Introduction

It makes reasonable to explore for alternatives to gasoline- and diesel-powered cars given the depletion of fossil fuels. Plug-in electric vehicles are becoming more and more popular since they are clean and environmentally friendly. They will eventually replace the traditional internal combustion engine (ICE) drive system because they produce no pollutants [1–3]. Future electric vehicle uses are anticipated to increase [4]. PEVs' size can be decreased and their power density can be increased by combining the motor system and charging system, according to the EDROC [5]. In driving and charging modes, EDROC can therefore function independently. By employing PMSM as a charger's charging inductors, the EDROC boosts power density while decreasing volume. Considering the quantity [6] of converters, EDROC are divided into three types: composite converter systems, single-stage converter systems, and double-stage converter systems. The composite converter system, which is widely utilized in four-wheel drive electric vehicles and series hybrid vehicles [7, 8], is made up of several motors and power converters. For EVs with a single motor, like PEVs, a composite converter is not the best option. One motor is often all that is required for electric vehicles and three-phase fast charging typically only requires the usage of double-stage converter technology. The single-stage converter system is better from a size and cost standpoint because it has more components than a multi-stage converter.

From the EDROC, three groups—composite converter systems, double-stage converter systems, and single-stage converter systems—can be derived [8]. Frequently used with a four-wheel-drive electric vehicle (EV) or a series hybrid electric vehicle (HEV) is the composite converter system, which consists of multiple motors and power converters [9, 10].

The charging inductor for the EDROC being exhibited is an auxiliary circuit that applies the PMSM's inductance. However, the PMSM that drives the PEV's conventional motor [9, 10] makes it difficult to convert the converter into a charger. In this work, some specially created motors are used as chargers in [11–17], providing a novel and simple electric-drive-reconstructed converter. Unlike the normal EDROC that was the focus of investigations [18, 19], the converter that is being offered does not require additional AC equipment. The converter that is being discussed uses a switching network to rebuild the typical PEV motor. The system has the advantages of a straightforward design, low cost, and the ability to use the existing drive system without special construction. The efficiency of the suggested EDROC and control mechanism is verified using simulation data.

Consider the traditional PEV electric motor, the EDROC with AC auxiliary gear, as proposed and studied in [20, 21]. There is no need for an insulating coating in order to accommodate the extra equipment. The system will need more room if it additionally includes a relay or an unregulated rectifier unit though. The study recommends an innovative and simple rebuilt onboard converter for electric drives. The suggested converter uses a switching network to rebuild the standard PEV motor. No additional AC inductor is required to implement the suggested converter's topology circuit. You can plug the suggested converter directly into a power outlet at home or at work. The system has the advantages of a straightforward design, low cost, and the ability to use the existing drive system without special construction. The effectiveness of the suggested EDROC is verified using simulation and experiment results.

10.2 Proposed Electric-Drive-Reconstructed Converter Topology

10.2.1 Circuit Diagram

The being exhibited electric-drive-reconstructed on-board converter (EDROC) is shown in Figure 10.1. By installing auxiliary circuits, the driver hardware can operate independently in driving and charging modes and increase power density. The EDROC that is being shown can function in both charging and driving modes. Different control strategies are adapted in two operational modes. The tactics for driving and charging modes differ from one another.

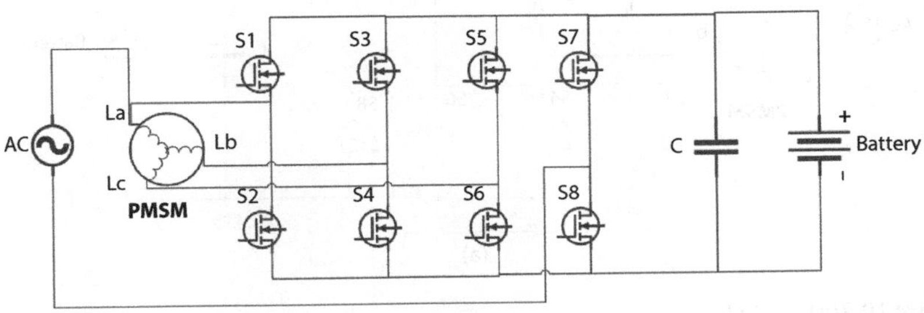

Figure 10.1 The topology of the onboard converter with electric drive reconstruction (EDROC).

10.2.2 Charging Mode

Switches S1 and S2 are off while switches S7 and S8 are turned on during charging mode. There are two portions with separate sets of four work states for the switching states.

10.2.2.1 States I–IV

As shown in Figures 10.2(a-d), the states I–IV are operating in the grid's positive cycle.

The state equations are expressed as:

$$L_s \frac{di_a{}^I}{dt} = \frac{2V_{ac} - V_B}{3}$$

$$L_s \frac{di_b{}^I}{dt} = \frac{-V_{ac} - V_B}{3}$$

$$L_s \frac{di_c{}^I}{dt} = \frac{-V_{ac} + 2V_B}{3}$$

In state I, the inductor Lb stores energy; the switch S5 causes the inductor Lc to discharge the energy to the battery.

According to Figure 10.2(b), while the system is in state II, the switches S3, S6, and S8 are switched on, while S4, S5, and S7 are off.

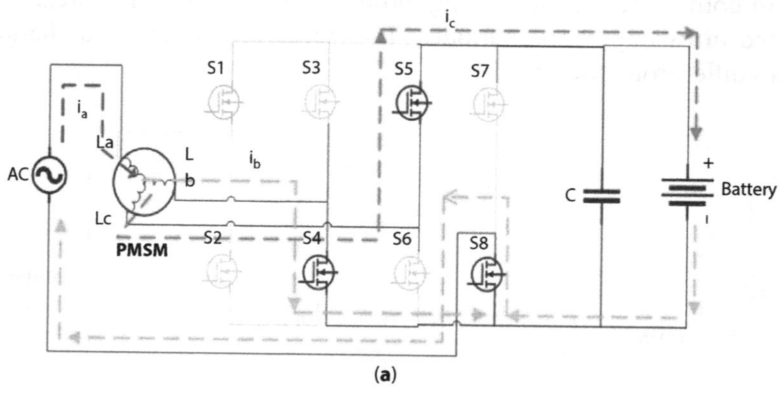

Figure 10.2(a) State I.

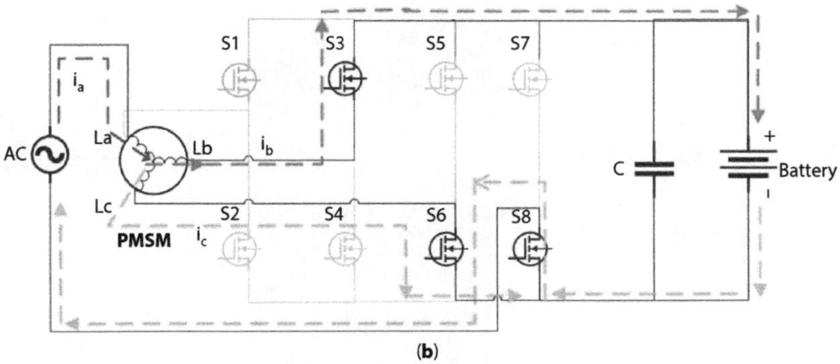

Figure 10.2(b) State II.

The state equations for state II are expressed as:

$$L_s \frac{di_a{}^{II}}{dt} = \frac{2V_{ac} - V_B}{3}$$

$$L_s \frac{di_b{}^{II}}{dt} = \frac{-V_{ac} + 2V_B}{3}$$

$$L_s \frac{di_c{}^{II}}{dt} = \frac{-V_{ac} - V_B}{3}$$

According to Figure 10.2(c), in state III, the switches S3, S5, and S7 are off, while S4, S6, and S8 are switched on.

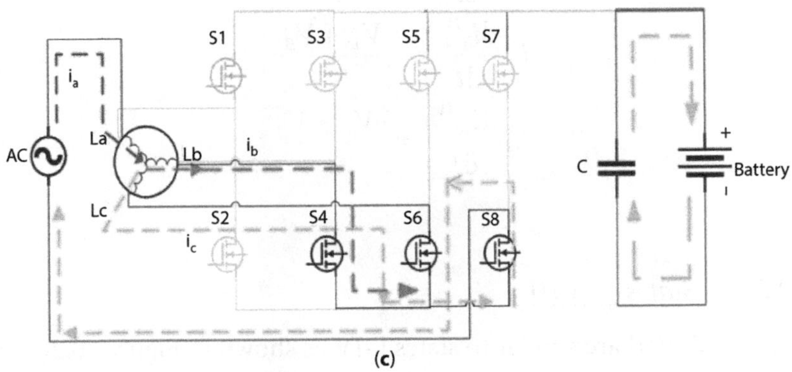

Figure 10.2(c) State III.

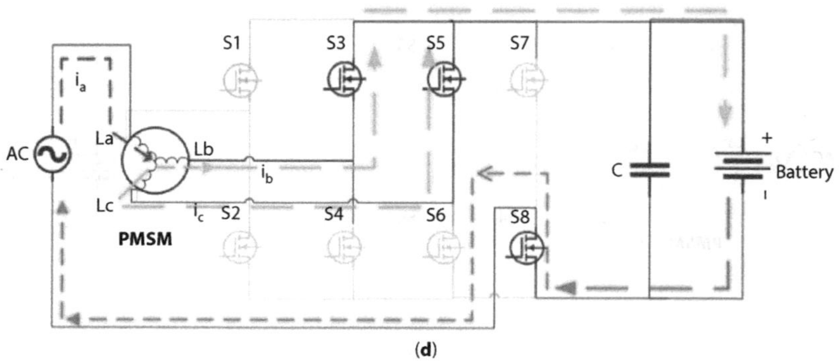

Figure 10.2(d) State IV.

The state equations for state III are expressed as:

$$L_s \frac{di_a^{III}}{dt} = \frac{2V_{ac}}{3}$$

$$L_s \frac{di_b^{III}}{dt} = \frac{-V_{ac}}{3}$$

$$L_s \frac{di_c^{III}}{dt} = \frac{-V_{ac}}{3}$$

The state equations for state IV are expressed as:

$$L_s \frac{di_a^{IV}}{dt} = \frac{2V_{ac} - 2V_B}{3}$$

$$L_s \frac{di_b^{IV}}{dt} = \frac{-V_{ac} + V_B}{3}$$

$$L_s \frac{di_{ac}^{IV}}{dt} = \frac{-V_{ac} + V_B}{3}$$

10.2.2.2 States V–VIII

The states V–VIII are similar to states I–IV as shown in Figure 10.2(e–h).

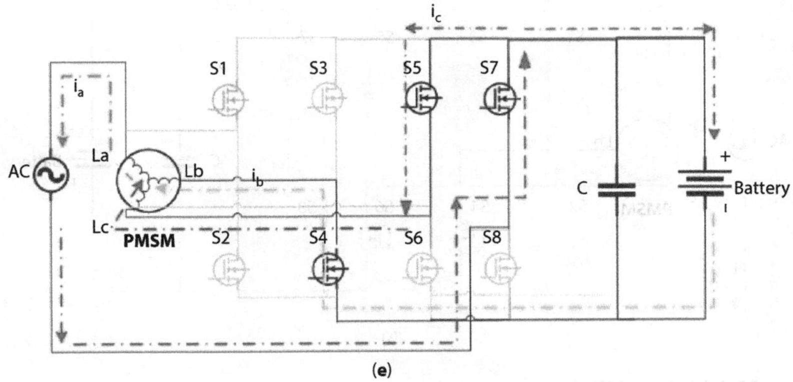

Figure 10.2(e) State V.

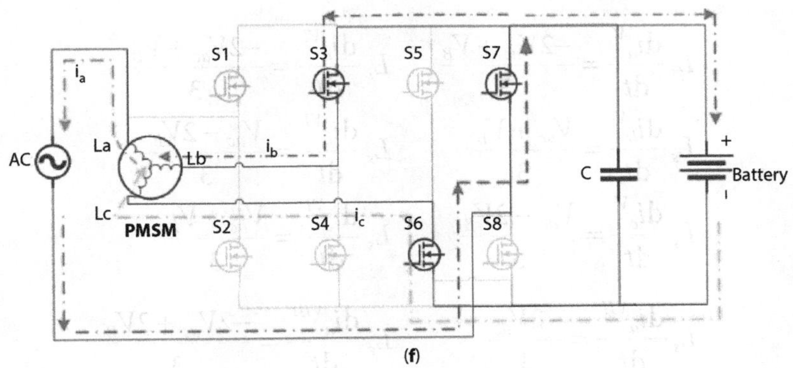

Figure 10.2(f) State VI.

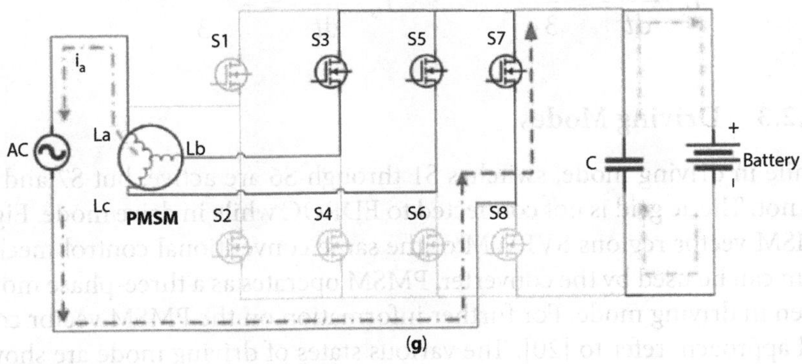

Figure 10.2(g) State VII.

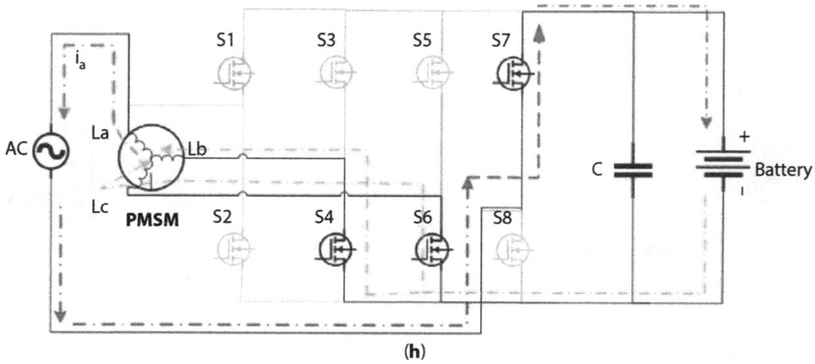

Figure 10.2(h) State VIII.

The state equations for the states V–VIII are given as:

$$L_s\frac{di_a{}^{V}}{dt}=\frac{-2V_{ac}+V_B}{3} \qquad L_s\frac{di_a{}^{VI}}{dt}=\frac{-2V_{ac}+V_B}{3}$$

$$L_s\frac{di_b{}^{V}}{dt}=\frac{V_{ac}+V_B}{3} \qquad L_s\frac{di_b{}^{VI}}{dt}=\frac{V_{ac}-2V_B}{3}$$

$$L_s\frac{di_c{}^{V}}{dt}=\frac{V_{ac}-2V_B}{3} \qquad L_s\frac{di_c{}^{VI}}{dt}=\frac{V_{ac}+V_B}{3}$$

$$L_s\frac{di_a{}^{VII}}{dt}=\frac{-2V_{ac}}{3} \qquad L_s\frac{di_a{}^{VIII}}{dt}=\frac{-2V_{ac}+2V_B}{3}$$

$$L_s\frac{di_b{}^{VII}}{dt}=\frac{V_B}{3} \qquad L_s\frac{di_b{}^{VIII}}{dt}=\frac{V_{ac}-V_B}{3}$$

$$L_s\frac{di_c{}^{VII}}{dt}=\frac{V_B}{3} \qquad L_s\frac{di_c{}^{VIII}}{dt}=\frac{V_{ac}-V_B}{3}$$

10.2.3 Driving Modes

While in driving mode, switches S1 through S6 are active, but S7 and S8 are not. The ac grid is not connected to EDROC while in drive mode. Eight PMSM vector regions SVPWM of the same conventional control mechanism can be used by the converter. PMSM operates as a three-phase motor when in driving mode. For further information on the PMSM vector control approach, refer to [20]. The various states of driving mode are shown in Figure 10.3.

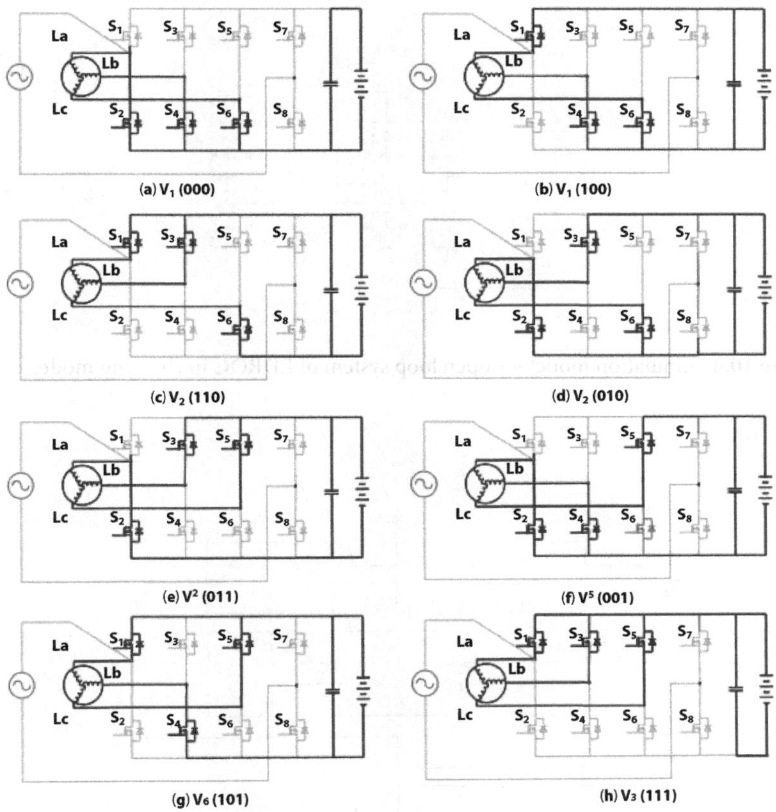

Figure 10.3 Driving mode switching states.

10.3 Operation of a Proposed System in Charging Mode

10.3.1 Open Loop Simulation in Charging Mode

Numerous simulations are run in both charging mode and driving mode to verify the provided technology. The inductance of PMSM is used to store energy during charging mode. Both the closed system and the open loop system simulations are run. The converter receives a firing pulse in an open loop system.

Figure 10.4 depicts the simulation model for the charging mode of the described EDROC's open loop system. The firing pulses for the open loop system are presented in Figure 10.5. The parameters used in the simulation model are given in Table 10.1.

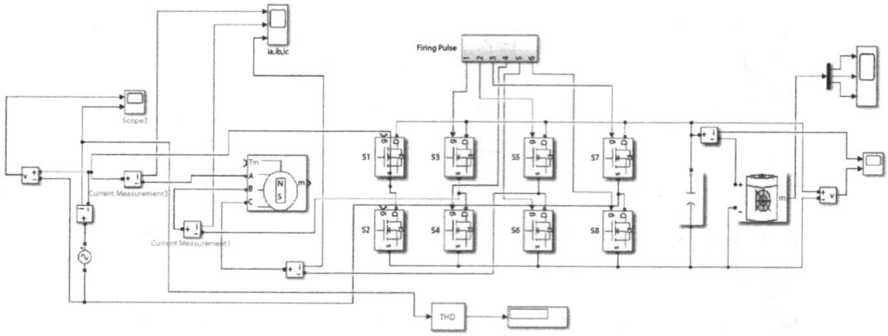

Figure 10.4 Simulation model for open loop system of EDROC in charging mode.

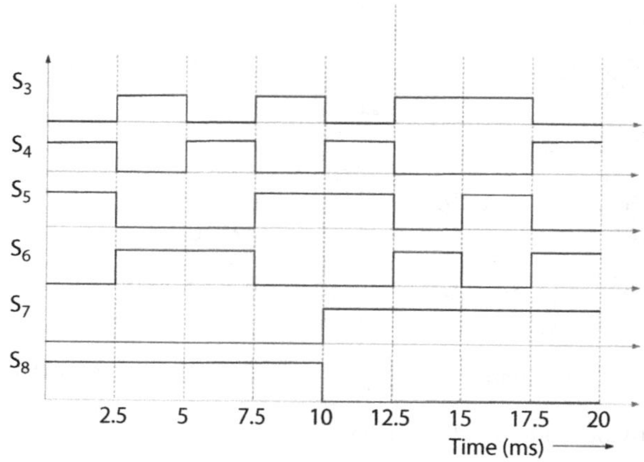

Figure 10.5 The firing pulses for switching circuit of presented EDROC in charging mode.

Table 10.1 Parameters in the simulation.

Description	Value
AC grid voltage	230 volts (r.m.s)
Stator inductance of PMSM	0.395 mH
Output capacitor	560 F
Nominal voltage of battery	72 volts
Rated capacity of battery	100 Ah
Initial state of charge	50%

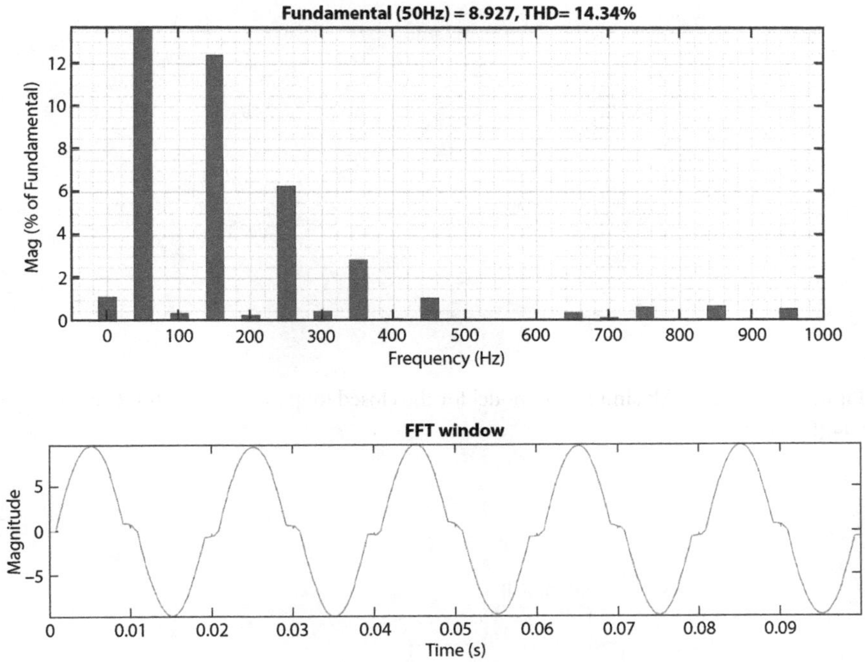

Figure 10.6 Simulation results of input current and its THD.

Figure 10.6 shows the input current of the presented EDROC in an open loop system. From the simulation results, the total harmonic distortion (THD) is 14.34%.

Since the THD of input current is very high, this is undesirable, so there is a need of closed loop charging system.

10.3.2 Closed Loop Simulation in Charging Mode

The simulation for closed loop system in charging mode of EDROC is shown in Figure 10.7. The additional parameters used for closed loop control are given in Table 10.2. The pulse width modulation (PWM) technique is used for switching circuits to reduce ripple in input current and obtain a desired voltage.

The PI controller is implemented in closed loop system in order to have efficient charging.

The proportional–integral controller makes use of both proportional and integral controller control actions. The drawbacks of each individual controller are eliminated when two distinct controllers are combined to create a more powerful controller as shown in Figure 10.8.

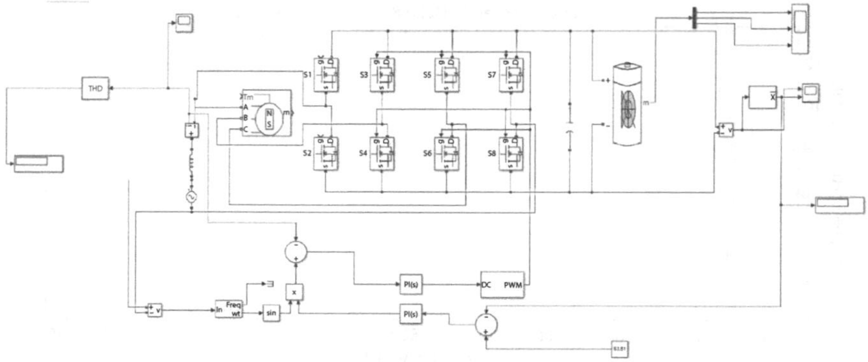

Figure 10.7 MATLAB simulation model for the closed loop system of EDROC in charging mode.

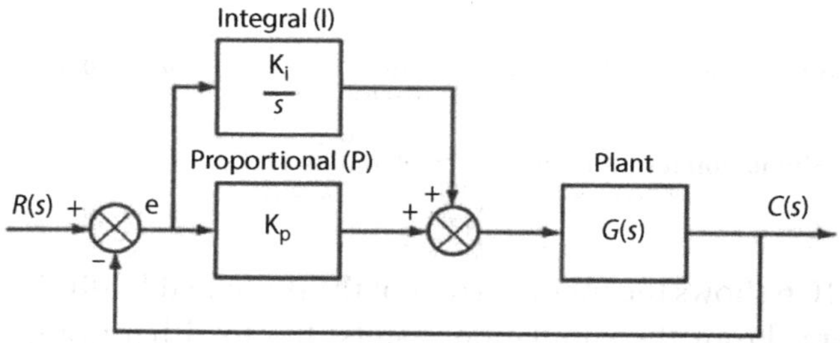

Figure 10.8 Block diagram of a PI controller.

Table 10.2 Parameters and the values used for simulation.

Parameter	Value
Rated frequency	50 Hz
Proportional constant, K_p	2
Integral constant, K_i	0.6
Switching frequency, f_s	15 kHz
Battery voltage	72 volt

It is vital to note that the circuit cannot be linearized to obtain the values of K_p and K_i by auto-tuning PI controller in MATLAB due to the intricacy of the electrical components inside the circuit route. Because there is no model, the settings are therefore determined through trial and error in order to achieve the desired voltage with the least amount of ripple. Finally, the PI controller's parameters for achieving the best performance are as follows:

$$K_p = 2 \text{ and } K_i = 0.6$$

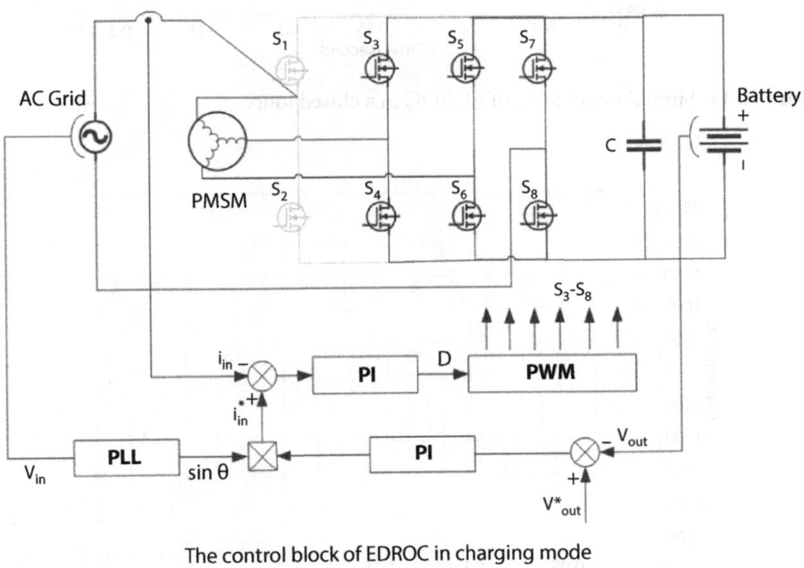

Figure 10.9 Closed loop charging of EDROC.

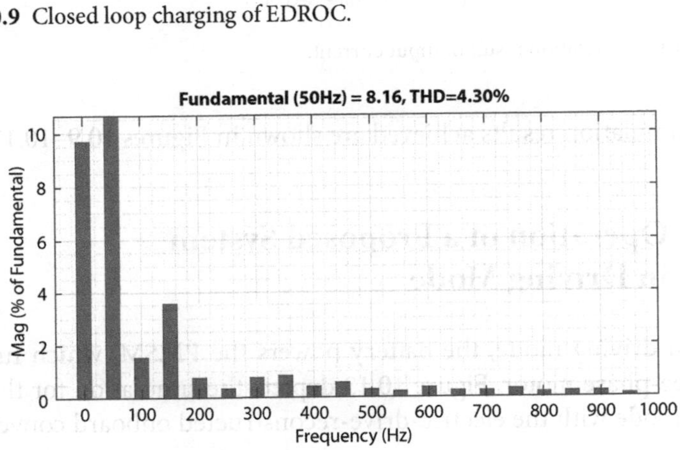

Figure 10.10 Simulation of THD of the presented EDROC in a closed loop.

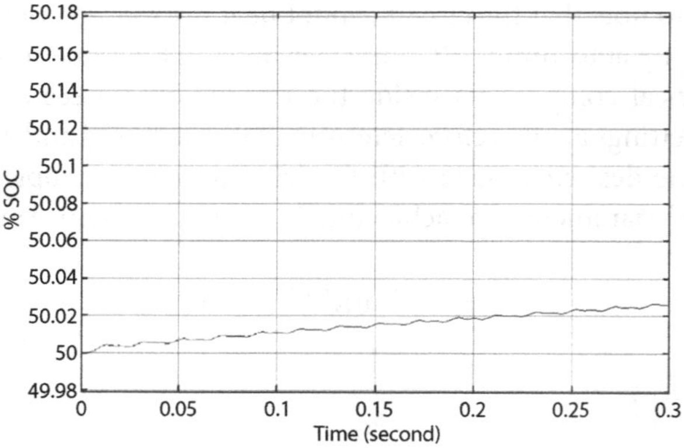

Figure 10.11 Simulation of SOC of EDROC in a closed loop.

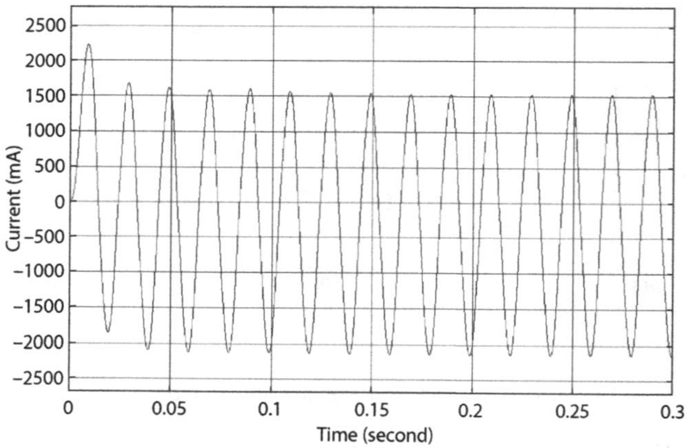

Figure 10.12 Simulation result of input current.

The simulation results achieved are shown in Figures 10.9–10.12.

10.4 Operation of a Proposed System in Driving Mode

When in driving mode, the battery powers the PMSM, which functions as a three-phase motor. Figure 10.13 depicts the simulation for the PEV's driving mode with the electric-drive-reconstructed onboard converter.

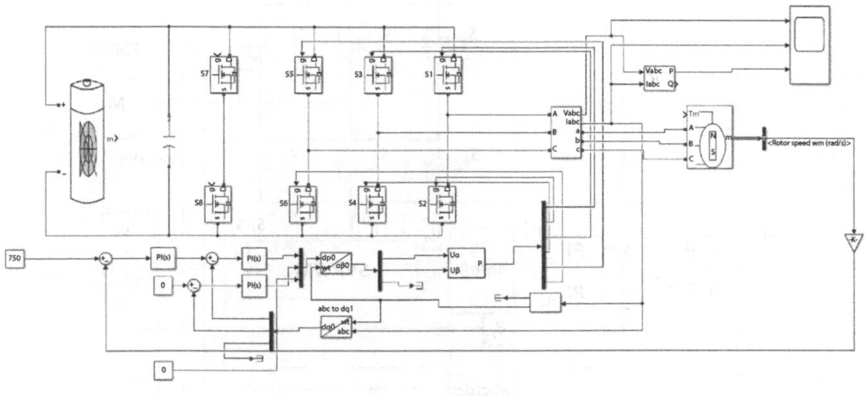

Figure 10.13 Simulation of EDROC in driving mode.

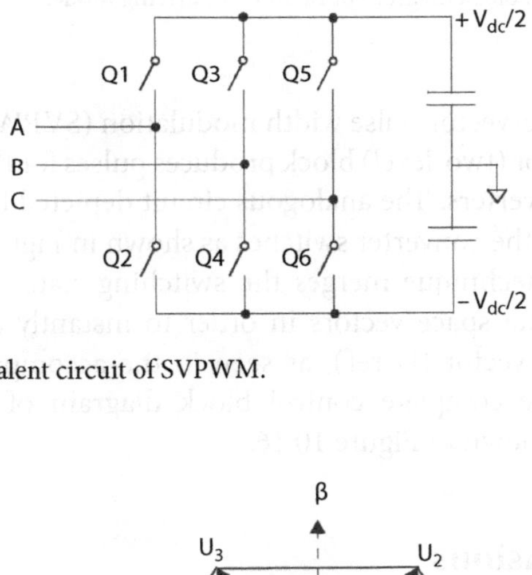

Figure 10.14 Equivalent circuit of SVPWM.

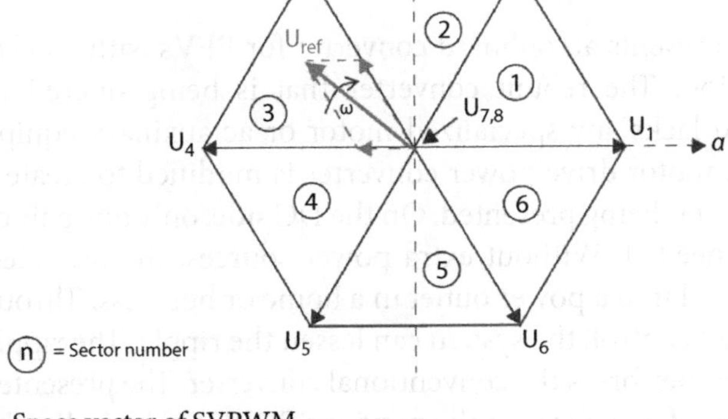

Figure 10.15 Space vector of SVPWM.

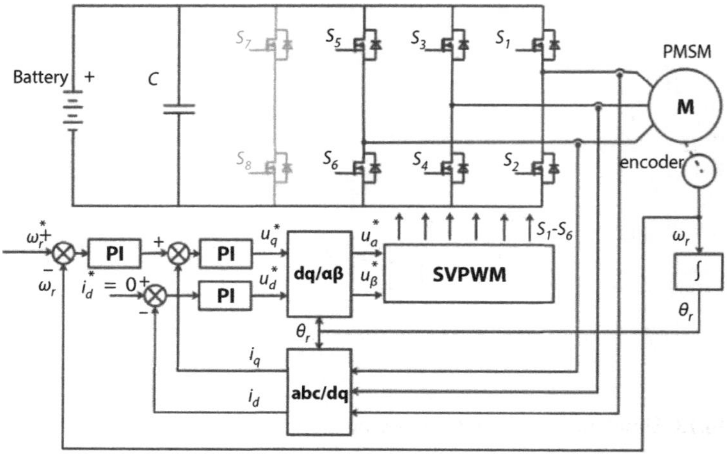

Figure 10.16 Control block diagram of EDROC in driving mode.

Using the space-vector pulse width modulation (SVPWM) method, the SVPWM generator (two-level) block produces pulses for three-phase, two-level DC/AC converters. The analogous circuit depicted below serves as a representation of the converter switches as shown in Figure 10.14.

The SVPWM technique merges the switching states that correspond to the fundamental space vectors in order to instantly approximate the reference voltage vector (U ref), as seen in the accompanying image of Figure 10.15. The complete control block diagram of EDROC under driving mode is shown in Figure 10.16.

10.5 Conclusions

This article presents an onboard converter for PEVs with an electric drive reconstruction. The rebuilt converter that is being offered is straightforward and lacks any specialized motor or ac auxiliary equipment. The three-phase motor drive power converter is modified to create the power converter that is being presented. On the DC side, only one pair of auxiliary switches is needed. Without extra power sources, the provided EDROC can be plugged into a power outlet in a home or business. Through the use of interleaved control, the system can lessen the ripple. The ripple suppression effect outperforms the conventional converter. The presented EDROC has a number of benefits over the previous EDROCs, including being more affordable and smaller in size. The motor driving and charger operations of the exhibited EDROC are realized on the workbench.

References

1. Husain, M. A., Rajput, R., Gupta, M. K., Tabrez, M., Ahmad, M. W., & Bakhsh, F. I. (2022). Design and Implementation of Different Drive Topologies for Control of Induction Motor for Electric Vehicle Application. *Distrib. Gener. Altern. Energy J.*, 37(4), 999–1026. https://doi.org/10.13052/dgaej2156-3306.3746
2. Amir, M., Zaheeruddin, Haque, A., Bakhsh, F.I., Kurukuru, V.S.B., Sedighizadeh, M.: Intelligent energy management scheme-based coordinated control for reducing peak load in grid-connected photovoltaic-powered electric vehicle charging stations. *IET Gener. Transm. Distrib.*, 1–18 (2023). https://doi.org/10.1049/gtd2.12772
3. Amir, M., Z., Haque, A., Kurukuru, V.S.B., Bakhsh, F.I., Ahmad, A.: Agent based online learning approach for power flow control of electric vehicle fast charging station integrated with smart microgrid. *IET Renew. Power Gener.*, 1–13 (2022). https://doi.org/10.1049/rpg2.12508
4. C. Chan, A. Bouscayrol, and K. Chen, "Electric, hybrid, and fuel-cell vehicles: Architectures and modeling," *IEEE Trans. Veh. Technol.*, vol. 59, no. 2, pp.589–598, Feb. 2010.
5. F. Yu, W. Zhang, Y. Shen, and J. Mao, "A nine-phase permanent magnet electric-drive-reconstructed onboard charger for electric vehicle," *IEEE Trans. Energy Convers.*, vol. 33, no. 4, pp. 2091–2101, Dec. 2018.
6. S. Mateen, M. Amir, A. Haque, F. I. Bakhsh, "Ultra-fast charging of electric vehicles: A review of power electronics converter, grid stability and optimal battery consideration in multi-energy systems, *Sustain. Energy Grids Netw.*, 4 (2023), doi: https://doi.org/10.1016/j.segan.2023.101112.
7. S. Haghbin, K. Khan, S. Lundmark, M. Alakla, O. Carlson, M. Leksell, and O. Wallmark, "Integrated chargers for EV's and PHEV's: Examples and new solutions," in *Proc. 19th Int. Conf. Elect. Mach.-ICEM*, Sep. 2010, pp. 1–6.
8. Dar, U., Siddiqui, A. S., Bakhsh, F. I. (2022). Design and Control of an Off Board Battery Charger for Electric Vehicles. *Distrib. Gener. Altern. Energy J.*, 37(4), 959–978. https://doi.org/10.13052/dgaej2156-3306.3744.
9. S. Morimoto, S. Ooi, Y. Inoue, and M. Sanada, "Experimental evaluation of a rare-Earth-free PMASynRM with ferrite magnets for automotive applications," *IEEE Trans. Ind. Electron.*, vol. 61, no. 10, pp. 5749–5756, Oct. 2014.
10. S. A J. Nerg, M. Rilla, V. Ruuskanen, J. Pyrhonen, and S. Ruotsalainen, "Direct-driven interior magnet permanent-magnet synchronous motors for

a full electric sports car," *IEEE Trans. Ind. Electron.*, vol. 61, no. 8, pp. 4286–4294, Aug. 2014.
11. L. De Sousa, B. Silvestre, and B. Bouchez, "A combined multiphase electric drive and fast battery charger for electric vehicles," in *Proc. IEEE Vehicle Power Propuls. Conf.*, Sep. 2010, pp. 1–6.
12. A. Bruyère, L. De Sousa, B. Bouchez, P. Sandulescu, X. Kestelyn, and E. Semail, "A multiphase traction/fast-battery-charger drive for electric or plug-in hybrid vehicles: Solutions for control in traction mode," in *Proc. IEEE VPPC*, Sep. 2010,pp. 1–7.
13. S. Lacroix, E. Laboure, and M. Hilairet, "An integrated fast battery charger forelectric vehicle," in *Proc. IEEE Vehicle Power Propuls. Conf.*, Sep. 2010, pp. 1–6.
14. S. Haghbin, S. Lundmark, M. Alakula, and O. Carlson, "An isolated high power integrated charger in electrified-vehicle applications," *IEEE Trans. Veh. Technol.*,vol. 60, no. 9, pp. 4115–4126, Nov. 2011.
15. S. Haghbin, S. Lundmark, M. Alakula, and O. Carlson, "Grid-connected integratedbattery chargers in vehicle applications: Review and new solution," *IEEE Trans. Ind. Electron.*, vol. 60, no. 2, pp. 459–473, Feb. 2013.
16. L. Solero, "Non conventional on-board charger for electric vehicle propulsion batteries," *IEEE Trans. Veh. Technol.*, vol. 50, no. 1, pp. 144–149, Jan. 2001.
17. L. Tang and G.-J. Su, "A low-cost, digitally-controlled charger for plugin hybrid electric vehicles," in *Proc. IEEE Energy Convers. Congr. Expo.*, Sep. 2009, pp.3923–3929.
18. J. Ye, C. Shi, and A. Khaligh, "Single-phase charging operation of a three phase integrated onboard charger for electric vehicles," in *Proc. IEEE Transp. Electrific. Conf. Expo. (ITEC)*, Long Beach, CA, USA, Jun. 2018, pp. 681–686.
19. C. Shi, Y. Tang, and A. Khaligh, "A three-phase integrated onboard charger for plug-in electric vehicles," *IEEE Trans. Power Electron.*, vol. 33, no. 6, pp.4716–4725, Jun. 2018.
20. J. Wu, J. Wang, C. Gan, Q. Sun, and W. Kong, "Efficiency optimization of PMSMdrives using field-circuit coupled FEM for EV/HEV applications," *IEEE Access*, vol. 6, pp. 15192–15201, 2018.
21. L. Solero, "Non conventional on-board charger for electric vehicle propulsion batteries," *IEEE Trans. Veh. Technol.*, vol. 50, no. 1, pp. 144–149, Jan. 2001.

11

Packed U-Cell Multilevel Inverter and Applications in Solar Photovoltaic System

Salman Ahmad[1]*, Tajamal Hayat Parray[2] and Farhad Ilahi Bakhsh[2]

[1]*Department of Electrical Engineering, Islamic University of Science and Technology, Awantipora, Pulwama, Jammu and Kashmir, India*
[2]*Department of Electrical Engineering, National Institute of Technology, Hazratbal, Srinagar, Jammu and Kashmir, India*

Abstract
This chapter focuses on the development of a multilevel inverter topology for power systems using a single DC source. The objective is to address the issues associated with large and expensive inverters that rely on isolated DC sources. The chosen approach involves the utilization of a packed U-cell (PUC) inverter to generate a symmetrical five-level stepped voltage waveform with low harmonic distortion. To simplify the control system, a sensorless voltage regulation method based on switching states with redundancies is implemented to maintain the DC capacitor voltage up to the permissible level. The simulation and experimental results are in very close agreement, making it highly attractive for solar photovoltaic applications.

Keywords: PUC, multilevel inverter, sensorless

11.1 Introduction

Traditional two-level inverters produce output voltage with high harmonic content, which is undesirable [1–3]. Current research in power electronic converters concentrates on employing additional switches and DC sources to generate higher output voltage levels [1, 2, 4]. This approach reduces harmonic content, decreases the need for extensive output filters, and

**Corresponding author*: salmanahmad19@gmail.com

Salman Ahmad, Farhad Ilahi Bakhsh, and P. Sanjeevikumar (eds.) Multilevel Converters, (203–232)

lowers manufacturing costs. Recent advancements in the realm of power electronics can be classified into three primary domains: topology, control, and application [5–17]. Power electronic converters fulfil diverse roles, including transforming AC to DC and performing DC-to-DC and DC-to-AC conversions. In terms of topology development for converters, the initial focus centers on alleviating stress on the switches employed in these devices while concurrently reducing the number of components [6]. This reduction in components is essential as excessive components may impose stress on power semiconductors and other passive elements such as inductors and capacitors. The ultimate goal of these efforts is to enhance overall efficiency. By reducing stress on these components and minimizing their quantity, overall efficiency can be improved [1]. The second aspect of topology development focuses on controlling parameters such as current, voltage, and power quality enhancement. The third facet of topology advancement deals with various applications, including standalone and grid integration. Recent advancements in semiconductor switch technology have made power electronics devices more economically feasible, resulting in a significant expansion of their applications. Multilevel inverters, characterized by multiple switches and DC supplies, are designed to produce output waveforms with minimal harmonic content [1, 17–20]. They are particularly suitable for energy conversion applications that efficiently transfer power from renewable sources like PV systems to loads. Multilevel inverters offer advantages such as reduced stress on switches, low total harmonic distortion (THD), and smaller passive filters, making them preferable over two-level inverters despite their bulkiness. These multilevel inverters find use in high-voltage, high-power scenarios. However, multilevel inverters require more independent DC sources, leading to the utilization of larger rectifiers and transformers. Alternately, employing DC capacitors in place of DC sources might be needed for precise voltage control strategies. For low- and medium-power applications, such as connecting solar panels to the local grid for domestic use or street lighting, researchers have been investigating various multilevel inverter configurations. These inverters are needed in order to convert the DC voltage produced by renewable energy sources into the appropriate AC waveform required by the load and the grid. In such scenarios, there is a preference for a single-phase transformerless inverter that utilizes a minimal number of DC sources. One notable topology in this category is the switched capacitor multilevel inverter (SCMIL), often referred to as the packed U-cell inverter (PUC). This name is derived from the arrangement of switches and capacitors. The recently introduced packed U-cell multilevel inverter (MLI) has gained popularity due to its advantages, including a reduced device count, lower total harmonic distortion, and enhanced power quality. The packed U-cell configuration requires fewer devices to achieve the same output voltage levels compared to other multilevel inverters with reduced device counts found in existing literature. The packed U-cell (PUC) inverter, initially introduced by Al-Haddad and colleagues, stands out as a noteworthy innovation. It achieves a seven-level voltage output while employing only six active switches, a single isolated DC source, and a second capacitor source [21–24]. Notably, the voltage of this second capacitor source must be carefully controlled to maintain a fixed ratio of 1/3 relative to the first DC source. However, it is important to acknowledge that the PUC inverter, despite its advantages in terms of component count compared to other seven-level inverters, comes with significant limitations [24]. The drawbacks include, among other things, the need for a complex and intricate controller equipped with numerous feedback sensors, operating at a high switching frequency, producing asymmetric output voltage cycles and levels, and requiring a large capacitor for voltage regulation under varying conditions. This chapter examines the PUC topology's modulation, control, and use inside a solar photovoltaic system. Constant voltage levels in the capacitor allow for a symmetrical five-level output voltage waveform with little harmonic distortion.

11.1.1 Inverter

The power electronic converters that transform an input DC voltage into a symmetric output AC voltage with a desired magnitude and frequency are called inverters. The voltage at output can be variable or fixed, either at a variable or fixed frequency. To achieve variable voltage at output, the input DC voltage can be adjusted while keeping the gain of inverter constant. Alternatively, if the input DC voltage is uncontrollable and fixed, the output voltage can be varied by adjusting the inverter gain using control techniques like pulse-width modulation.

11.1.1.1 Evolution of Multilevel Inverter (MLI) Configurations

Figure 11.1 presents an overview of the evolution of various multilevel inverter topologies. Over time, researchers have introduced numerous novel multilevel inverter configurations with application-based approaches, emphasizing the creation of reduced switch count designs.

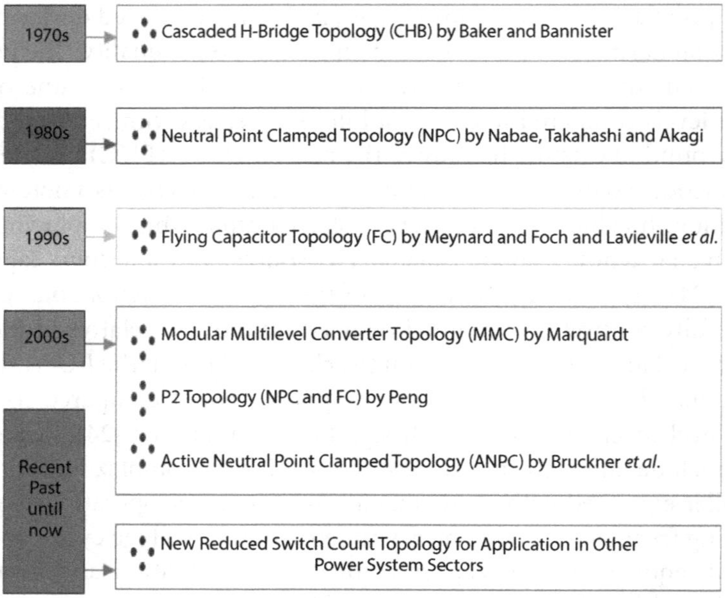

Figure 11.1 An overview of the evolution of different multilevel inverter topologies.

11.1.1.2 Two-Level Inverter

Two-level inverters generate two levels in the output voltage or current. The two levels are either 0 or +Vdc,-Vdc. However, when operating at high frequencies, it possesses limitations in high-voltage and high-power applications. The primary reasons behind these limitations are switching losses and device rating constraints. A detailed comparison of two-level and multilevel topologies is shown in Table 11.1.

11.1.1.3 Need for Higher Level for Producing Output Voltage

Three-level converters were the first multilevel inverter (MLI) used in 1975, and over time, different MLI topologies emerged. An alternate method for integrating MLIs into grid-tied systems is to send a waveform that is entirely sinusoidal to the grid as opposed to one that has less harmonics in either the current or voltage. The grid frequencies must match the implanted signal frequencies exactly. An LC filter is often used in an MLI arrangement to create sinusoidal three-phase current and voltage with decreased harmonics. Initially, the industry's top option for connecting renewable energy sources to the electrical grid was a three-level inverter. These inverters had a number of shortcomings though, which called for more topological

Table 11.1 Comprehensive overview of merits and demerits of different MLI configurations.

Properties	MLI	Two-level inverter
Structure	Modular	Complicated
High voltage and current operation	Can operate	Can operate (parallelized structures)
Application	High voltage	Low voltage
Stress on power electronics switches	Less	More
Performance with excellent power	High	Low
Harmonics	High	Low
Ability to operate without a transformer	Yes	No
Generating common mode voltage	Lower	Higher
Switching losses	Low	High
Efficiency	High	Low
Low frequency operation	Can operate	Fails
dv dt	Low	High
Input current distortions	Low	High
Fault-tolerant operation	Possible	Impossible

changes. These drawbacks included the need for high-power-rated components for applications demanding high power levels, thereby increasing inverter costs and resulting in higher switching losses. Additionally, conventional inverters produced an output limited to two levels (0 V and Vdc), resulting in low-quality waveforms with substantial harmonic content. In response to these limitations, MLIs were introduced and have since gained widespread adoption within the industry. The MLIs have a number of benefits over conventional inverters, including improved harmonic properties, operational performance, decreased stress on MLIs under load, and less electromagnetic interference (EMI). A PUC inverter technique's improved algorithm keeps the capacitor voltage constant despite changes in load or starting. Recent studies confirm that this method substantially improves system speed compared to the preceding controllers. Notably, it operates without the requirement for feedback sensors, modulation index, switching frequency, switching index, or system modeling. It can work effectively across a large voltage range, from zero to infinite amplitudes, even when the DC supply voltage changes.

11.1.2 Multilevel Inverters

In recent times, multilevel inverters have gained significant recognition as state-of-the-art technology in various domains for conversion of power from DC to AC. These domains include power generation, transmission, distribution, and energy utilization. MLIs are being innovatively employed in various areas such as FACTS, UPFCs, HVDC transmission, STATCOMs, active filters (AFs), dynamic voltage restorers, drives for traction and transportation systems, marine propulsion, mine hoists, conveyors, and induction heating power supplies. Their adaptability and efficiency make them valuable assets in modern power systems across diverse industries. Multilevel inverters have gained immense popularity in the power industry, transportation, and renewable energy sectors due to their unique features that make them suitable for reactive power compensation. Their design enables the production of high-power, high-voltage inverters with better control over device voltage stresses. This characteristic allows for increasing the number of voltage levels without requiring devices of higher ratings, thus enhancing the power rating of the inverter. A notable advantage of multilevel inverters is their capability to obtain high output voltages with reduced harmonics, all without the need for series-connected synchronized-switching devices or transformers. The input of MLI is direct current (dc), and its ideal output should be a sine wave, making it an attractive option for various applications in power conversion and energy control. The detailed classification of multilevel inverter topologies is shown in Figure 11.2. The main multilevel inverters (MLIs) can be classified into various categories based on the following important factors:

1. Symmetric type—utilizing equal DC sources
2. Asymmetric type—employing unequal DC sources
3. With inherent negative levels
4. Without inherent negative levels
5. Utilizing capacitor links with single sources

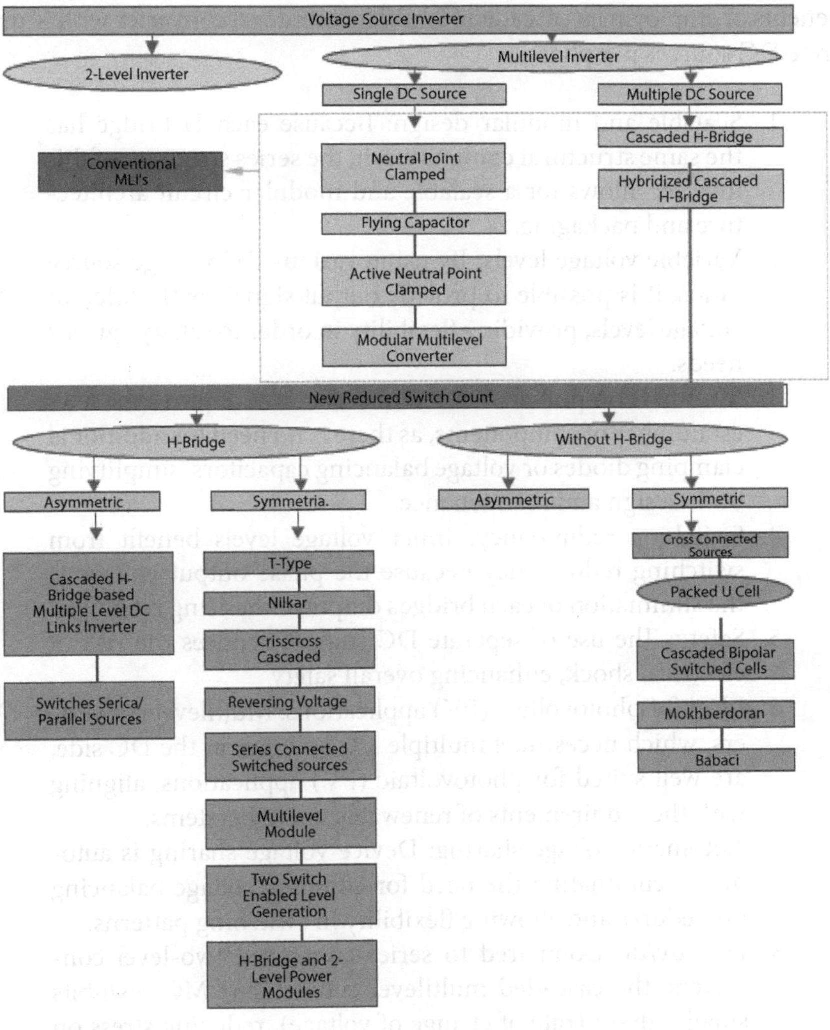

Figure 11.2 Overall classification of MLI.

6. Regenerative configuration operating as both rectifier and an inverter
7. Hybrid approach of FC, NPC, and CHB

11.1.3 Cascaded H-Bridge

The cascaded H-bridge multilevel inverter proves most advantageous in scenarios where multiple DC sources are concurrently active. Here are the benefits of employing a cascaded H-bridge multilevel converter with 's' discrete DC sources per phase:

1. Scalable and modular design: Because each H-bridge has the same structural configuration, the series structure of this inverter allows for a scalable and modular circuit architecture and packaging.
2. Variable voltage levels: By using various DC voltage source ratios, it is possible to provide output signals with different voltage levels, providing flexibility in order to satisfy various needs.
3. Minimal component count: This converter requires the fewest number of components, as there is no need for additional clamping diodes or voltage balancing capacitors, simplifying both design and maintenance.
4. Switching redundancy: Inner voltage levels benefit from switching redundancy because the phase output voltage is the summation of each bridge's output, enhancing reliability.
5. Safety: The use of separate DC sources reduces the risk of electrical shock, enhancing overall safety.
6. Ideal for photovoltaic (PV) applications: Multilevel converters, which necessitate multiple DC sources on the DC side, are well suited for photovoltaic (PV) applications, aligning with the requirements of renewable energy systems.
7. Automatic voltage sharing: Device voltage sharing is automatic, eliminating the need for complex voltage balancing procedures and allowing flexibility in switching patterns.
8. Low dv/dt: Compared to series-connected two-level converters, the cascaded multilevel converter (CMC) exhibits smaller dv/dt (rate of change of voltage), reducing stress on components and enhancing performance.
9. Modularity and control: The modularity inherent in this topology simplifies control strategies, making it relatively straightforward to manage.
10. Efficiency: When compared to other multilevel topologies, CMC stands out for its efficiency, requiring the fewest components due to the absence of clamping diodes and flying capacitors.
11. Redundancy and scalability: The CMC offers redundancy in inverter operation and can be easily scaled up by incorporating more cells per phase than initially required, enhancing

system resilience. However, it is essential to note that the main limitation of the cascaded H-bridge multilevel converter is its dependence on having separate DC sources available, limiting its application to specific scenarios where such sources are accessible.

11.1.4 Flying Capacitor Multilevel Inverter

The structure of the capacitor-clamped inverter closely resembles that of the diode-clamped converter, with the primary distinction being the replacement of diodes with capacitors for clamping. The most common application of this topology is static VAR generation. The key advantages and disadvantages are discussed in the following subsections.

11.1.4.1 Advantages

Following are the main advantages of FCMLI:

1. Control of real and reactive power: The capacitor-clamped inverter enables precise control over both real and reactive power transmission, making it suitable for applications requiring power regulation.
2. Voltage handling capability: The inclusion of a substantial number of capacitors equips the inverter to effectively manage voltage drops, spikes, and short outages, enhancing grid stability and reliability.
3. Redundancy for voltage balancing: Redundant switching states are available to balance the voltages across the capacitors, promoting consistent performance and reducing the risk of voltage imbalances.

11.1.4.2 Disadvantages

Following are the main disadvantages of FCMLI:

1. Reduced efficiency for real power transmission: The efficiency of the capacitor-clamped inverter tends to be lower when transmitting real power compared to other inverter topologies, which may limit its suitability for applications prioritizing efficiency.

2. Cost and bulk: The large number of capacitors required in this topology can be more expensive and bulkier than the diodes used in the neutral point-clamped inverter. This aspect can pose challenges, especially in high-level converters, both in terms of cost and physical packaging.
3. Complex initialization: Initiating the system and precharging the capacitor voltages can be a complex and critical procedure, potentially requiring additional control and safety measures to ensure proper operation.

In summary, the capacitor-clamped inverter offers advantages in terms of power control, voltage handling, and redundancy for voltage balancing, making it suitable for specific applications like static VAR generation. However, it comes with drawbacks related to efficiency, cost, bulkiness, and complex initialization, which must be carefully considered when selecting this topology for a particular.

11.2 Packed U-Cell Inverter

Researchers have been actively examining a number of multilevel inverter topologies suitable for moderate and medium power applications, such as connecting solar panels to the local grid for domestic usage or street lighting. For the conversion of the DC voltage generated from renewable energy sources into the appropriate AC waveform that can be utilized by loads and the grid, these inverters are needed installed. Single-phase transformer-free inverters with the required number of DC sources are given preference in such scenarios. One innovative method in that particular field is the packed U-cell (PUC) inverter, which has been demonstrated for the first time by Al-Haddad *et al.* [21]. This inverter method may also offer a seven-level voltage using just six active switches, an isolated DC source, and a second capacitor as a backup supply while using the same amount of isolated DC sources. The PUC topology has fewer components than other seven-level inverters, but it also has some disadvantages that must be taken into account, including a high switching frequency, an asymmetry in the output voltage cycles and levels, the requirement for a quick and difficult controller with numerous feedback sensors, and the use of a large capacitor to regulate voltage under changing conditions. The introduction of the forefront five-level PUC inverter using sensor-less PUC5 technological advances, which autonomously balances voltage, is of the utmost significance [22]. The PUC5 inverter maintains its capacitor voltage at half of the DC source magnitude by employ of a self-voltage-balancing mechanism incorporated into multi-carrier pulse width modulation (PWM). Therefore, complicated controls and voltage or current sensors are not essential. The five-level, symmetric output voltage waveform has reduced harmonic distortion as a result of maintaining a steady capacitor voltage at an appropriate level. The component counts of various five-level typologies by considering different aspects such as component counts and application domains for various multilevel converter typologies are shown in Tables 11.2, 11.3, and 11.4, respectively. The literature [21–32] contains a substantial amount of topological modification and control techniques as well as implementations for these approaches.

Table 11.2 Component comparison of different five-level MLI.

Topology	Switches	Capacitors	Size	Cost
Flying capacitor	8	4	Complex	300% cost compared to PUC
H-bridge	8	2	Moderate	275% cost compared to PUC
PUC	6	2	Small	Less cost compared to others

Table 11.3 Component count of MLI.

Parameters	CHB	FC	DC
Switch count	$2(N-1)$	$2(N-1)$	$2(N-1)$
DC source	$N-1/2$	1	1
Diodes	0	0	$(N-1)(N-2)$
Capacitors	0	$N(N-1)$	$N-1$

Table 11.4 Applications of different multilevel inverter (MLI) configurations in various power system areas.

MLI topologies	Applications
FC-MLI	- Renewable energy - Motor drives - FACTS
CHB-MLI	- Drives
	- Renewable energy
PUC-MLI	- Motor drives - Renewable energy

11.3 Comparison of MLI Topologies

Among the different MLI (Multilevel Inverter) topologies analyzed in the literature, the 3-phase Flying Capacitor topology exhibited THD of 38.76%, while the Cascaded H-Bridge topology showed a THD of 39.86%. However, our proposed PUC5 inverter demonstrated significantly reduced harmonic content, presenting an even lower THD value. Also PUC5 inverter has low component count compared to other MLIs as shown in Figure 11.3.

The Switched Capacitor Multilevel Inverter (SCMLI), known as the Packed U-Cell Inverter, is a widely recognized topology in power electronics. It is renowned for its ability to generate a seven-level output waveform using minimal components, thus achieving maximum voltage levels. However, the complexity of the controller required for its operation and the need for numerous sensors to provide state feedback for controller calculations pose challenges for widespread acceptance in industries and the market. The name "packed U-cell" is derived from the arrangement of switches and capacitors in the inverter. Recently, the packed U-cell MLI has gained significant attention due to its advantages, such as lower THD, reduced device count, and improved power quality. In comparison to other low device count MLIs mentioned in the literature, the packed U-cell requires fewer devices so as to obtain the same number of output voltage levels. Figure 11.4 illustrates the single-phase PUC inverter topology, while Figure 11.5 presents a comprehensive list of the associated switching states. It comprises a single DC source and six active switches, effectively

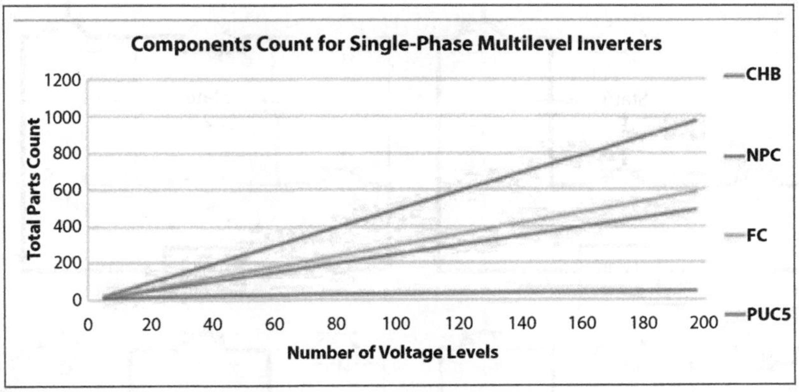

Figure 11.3 Component count chart in multilevel inverters.

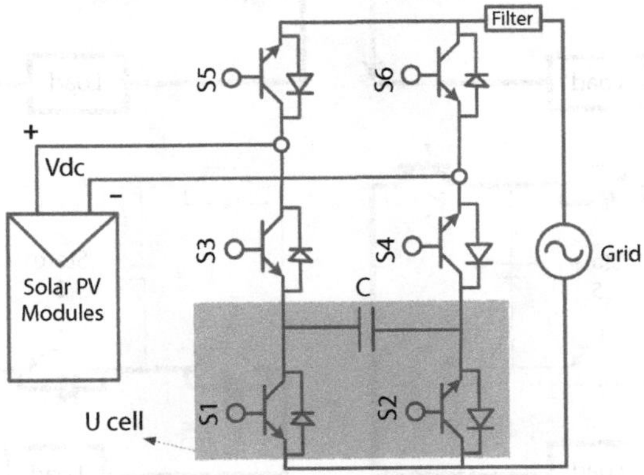

Figure 11.4 Circuit topology.

producing a five-level voltage waveform. Despite its promising features, further research and development are necessary to overcome the challenges posed by complex controllers and asymmetric voltage levels, making it more attractive for industrial adoption. A single-phase PUC inverter consists of single DC source and six active switches, which give a five-level voltage waveform.

11.3.1 Operation of Packed U-Cell Inverter

Moreover, it is apparent that the system benefits from eight distinct switching states, providing a variety of pathways for electric current to traverse

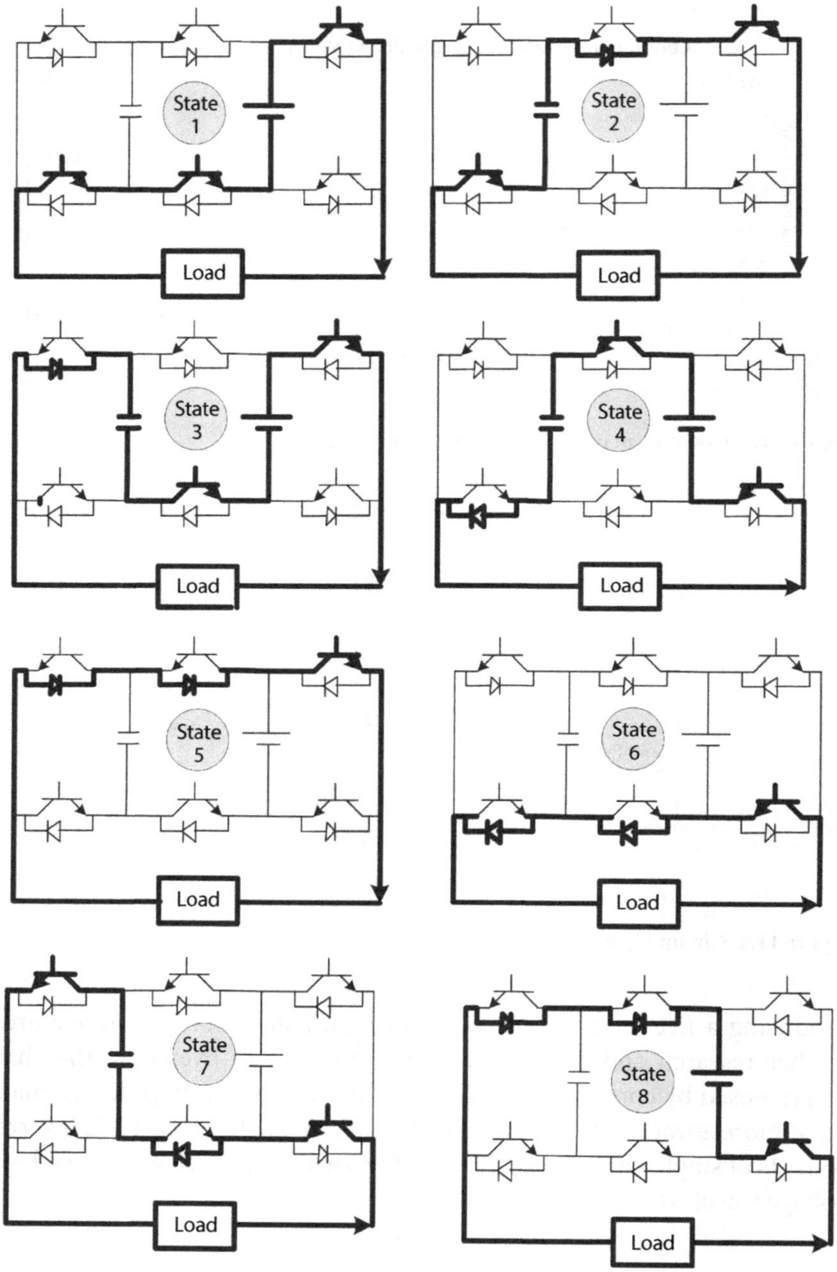

Figure 11.5 Circuit diagram of switching states.

Table 11.5 Capacitor states.

State	Voltage states of capacitor
1	No change
2	Discharging
3	Discharging
4	Charging
5	No change
6	No change
7	Discharging
8	No change

the entire setup, including the DC sources and loads. Importantly, the number of achievable output voltage levels hinges on the amplitudes of the DC sources. To maximize the number of output levels, it is crucial to set V2's amplitude to one-third of V1's. Assuming V1=3V2=3E, this configuration yields seven levels: 0, ±E, ±2E, ±3E. However, the seven-level PUC inverter presents certain drawbacks, mainly arising from the complexities associated with voltage balancing. The charging and discharging states of the capacitor is shown in Table 11.5.

It is evident that the PUC inverter can alternatively function as a five-level inverter when V1=2V2=2Vdc. Consequently, the resulting five-level output voltage waveform encompasses the levels 0, ±Vdc, and ±2Vdc. In this particular scenario, the capacitor voltage (V2) is consistently maintained at half of the amplitude of the DC source (V1). The presence of six switching states capable of producing three levels, which include -Vdc, 0, and +Vdc. This observation signifies the existence of redundant switching states that can facilitate the establishment of diverse current pathways through the load. These redundant switching states assume a pivotal role in overseeing the charging and discharging of the capacitor to preserve voltage equilibrium at half of the DC source voltage. To seamlessly integrate these redundant switching states into the requisite PWM technique for the design of the PUC5 inverter, a comprehensive analysis of all switching states and their effects on capacitor voltage has been conducted, as illustrated in Figure 11.4 We need to generate gate pulses for three switches S1, S3, and S5 only as the other three switches S2, S4, and

S6 operate in complementary manners. The different switching states are given in Figure 11.5. The switching state and output voltage across are discussed in the subsection that follows.

11.3.2 State 1

In this state, switches S2, S4, and S5 are on, while switches S1, S3, and S6 are off. Thus, the circuit gets completed as S2, S4, and S5 are involved to complete the circuit, and the capacitor is disconnected in this particular state. The battery voltage V1 appears across the output.

11.3.3 State 2

In this state, switches S2, S3, and S5 are on, while switches S1, S2, and S6 are off. Thus, the circuit gets completed as S2, S4, and S5 are involved to complete the circuit along with the capacitor, and the main source of voltage is disconnected in this particular state.

11.3.4 State 3

In this state, switches S1, S4 and S5 are on, while switches S2, S3, and S6 are off. The circuit gets completed through S1, S4, and S5 which are involved to complete the circuit, and the capacitor is connected in this particular state along with the main input source.

11.3.5 State 4

In this state, switches S2, S3, and S6 are on, while switches S1, S2, and S5 are OFF. The circuit gets completed *via* S1, S4, and S5 which are involved to complete the circuit, and the capacitor is connected in this particular state along with the main input source as in state 3.

11.3.6 State 5

In this state, switches S1, S3, and S5 are on, while as switches S2, S4, and S6 are off. Thus, the circuit gets completed as S1, S3, and S5 are involved to complete the circuit. The battery and capacitor are disconnected in this particular state, and the output voltage is zero.

11.3.7 State 6

In this state, switches S2, S4, and S6 are on, while switches S1, S3 and S5 are off. The battery and capacitor are disconnected in this particular state, and the output voltage is zero like in state 5.

11.3.8 State 7

In this state, switches S1, S4, and S6 are on, while switches S2, S3, and S5 are off. Thus, the circuit gets completed as S1, S4, and S6, and the capacitor involved to complete the circuit while the battery remain disconnected in this particular state.

11.3.9 State 8

In this state, switches S1, S3, and S6 are on, while switches S2, S4, and S5 are off. Thus, the circuit gets completed as S2, S4, and S5. Battery voltage is involved to complete the circuit, and the capacitor is disconnected in this particular state.

11.4 Output Equation

From the discussion above, we can notice that there are redundancies at certain states. So, we can exploit this redundancy and serve our purpose to keep the output voltage of the capacitor maintained at the desired level. The output voltage and current can be represented by Equations 11.1 and 11.2, respectively.

$$y_l(t) = y_m \sin(\omega t) \tag{11.1}$$

$$x_l(t) = x_m \sin(\omega t - \theta_0) \tag{11.2}$$

where y_m and x_m are the maximum values of the output voltage and current waveforms. In addition, θ_0 represents the phase difference between the output voltage and current. The equations mentioned below can be formulated on the basis of the energy absorbed or delivered by a dc capacitor to the load:

$$x = \frac{dq}{dt}$$
$$dE = y\,dq = y\,x\,dt$$
$$E = \int y\,x\,dt \tag{11.3}$$

Here the variables x, y, q, and E represent the current, voltage, charge, and capacitor energy, respectively. Referring to Figure 11.12 and substituting Equation 11.2 into Equation 11.3, the energy absorbed or delivered from the capacitor in PUC5 inverter can be computed over the positive and negative halves of the output voltage cycle. It is worth noting that the voltage of capacitor remains constant at value V.

$$E_+ = \int_0^{\pi} yVx_m \sin(\omega t)d(\omega t)$$
$$= x_m \int_0^{\pi} yV \sin(\omega t - \theta_0)d(\omega t)$$
$$= x_m\left[\int_0^{\alpha_1} 0 \times \sin(\omega t - \theta_0)d(\omega t)\right.$$
$$+\int_{\phi_1}^{\phi_2} V \times \sin(\omega t - \theta_0)d(\omega t)$$
$$+\int_{\phi_2}^{\phi_3} 0 \times \sin(\omega t - \theta_0)d(\omega t)$$
$$\left.+\int_{\alpha_3}^{\pi} V \times \sin(\omega t - \theta_0)d(\omega t)\right]$$
$$= -V\,xm\cos(\omega t - \theta_0)\Big|_{\phi_1}^{\phi_2} - V\,xm\cos(\omega t - \theta_0)\Big|_{\phi_3}^{\phi_4}$$
$$= V\,xm[\cos(\phi_1 - \theta_0) - \cos(\phi_2 - \theta_0) + \cos(\phi_3 - \theta_0) - \cos(\phi_4 - \theta_0)]$$

Similar calculations are carried out to determine the capacitor energy during the negative half-cycle, resulting in the following equation for U-:

$$E^{-} = \int_{\pi}^{2\pi} y_V x_m \sin(\omega t - \theta_0) d(\omega t)$$

$$= x_m \int_{\pi}^{2\pi} y_V \sin(\omega t - \theta_0) d(\omega t)]. \quad (11.5)$$

$$= V\, xm \cos(\omega t - \theta_0) \Big|_{\phi_6}^{\phi_6} + V\, xm \cos(\omega t - \theta_0) \Big|_{\phi_7}^{\phi_8}$$

$$= V\, xm [\cos(\phi_6 - \theta_0) - \cos(\phi_5 - \theta_0) + \cos(\phi_8 - \theta_0) - \cos(\phi_8 - \theta_0)]$$

After recognizing the symmetry between the two half cycles of the output voltage, it is reasonable to assume that

$$\begin{bmatrix} \phi_5 = \pi + \phi_2 \\ \phi_6 = \pi + \phi_2 \\ \phi_7 = \pi + \phi_3 \\ \phi_8 = \pi + \phi_4 \end{bmatrix} \quad (11.6)$$

Then, the energy in half cycles would then be equal in magnitude but opposite in sign, i.e.:

$$U- = -U_{+} \quad (11.7)$$

Due to the periodic structure of the reference waveform, this ensures that the energy stored in the capacitor remains constant and in balance during an entire cycle, maintaining the capacitor's voltage consistency under all circumstances. The time it takes for the capacitor to charge and discharge is efficiently decreased by raising the switching frequency, which results in more frequent switching pulses. Voltage balance is thus improved by this improvement in charging and discharging symmetry. Sensorless voltage balancing is resilient even in the presence of grid voltage distortion or unbalanced voltage situations, such as when the presence of third or fifth harmonics alters the complete cycle. This is because sensorless voltage balancing relies on the symmetry between charge and discharge periods. In these situations, the positive and negative half cycles retain their identical and symmetrical characteristics, ultimately resulting in the capacitor's self-voltage balancing.

11.4.1 Switching Techniques Used in PUC5

Multilevel inverters possess a higher number of switching components, contributing to increased switching losses and complex modulation technique design. In high-power applications of multilevel inverters, opting for modulation techniques with lower switching frequencies is preferred to mitigate these losses. Various modulation techniques have been developed for multilevel inverters, but only a select few have gained industrial traction. Techniques like PWM, SHE, and SVM are suitable for generating pulses for multilevel inverter. Among these, PWM is primarily favored in industrial applications. The proposed packed U-cell multilevel inverter employs the techniques discussed in the following subsection.

11.4.1.1 Pulse Width Modulation

In this switching technique, the width of pulse is controlled by tunning switches between supply and load. The pulse width is varied so as to maintain the instantaneous magnitude same as that of input. The aim of PWM is to control the voltage at the output of the inverter and also reduce its harmonic content. In PWM technique, there are two signals: one signal is reference signal and another one is carrier signal. The switching pulses needed to change the mode of the inverter can be produced through a comparison of these two signals.

1. Level-shifted PWM: In this technique, different carrier waveforms are compared with fundamental reference waveforms to obtain stepped multilevel output voltage from the inverter.
2. Phase-shifted PWM: In this approach, the carrier signals are phase-shifted, while the modulating wave for each cell remains the same. As a result, fundamental voltages get added.
3. Space vector modulation: It is mainly used for three-phase configuration. In SVM technique, a vector-based approach is employed to generate switch control pulses. In this method, voltage vectors generated by different sets of switching states are computed, and the phase and magnitude of the reference voltage vector are instantaneously determined. A combination of nearest vectors is chosen based on the location of the reference voltage vector. Consequently, the corresponding switches responsible for generating these voltage vectors are fired within specified intervals.

11.5 Simulation Model

Sinusoidal reference waveform and four triangular carrier waves: These four carrier waveforms—abbreviated as Cr1, Cr2, Cr3, and Cr4—are vertically adjusted to entirely modify the reference waveform, according to [15] and [24]. The firing pulses associated with switching states 1, 2, 4, 5, 6, and 8 (as given in Table 11.1) are generated by comparing Vref to these carrier waves. The gate pulses for all the six semiconductor switches are shown in Figure 11.6. Reliable switching states 4 and 5 are also utilized to reduce switching frequency. When Vref is positive, state 4 is activated to provide a zero-level output voltage. When Vref is negative, on the other hand, state 5 is employed to provide the output voltage at zero. With a low switching frequency and no feedback sensors, the method demonstrated in Figure 11.7 demonstrates how to generate a five-level voltage

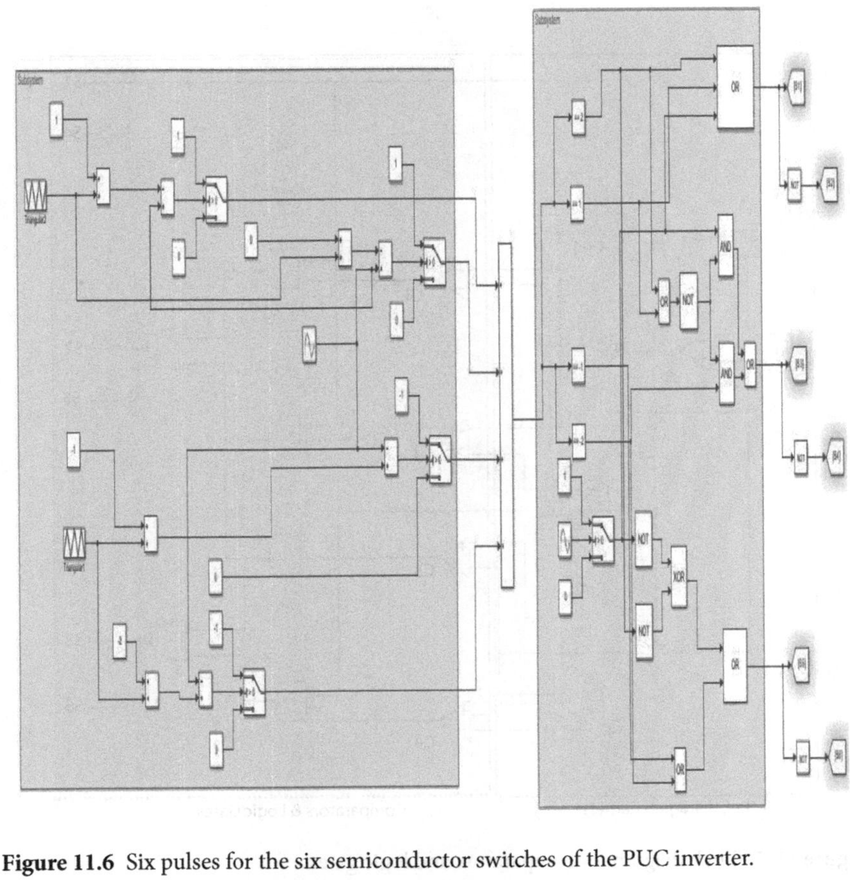

Figure 11.6 Six pulses for the six semiconductor switches of the PUC inverter.

waveform at the output while keeping an appropriate capacitor voltage. As already stated, by implementing this new approach, the PUC5 inverter has the ability to produce a five-level voltage waveform at the output without the use of voltage sensors or tedious computations in the controller. The capacitor voltage remains constant regardless of the load fluctuates and during starting. In comparison to the controller that was earlier employed in the PUC inverter, this simplified algorithm substantially boosts system speed, according to the most recent studies. Furthermore, this technique operates without the need for feedback sensors, modulation index, switching frequency, or grid frequency. Additionally, it does not depend on system modeling (like average modeling). It can run the system effectively throughout a wide voltage range, from zero to unlimited amplitudes, even when the DC supply voltage varies.

The capacitor voltage remains constant not only during start-up but also during the load change condition. The light algorithm significantly

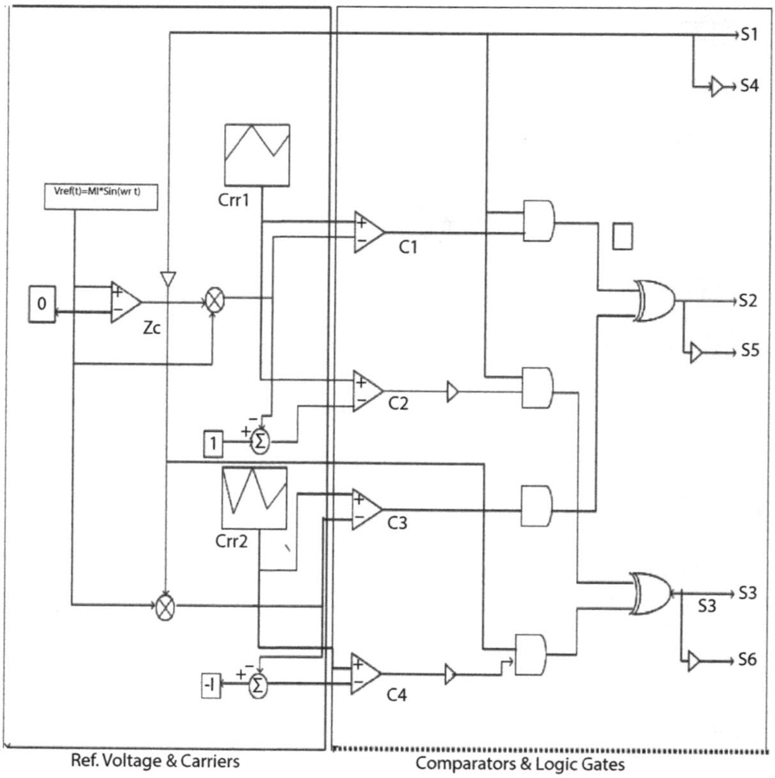

Figure 11.7 Block diagram of comparators and logic gates.

enhances system speed compared to the previously employed controller for the PUC inverter, as mentioned in existing literature. The technique is independent of system models, feedback sensors, modulation index, grid and switching frequency and can effectively operate the system from zero voltage to any amplitude, even in the situation when DC source voltage varies. The gate pulses, output voltage, and harmonics spectrum have been shown in Figures 11.8, 11.9, and 11.10, respectively.

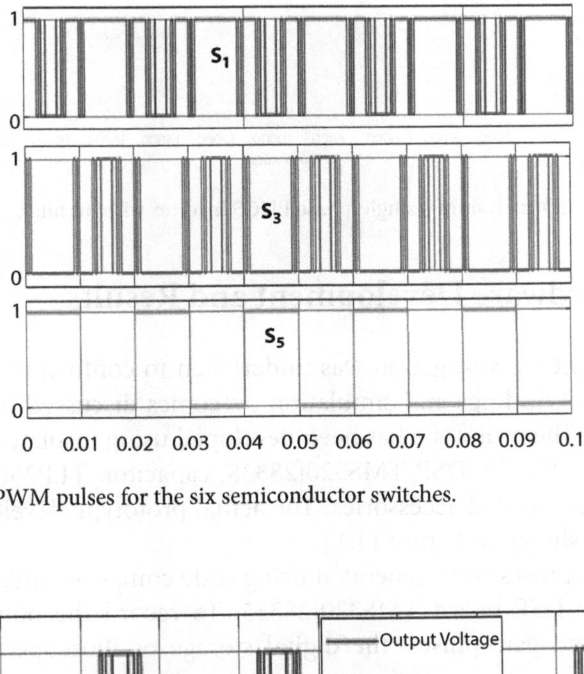

Figure 11.8 PWM pulses for the six semiconductor switches.

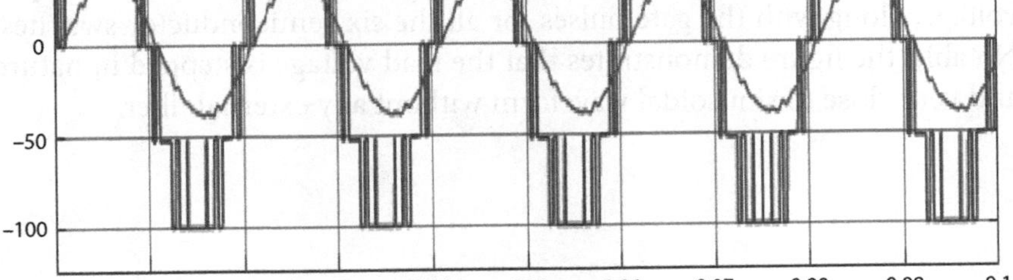

Figure 11.9 Output voltage of a single-phase PUC5 inverter without filter.

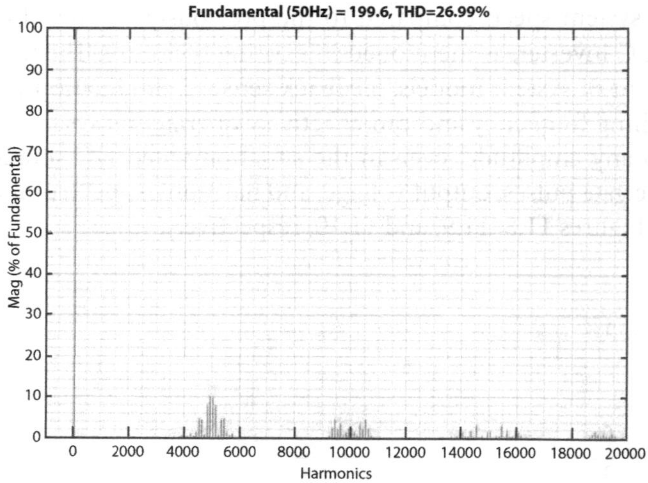

Figure 11.10 THD analysis of a single-phase PUC5 inverter without filter.

11.6 Hardware Development and Results

An experimental investigation was undertaken to confirm the validity of the theoretical findings and simulation outcomes discussed in the previous sections. The prototype has been developed using insulated gate bipolar transistor (IGBT), DSP TMS320f28335, capacitor, TLP250 gate drive boards, and associated accessories. The actual prototype developed in the laboratory is shown in Figure 11.11.

The PWM pulses were generated using code composer studio and executed on the DSP board TMS320f28335. To record the output voltage waveforms and gate pulses, the digital storage oscilloscope (DSO) was employed.

Figure 11.12 displays the experimental results, illustrating a five-level waveform and the sinusoidal fundamental of the PUC inverter's output voltage, along with the gate pulses for all the six semiconductor switches. Notably, the figure demonstrates that the load voltage is stepped in nature and very close to sinusoidal waveform without any external filter.

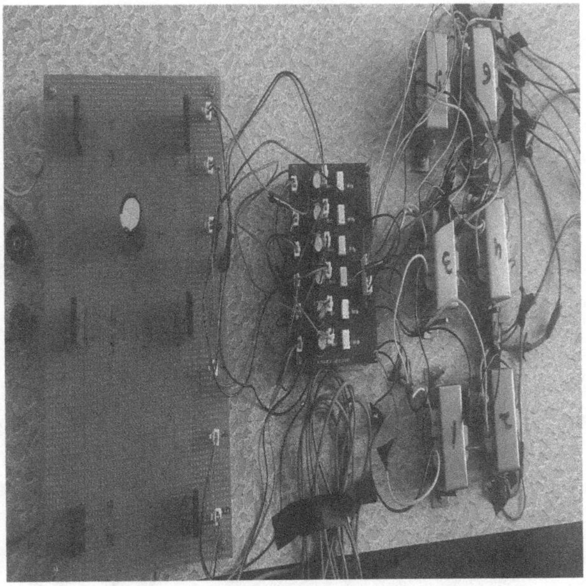

Figure 11.11 Hardware setup for PUC MLI with six semiconductor switches.

11.7 Conclusion

This chapter introduces reduced device count multilevel power converter topology known as the packed U-cells (PUC). Unlike conventional topologies, PUC utilizes a configuration comprising two power switches and one capacitor within each U-cell. This design delivers superior energy conversion efficiency while minimizing the need for capacitors and power devices, resulting in a cost-effective solution. The chapter presents a detailed averaged model of this topology and thoroughly analyzes the operational principles of the transformerless five-level inverter. To harness the advantages of the PUC-based structure, multilevel sinusoidal modulation is employed, and a control strategy is devised to significantly reduce load voltage harmonic content.

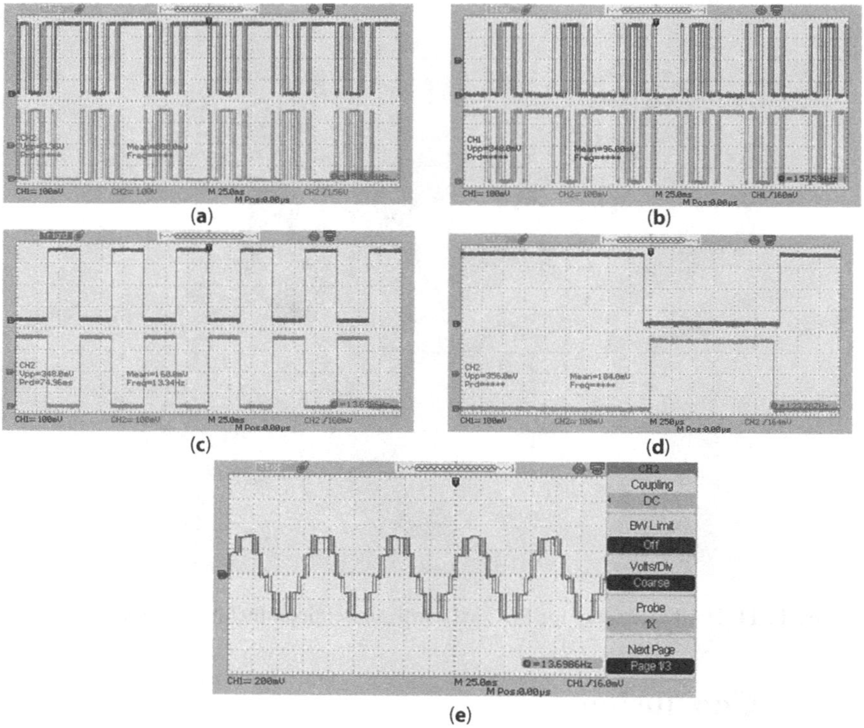

Figure 11.12 Hardware results: (a) PWM pulses for switches S1 and S2, (b) PWM pulses for switches S3 and S4, (c) PWM pulses for switches S5 and S6, (d) dead time between two complementary switch pulses, and (e) output voltage waveform reliance on filters.

The paper conducts a comparative study to underscore the benefits of the packed U-cells' topology and validates its operation through simulations and experimental tests utilizing the DSP board.

References

1. D. G. Holmes and T. A. Lipo, *Pulse width modulation for power converters: principles and practice*. John Wiley Sons, 2003, vol. 18.
2. J. Rodriguez, L. G. Franquelo, S. Kouro, J. I. Leon, R. C. Portillo, M. A. M. Prats, and M. A. Perez, "Multilevel converters: An enabling technology for high-power applications," *Proc. IEEE*, vol. 97, no. 11, pp. 1786–1817, 2009.
3. A. M. Trzynadlowski, "An overview of modern pwm techniques for three-phase, voltage-controlled, voltage-source inverters," in *Proceedings of IEEE International Symposium on Industrial Electronics*, vol. 1. IEEE, 1996, pp. 25–39.
4. Hussan, M.R., Sarwar, A., Tayyab, M., Tariq, M., Bakhsh, F.I., Ahmad, A.: A single-source switched-capacitor based 13-level sextuple boost inverter with

reduced component count and low THD. *IET Gener. Transm. Distrib.* 17, 2587–2599 (2023).

5. Islam, S., Bakhsh, F.I., Islam, Q.U. (2019). Modeling and Analysis of the Photovoltaic Array Feeding a SPWM Inverter. In: Malik, H., Srivastava, S., Sood, Y., Ahmad, A. (eds) Applications of Artifcial Intelligence Techniques in Engineering. *Adv. Intell. Syst. Comput.*, vol 698. Springer, Singapore.
6. M. Kavitha , A. Arunkumar , N. Gokulnath , S. Arun 4 .New Cascaded H-Bridge Multilevel Inverter Topology with Reduced Number of Switches and Sources. Final year students / Dept. of EEE / DR.S.J.S Paul Memorial College of Engineering Technology / Pondicherry / India. Volume 2, Issue 6 (Sep-Oct. 2012), Pp 26-36.
7. M. Zaid, M. Tayyab, A. R. Moonis, F. I. Bakhsh, A. Sarwar and A. Iqbal, "A Switched Capacitor Triple Boost Single-Source Thirteen-level Inverter," *2023 IEEE IAS Global Conference on Renewable Energy and Hydrogen Technologies (GlobConHT)*, Male, Maldives, 2023.
8. M. Al-Hitmi, S. Ahmad, A. Iqbal, S. Padmanaban, and I. Ashraf, "Selective harmonic elimination in a wide modulation range using modified newton-raphson and pattern generation methods for a multilevel inverter," *Energies*, vol. 11, no. 2, p. 458, 2018.
9. S. Ahmad, M. Meraj, A. Iqbal, and I. Ashraf, "Selective harmonics elimination in multilevel inverter by a derivative-free iterative method under varying voltage condition," *ISA Trans.*, vol. 92, pp. 241–256, 2019.
10. S. Ahmad, I. Khan, A. Iqbal, and S. Rahman, "A novel pulse width amplitude modulation for elimination of multiple harmonics in asymmetrical multilevel inverter," in *2021 IEEE Texas Power and Energy Conference (TPEC)*. IEEE, 2021, pp. 1–6.
11. S. Ahmad, I. Ashraf, A. Iqbal, and M. Fatimi, "She pwm for multilevel inverter using modified nr and pattern generation for wide range of solutions," in *2018 IEEE 12th International Conference on Compatibility, Power Electronics and Power Engineering (CPE-POWERENG 2018)*. IEEE, 2018, pp. 1–6.
12. S. Ahmad, M. Al-Hitmi, A. Iqal, I. Ashraf, M. Meraj, and S. Padmanaban, "Low-order harmonics control in staircase waveform useful in high-power application by a novel technique," *Int. J. Electr. Power Energy Syst.*, vol. 29, no. 3, p. e2769, 2019.
13. S. Ahmad, A. Iqbal, R. Al-Ammari, and H. Abu-Rub, "Selected harmonics elimination in multilevel inverter using improved numerical technique," in *2018 IEEE 12th International Conference on Compatibility, Power Electronics and Power Engineering (CPE-POWERENG 2018)*. IEEE, 2018, pp.1–6.
14. S. Ahmad, M. Al-Hitmi, A. Iqbal, K. Rahman, and I. Ashraf, "Low switching frequency modulation of a 3× 3 matrix converter in upfc application using differential evolution method," *Int. Trans. Electr. Energy Syst.*, vol. 30, no. 1, p. e12179, 2020.
15. S. Ahmad, A. Iqbal, M. Ali, K. Rahman, and A. S. Ahmed, "A fast convergent homotopy perturbation method for solving selective harmonics elimination

pwm problem in multi level inverter," *IEEE Access*, vol. 9, pp. 113 040–113 051, 2021.
16. S. Ahmad and A. Iqbal, "Switching angles computations using pso in selective harmonics minimization pwm," in *Metaheuristic and Evolutionary Computation: Algorithms and Applications*. Springer, 2021, pp. 437–461.
17. Z. A. Ganie, A. H. Bhat, and S. Ahmad, "Harmonics minimization in wide modulation range using improved differential evolution optimization technique and hardware validation," in *2022 2nd Asian Conference on Innovation in Technology (ASIANCON)*. IEEE, 2022, pp. 1–6.
18. S. Ahmad, A. Iqbal, I. Ashraf, and Z. A. Ganie, "Harmonics minimization in multilevel inverter by continuous mode aco technique," in *Innovations in Electrical and Electronic Engineering*. Springer, 2021, pp. 95–104.
19. S. Ahmad, A. Iqbal, I. Ashraf, and M. Meraj, "Improved power quality operation of symmetrical and asymmetrical multilevel inverter using invasive weed optimization technique," *Energy Rep.*, vol. 8, pp. 3323–3336, 2022.
20. S. Ahmad, Z. A. Ganie, I. Ashraf, and A. Iqbal, "Harmonics minimization in 3-level inverter waveform and its fpga realization," in *2018 3rd International Innovative Applications of Computational Intelligence on Power, Energy and Controls with their Impact on Humanity (CIPECH)*. IEEE, 2018, pp. 1–5.
21. Y. Ounejjar, K. Al-Haddad, and L. A. Grégoire, "Packed U cells multilevel converter topology: theoretical study and experimental validation," *IEEE Trans. Ind. Electron.*, vol. 58, no. 4, pp. 1294- 1306, 2011.
22. H. Vahedi and K. Al-Haddad, "Real-Time Implementation of a Packed U-Cell Seven-Level Inverter with Low Switching Frequency Voltage Regulator," *IEEE Trans. Power Electron.*, 2015.
23. Y. Ounejjar, K. Al-Haddad, and L. A. Dessaint, "A Novel Six-Band Hysteresis Control for the Packed U Cells Seven-Level Converter: Experimental Validation," *IEEE Trans. Ind. Electron.*, vol. 59, no.10, pp. 3808-3816, 2012.
24. H. Vahedi, K. Al-Haddad, and H. Y. Kanaan, "A New Voltage Balancing Controller Applied on 7-Level PUC Inverter," in *IECON 2014-40th Annual Conference on IEEE Industrial Electronics Society, USA*, 2014, pp. 5082-5087.
25. M. Abarzadeh, S. Peyghami, K. Al-Haddad, N. Weise, L. Chang, and F. Blaabjerg, "Reliability and performance improvement of PUC converter using a new single-carrier sensor-less PWM method with pseudo reference functions," *IEEE Trans. Power Electron.*, vol. 36, no. 5, pp. 6092–6105, May 2021.
26. H. Makhamreh, M. Trabelsi, O. Kükrer, and H. Abu-Rub, "A Lyapunovbased model predictive control design with reduced sensors for a PUC7 rectifer," *IEEE Trans. Ind. Electron.*, vol. 68, no. 2, pp. 1139–1147, Feb. 2021.
27. R. Vasu, S. K. Chattopadhyay, and C. Chakraborty, "Seven-level packed U-cell (PUC) converter with natural balancing of capacitor voltages," *IEEE Trans. Ind. Appl.*, vol. 56, no. 5, pp. 5234–5244, Sep./Oct. 2020.
28. K.R. Sorto-Ventura, M. Abarzadeh, K. Al-Haddad, and L. A. Dessaint, "23-level single DC source hybrid PUC (H-PUC) converter topology with

reduced number of components: Real-time implementation with model predictive control," *IEEE Open J. Ind. Electron. Soc.*, vol. 1, pp. 127–137, 2020.

29. A. Sahli, F. Krim, A. Laib, and B. Talbi, "Energy management and power quality enhancement in grid-tied single-phase PV system using modified PUC converter," *Inst. Eng. Technol. Renew. Power Gener.*, vol. 13, no. 14, pp. 2512–2521, Oct. 2019.
30. I. Abari, M. Hamouda, M. Sleiman, J. B. H. Slama, H. Y. Kanaan, and K. Al-Haddad, "Open-circuit fault detection and isolation method for five-level PUC inverter based on the wavelet packet transform of the radiated magnetic field," *IEEE Trans. Instrum. Meas.*, vol. 71, 2022, Art. no. 3502611.
31. S. Arazm and K. Al-Haddad, "A novel single DC source three-phase wye packed U-cell (Y-PUC) converter," *IEEE Open J. Ind. Electron. Soc.*, vol. 3, pp. 1–13, 2022.
32. D. A. Schuetz, F. B. Grigoletto, F. Carnielutti, and H. Pinheiro, "Discontinuous space vector modulation for three-phase five-levels packed-U-cell converter," *IEEE Trans. Power Electron.*, vol. 36, no. 12, pp. 14353–14365, Dec. 2021

12

Unified Power Quality Conditioner (UPQC) Based on Multilevel Configurations

Javeed Bashir*, Salman Ahmad and Ahmed Sharique Anees

Department of Electrical Engineering, Islamic University of Science and Technology, Awantipora, Pulwama, Jammu and Kashmir, India

Abstract
This chapter explores the development, regulation, and deployment of a photovoltaic-interfaced Unified Power Quality Conditioner (PV-UPQC). The PV-UPQC serves to uphold power quality amidst various current and voltage irregularities, while integrating photovoltaic generation with existing infrastructure. Acting as a versatile power conditioning apparatus, the PV-UPQC combines shunt and series voltage source converters through a common DC-link. Moreover, the solar system connects to the UPQC DC-link to supply power to the load. During typical grid conditions, the series voltage source converter (VSC) shares a fraction of the load's reactive power, enhancing the utilization of the series VSC and reducing the rating of the shunt VSC. Additionally, the PV array integrated into the UPQC's DC-bus contributes active load power, consequently alleviating the demand on the supply system

Keywords: Unified power quality conditioner, multilevel converter, solar photovoltaic, power quality

12.1 Introduction

In the present-day world, which is civilized and developed, the essentiality of electricity is well known to everyone. The ever-increasing demand for the use of electricity with social, economic, and industrial development

**Corresponding author*: javeedbashir001@gmail.com

Salman Ahmad, Farhad Ilahi Bakhsh, and P. Sanjeevikumar (eds.) Multilevel Converters, (233–252)

stipulates the augmentation of existing generating stations and creation of new transmission systems [1–3]. However, environmental and social issues constrain the room for such expansion. Besides, apprehensions about global warming, pollution due to the use of conventional fuel, and the depletion of fossil fuels are increasing. So, with the difficulty of constructing new transmission lines, power stations, and their cost concerns, the whole attention is focused on enhancing the power transmission capacity of the existing system without altering the physical structure of the present power system [4, 5]. The solution for such a problem is successfully provided by the FACTS technology by not only enhancing the power transfer capacity of the existing system but also improving the voltage profile, stability, and line impedance control, etc. FACTS make the system operation flexible and the system utilization better. Different FACTS controllers are used to achieve the above-mentioned objectives. But most of the controllers have one or more disadvantages and particularly are not able to control the real (P) and reactive power (Q) independently. Such kind of diversity is provided by the FACTS controller known as Unified Power Quality Conditioner (UPQC) [6–8]. It is one of the most innovative devices among the FACTS controllers which have a great potential of voltage stability and control of P flow in a transmission system. It is capable of regulating the multi-electrical parameters and hence improves the capacity of power transfer and the stability of the power system. It also enables the transmission system to operate near the maximum rating (thermal limit) by changing the power flow through the well-defined transmission lines without affecting the stability limits along with controlling the P and Q independently [9–11]. UPQC comprises of two voltage source converters (VSCs), linked through a dc-link. One of them is connected in shunt and the other is connected in series with the transmission line through converter transformers. Regulating the voltage is performed by the former VSC and hence controls the P exchange between the two VSCs. The P and Q flow through the transmission system is controlled by series VSC which injects controllable ac quadrature voltage, with respect to the current in the line, to the primary side of the transformer. Traditional UPQCs are two- or three-level VSCs. Zig-zag transformers are used to increase the voltage levels so as to decrease the harmonic content, but it results in increased cost, lifetime reduction, and additional power losses. However, due to their high-power density, dependability, and notably flexibility, modular multilevel converters (MMC) have made a substantial contribution to medium and high voltage systems during the last decade. As a result, the use of UPQC based on MMC has drawn a lot of attention [12, 13].

12.2 Basic Principle of Operation

As indicated for the two-machine model in Figure 12.1, UPQC is a generalized synchronous voltage source (SVS), best represented at the fundamental frequency by a V_{pq} (voltage phasor) in series with the line and with adjustable magnitude V_{pq} ($0 < V_{pq} < V_{pqmax}$) and phase angle ρ ($0 < \rho < 2\pi$). An SVS normally communicates both P and Q with the transmission network in this operationally unrestricted operation, which obviously regulates the voltage and the phase angle. As stated previously, an SVS can produce only the Q; the P has to be provided to it, or consumed from it, by an appropriate power source or load. As shown in Figure 12.1, the P interchange in the UPQC structure is presented by one of the terminal buses (for example, sending-end bus).

Figure 12.2 shows a UPFC that comprises two converters (VSCs). These head-to-head converters, "Controller 1" and "Controller 2" in the figure, are powered by a dc storage capacitor and share the same dc link. As previously stated, this topology acts as a complete ac-to-ac controller in which P may flow freely in either way between the two controllers' ac terminals and each controller can independently create (or consume) Q at its own ac output of the inverter.

The major task of the UPQC is done by Converter 2 by inserting a voltage V_{pq} with adjustable strength at a phase angle ρ with respect to V_s connected in series with the line through a series transformer. This inserted voltage V_{pq} acts essentially as an ac synchronous voltage source. This voltage source interchange P and Q with the ac system when the line current runs through it. The Q exchanged at the ac terminal is created internally by

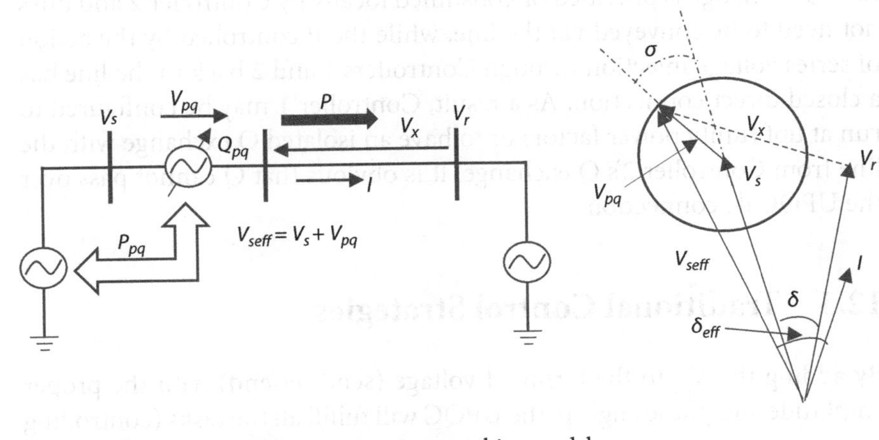

Figure 12.1 UPQC representation as a two-machine model.

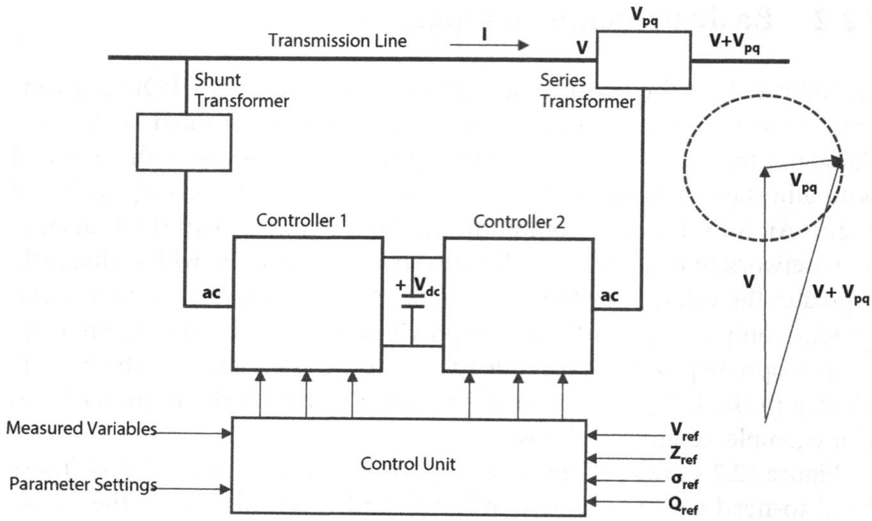

Figure 12.2 Implementation of the UPFC by two back-to-back VSCs.

the converter. The exchange of the P at the ac terminals is transformed to dc power and it seems to be the positive or negative power requirement at the dc link. Controller 1's primary role is to deliver or consume the P required by Controller 2 at the common dc link in order to enable the P exchange caused by the injection of voltage in series. Controller 1 transforms back the dc link power requirement of Controller 2 to ac and connects it to the transmission line bus through a shunt-connected transformer. Controller 2 fulfills the requirement of real power demand in addition to generation or consumption of adjustable reactive power, for Converter 2. The necessary Q exchange is provided or consumed locally by Controller 2 and does not need to be conveyed *via* the line, while the P controlled by the action of series voltage injection through Controllers 1 and 2 back to the line has a closed direct connection. As a result, Controller 1 may be configured to run at upf (unity power factor) or to have an isolated Q exchange with the line from Controller 2's Q exchange. It is obvious that Q cannot pass over the UPQC dc connection.

12.3 Traditional Control Strategies

By adding the V_{pq} to the terminal voltage (sending-end) with the proper amplitude and phase angle ρ, the UPQC will fulfill all the tasks (controlling magnitude of V, δ and X) and achieve many control goals. Figure 12.3 shows an illustration of the fundamental UPQC control operations by phasor representation. Figure 12.3(a) illustrates voltage control with variable in-phase/out-of-phase voltage insertion for voltage adjustments $V_{pq} = \pm\Delta V$ ($\rho = 0$). Technically speaking, this is comparable to what is possible with a transformer tap changer that has infinite tiny steps. Figure 12.3(b) illustrates series reactance compensation wherein Vpq =Vc is inserted perpendicular to the current I. Hence, this is equivalent to series reactive (inductive and capacitive) compensation. Figure 12.3(c) illustrates phase angle regulation by injecting Vpq = Vσ at an angle with respect to sending end voltage, V_s, that results in the appropriate phase change (forward or backward) without any variation in magnitude. Because of its internal VAr production, the UPFC can therefore provide the reactive power required for transmission angle control and behave as a complete phase angle regulator. Figure12.3(d), where Vpq= ΔV + Vc + Vσ, illustrates multi-functional power flow management that is carried out by simultaneous terminal voltage control, series reactance, and phase angle control. The UPQC alone has such functional capacity. No other conventional piece of technology has the same level of multifunctionality. The P and Q vs. δ curves of the straightforward two-machine model, depicted in Figure 12.1, provide the clear illustration of the overall power flow control capabilities of the UPQC when viewed from the perspective of traditional transmission control. The P (transmitted power) and the $-jQ_r$, provided by the receiving end, may be explained as in Equation 12.1.

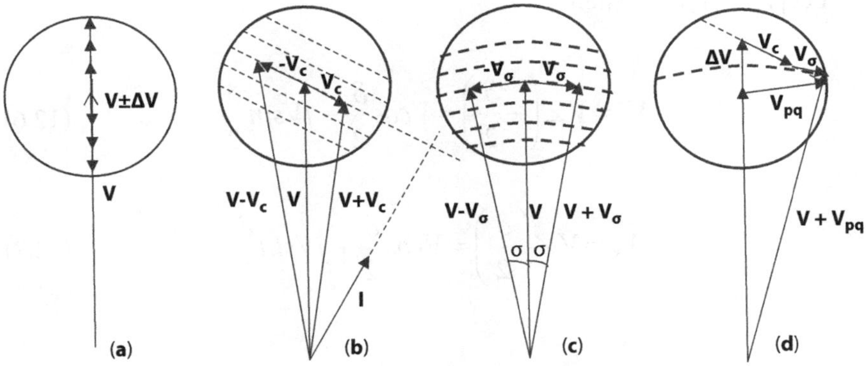

Figure 12.3 Phasor diagrams illustrating the conventional transmission control capabilities of the UPFC.

$$S = P + jQ = \overline{V_r}I^* = \overline{V_r} \times \left(\frac{\overline{V_s} + \overline{V_{pq}} - \overline{V_r}}{jX} \right)^* \tag{12.1}$$

If $V_{pq} = 0$, then (12.1) will describe the uncompensated system as derived in Equations 12.2–12.21,

$$P + jQ = \overline{V_r} \times \left(\frac{\overline{V_s} - \overline{V_r}}{jX} \right)^* \tag{12.2}$$

When V_{pq} is non-zero, the net P and Q can be expressed as:

$$P + jQ = \overline{V_r} \times \left(\frac{\overline{V_s} - \overline{V_r}}{jX} \right)^* + \overline{V_r} \times \left(\frac{\overline{V_{pq}}}{jX} \right)^* \tag{12.3}$$

Substituting,

$$\overline{V_r} = V_r \angle \left(-\frac{\delta}{2} \right) = V_r cos\frac{\delta}{2} - jV_r sin\frac{\delta}{2} \tag{12.4}$$

$$\overline{V_s} = V_s \angle \left(\frac{\delta}{2} \right) = V_s cos\frac{\delta}{2} + jV_s sin\frac{\delta}{2} \tag{12.5}$$

Let $|V_s| = |V_r| = V$, then

$$\overline{V_r} = V \angle \left(-\frac{\delta}{2} \right) = Vcos\frac{\delta}{2} - jVsin\frac{\delta}{2} \tag{12.6}$$

$$\overline{V_s} = V \angle \left(\frac{\delta}{2} \right) = Vcos\frac{\delta}{2} + jVsin\frac{\delta}{2} \tag{12.7}$$

and

$$\overline{V_{pq}} = V_{pq}\angle\left(\rho + \frac{\delta}{2}\right) = V_{pq}cos\left(\rho + \frac{\delta}{2}\right) + jV_{pq}sin\left(\rho + \frac{\delta}{2}\right) \quad (12.8)$$

$$I = \frac{\overline{V_s} + \overline{V_{pq}} - \overline{V_r}}{jX} \quad (12.9)$$

$$S = P + jQ = \overline{V_r}I^* = \overline{V_r} \times \left(\frac{\overline{V_s} + \overline{V_{pq}} - \overline{V_r}}{jX}\right)^* \quad (12.10)$$

$$P + jQ = \overline{V_r} \times \left(\frac{\overline{V_s} - \overline{V_r}}{jX}\right)^* + \overline{V_r} \times \left(\frac{\overline{V_{pq}}}{jX}\right)^* \quad (12.11)$$

Now,

$$\left(\frac{\overline{V_s} - \overline{V_r}}{jX}\right)^* = \left(\frac{2jVsin\frac{\delta}{2}}{jX}\right)^* = \left(\frac{2Vsin\frac{\delta}{2}}{X}\right) \quad (12.12)$$

$$\therefore \bar{V_r} \times \left(\frac{\bar{V_s} - \bar{V_r}}{jX}\right)^* = \frac{2V^2}{X}sin\frac{\delta}{2}cos\frac{\delta}{2} - j\frac{2V^2}{X}sin^2\frac{\delta}{2} \quad (12.13)$$

$$P_0(\delta) + jQ_0(\delta) = \overline{V_r}I^* = \frac{V^2}{X}sin\delta - j\frac{V^2}{X}(1 - cos\delta) \quad (12.14)$$

$$\left(\frac{\overline{V_{pq}}}{jX}\right)^* = \frac{V_{pq}cos\left(\rho + \frac{\delta}{2}\right) - jV_{pq}sin\left(\rho + \frac{\delta}{2}\right)}{-jX} = \frac{jV_{pq}cos\left(\rho + \frac{\delta}{2}\right) + V_{pq}sin\left(\rho + \frac{\delta}{2}\right)}{X} \quad (12.15)$$

$$\overline{V_r} \times \left(\frac{\overline{V_{pq}}}{jX}\right)^* = \frac{\begin{matrix} VV_{pq} \times (\cos\frac{\delta}{2} \times \sin\left(\rho + \frac{\delta}{2}\right) + \sin\frac{\delta}{2} \times \cos\left(\rho + \frac{\delta}{2}\right)) \\ +jVV_{pq} \times (\cos\frac{\delta}{2} \times \cos\left(\rho + \frac{\delta}{2}\right) - \sin\frac{\delta}{2} \times \sin\left(\rho + \frac{\delta}{2}\right)) \end{matrix}}{X} \tag{12.16}$$

$$\overline{V_r} \times \left(\frac{\overline{V_{pq}}}{jX}\right)^* = \frac{VV_{pq} \times \sin(\rho+\delta) + jVV_{pq} \times \cos(\rho+\delta)}{X} = P_{pq} + jQ_{pq} \tag{12.17}$$

$$S = P + jQ = \overline{V_r} I^* = \overline{V_r} \times \left(\frac{\overline{V_s} + \overline{V_{pq}} - \overline{V_r}}{jX}\right)^* = P_0(\delta) + jQ_0(\delta) + P_{pq} + jQ_{pq} \tag{12.18}$$

$$= \frac{V^2}{X}\sin\delta - j\frac{V^2}{X}(1-\cos\delta) + \frac{VV_{pq} \times \sin(\rho+\delta) + jVV_{pq} \times \cos(\rho+\delta)}{X} \tag{12.19}$$

Therefore,

$$P = P_0(\delta) + P_{pq} = \frac{V^2}{X}\sin\delta + \frac{VV_{pq} \times \sin(\rho+\delta)}{X} \tag{12.20}$$

$$Q = Q_0(\delta) + Q_{pq} = -\frac{V^2}{X}(1-\cos\delta) + \frac{VV_{pq} \times \cos(\rho+\delta)}{X} \tag{12.21}$$

are the P and Q respectively that define the uncompensated system's power transfer as a function of phase angle δ. At any specified phase angle δ ($0 < \delta < \pi$), angle ρ is freely modifiable between 0 to 2π, hence it follows that Ppq(ρ) and Qpq(ρ) are adjustable between -VVpq/X and +VVpq/X regardless of angle. Consequently, it is possible to regulate the transmitted real power P between as given in (12.22)

$$P_0(\delta)-\frac{VV_{pq}}{X}\le P(\delta,\rho)\le P_0(\delta)+\frac{VV_{pq}}{X} \tag{12.22}$$

And the Qr as given in (12.23)

$$Q_0(\delta)-\frac{VV_{pq}}{X}\le Q(\delta,\rho)\le Q_0(\delta)+\frac{VV_{pq}}{X} \tag{12.23}$$

Irrespective of the phase angle δ, as seen in Figure 12.4. The Figure 12.5 shows a broad range of control for the P which is not dependent of the phase angle δ, which points to the UPFC's better performance in power flow applications as well as its potent potential for improving transient stability and dampening power oscillations. Consider Figure 12.1 again to further clarify the meaning of (12.20) and (12.21). Figure 12.6(a) shows a phasor diagram that gives the relation between the voltage phasor across X, Vs, Vr, and Vx, as well as the inserted voltage Vpq, with adjustable value (0 ≤ V_{pq} ≤ Vpqmax) and phase angle angle (0 ≤ ρ_{pq} ≤ 360°). (For example, V_{pq} = 0.25 p.u, δ=30°, X = 0.5 p.u. and V_s = V_r = 1 p.u. values were taken.). The resultant sending-end voltage $V_{seff} = V_s + V_{pq}$ is formed as shown by adding variable voltage V_{pq} to the constant voltage V_s. V_x (compensated voltage) across X, is given by the V_{seff} - V_r. The tip of phasor V_{pq} rotates in a circular

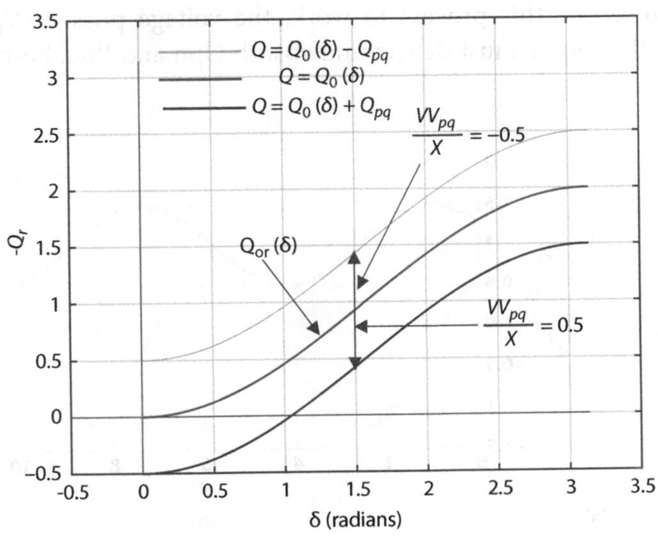

Figure 12.4 Range Q (at receiving end) vs. δ of a UPQC controller.

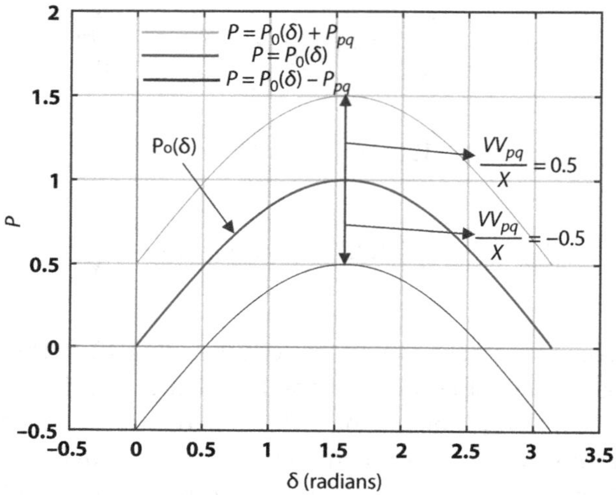

Figure 12.5 Range P vs. δ of a UPQC controller.

path whose centre is at the top of phasor V_s as angle ρ_{pq} is changed across its entire 360° span.

The region included within this circle, as determined by V_{pq}, specifies the phasor V_{pq}'s operational range and, thus, the line's practicable compensation. As shown in Figure 12.6(b), the transmitted P, and Qr fluctuate with ρ_{pq} in a sinusoidal fashion because the circular motion of V_{pq} with the change in phase angle ρ_{pq} modifies both the amplitude and phase angle of V_x.

Of course, for this process to work, the voltage phasor (Vpq) produced by the source must deliver and absorb Qpq and Ppq, both varying

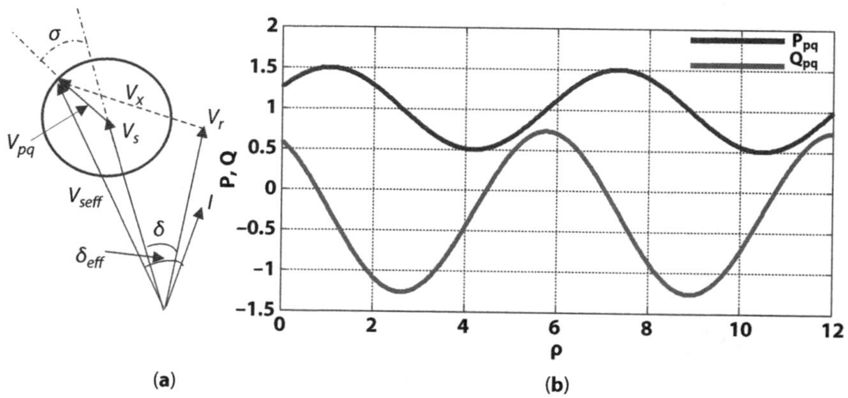

Figure 12.6 Variation of P and Q (supplied by UPQC) vs. ρ.

sinusoidally as the functions of angle ρ_{pq} as seen in the above-mentioned figure.

A global power flow controller that is capable of maintaining specified, independently adjustable P and Q in the line can be built using the robust, previously unachievable abilities of the UPQC that were described above in terms of classic transmission control ideas. By controlling the angle and the magnitude V_{pq}, the UPQC can regulate or change the P and Q flow in the line to fulfill load requirement and system operational conditions. Within this idea, the terms like series line compensation, phase angle regulation, etc. will be meaningless.

12.4 UPQC's P and Q Independent Control

Refer to Figure 12.6(a) to learn more about the UPQC's ability to regulate P and Q flow in the transmission line. Let us begin by assuming that V_{pq} = 0. Then, the basic two-machine model power system is recreated with V_s (sending-end voltage), V_r (receiving-end voltage), δ (transmission angle), and X (line reactance). After this the transmitted P and Q respectively will be normalized as, $P_0(\delta) = \{V^2/x\} \sin \delta = \sin \delta$, and $Q_0(\delta) = Q_{0s}(\delta) = -Q_{0r}(\delta) = \{V^2/x\}(1- \cos \delta) = 1-\cos \delta$, provided at the line's termini, are illustrated in Figure 12.7(a). With $V^2/X = 1$, the following connection between Ps (δ) and Qs (δ) at the source end may be easily expressed as in (12.24)-(12.31).

If $V_{pq} = 0$,

$$\text{Then, } P_0(\delta) = \frac{V^2}{x} sin\delta \text{ and } Q_{0r}(\delta) = \frac{V^2}{x}(1 - cos\delta) \tag{12.24}$$

$$\text{Since we know that } Q_{0s} = -Q_{0r} = -\frac{V^2}{x}(1 - cos\delta) \tag{12.25}$$

After normalizing: i.e., let $\frac{V^2}{x} = 1$,
then,

$$P_0(\delta) = sin\delta \text{ and} \tag{12.26}$$

$$Q_{0r}(\delta) = -(1 - cos\delta) = -1 + cos\delta = -1 + \sqrt{1 - sin^2\delta} = -1 + \sqrt{1 - (P_0(\delta))^2} \tag{12.27}$$

Again,
Let us replace $Q_{0r}(\delta)$ by $Q_0(\delta)$, then

$$Q_0(\delta) = -1 + \sqrt{1-(P_0(\delta))^2} \tag{12.28}$$

$$Q_0(\delta) + 1 = \sqrt{1-(P_0(\delta))^2} \tag{12.29}$$

$$(Q_0(\delta)+1)^2 = 1-(P_0(\delta))^2 \tag{12.30}$$

$$(Q_0(\delta)+1)^2 + (P_0(\delta))^2 = 1 \tag{12.31}$$

which gives us the equation of a circle with variables $P_0(\delta)$ on x-axis and $Q_0(\delta)$ on y-axis and with radius 1 and center (-1, 0).

For +ve values of P in Figure 12.7(a), (12.31) is a circle with a radius of 1.0 and the center P = 0 and Q_r=-1 in a $\{Q_r, P_0\}$ plane. With respect to a specified transmission angle δ, every point on this circular path provides the equivalent P_0 and Q_0 values for a system without any compensated. As an example, P_0=0 and Q_{0r}=0 for δ=0; P_0=0.5 and Q_{0r}=-0.134 for δ=30°; P_0 = 1.0 and Q_{0r} = -1.0 for δ=90° etc.

Observe Figure 12.6(a) once again and now suppose V_{pq} is non-zero. The P and Q change from their $P_0(\delta)$ and $Q_{0r}(\delta)$ (without compensation), as functions of the amplitude V_{pq} and its phase angle ρ, is evident from

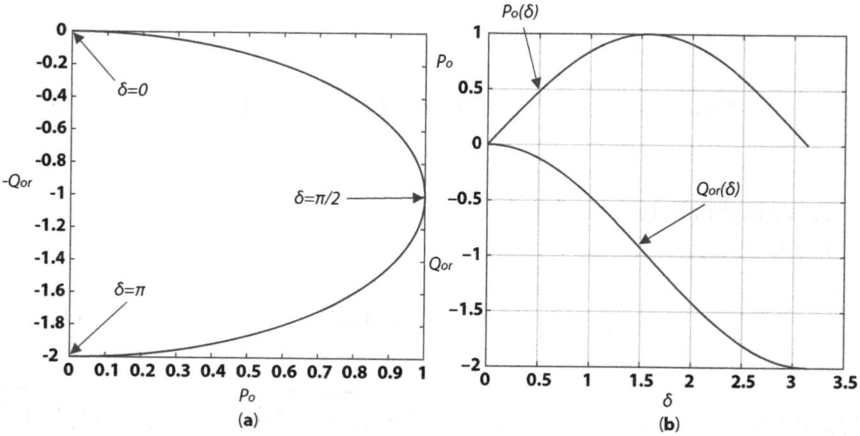

Figure 12.7 (a) Loci of Q_{0r} vs P_0. (b) $P_0(\delta)$ and $Q_{0r}(\delta)$ vs δ.

Equation 12.3, or Equations 12.20 and 12.21, and shown in the Figure 12.6(b). The limit of the control zone for P (δ, ρ) and Q_r (δ, ρ) is determined from a full revolution of Vpq with its maximum amplitude of V_{pqmax}, since angle is an unconstrained parameter (0 < ρ < 2π). The aforementioned equations prove that this control zone is a circle with a radius of $V_r V_{pq}/X$ and a center determined by the coordinates $P_0(\delta)$ and $Q_{0r}(\delta)$, the boundary of the circle is defined by the following equation with $V_s = V_r = V$, we can write Equation 12.32.

$$\{P(\delta,\rho)-P_0(\delta,\rho)\}^2+\{Q_r(\delta,\rho)-Q_{r0}(\delta,\rho)\}^2=\{\frac{VV_{pqmax}}{X}\}^2 \tag{12.32}$$

Figure 12.8 depicts the circular control zones described by Equation 12.32 with centers on the circular curve defined by (12.31) for phase angles δ = 0°, 30°, 60°, and 90° with V = 1.0, V_{pqmax} = 0.5, and X=1.0 (all in per unit or p.u. values). In other words, the $P_0(\delta)$ and $Q_{0r}(\delta)$ coordinates at angles = 0°, 30°, 60°, and 90° in the {Qr, P} plane serve as the definition of the centers of the control areas.

The system will at a halt at the origin of the Qr, P plane as shown in Figure 12.8, which shows the scenario when the δ is zero and V_{pq} is zero, P,

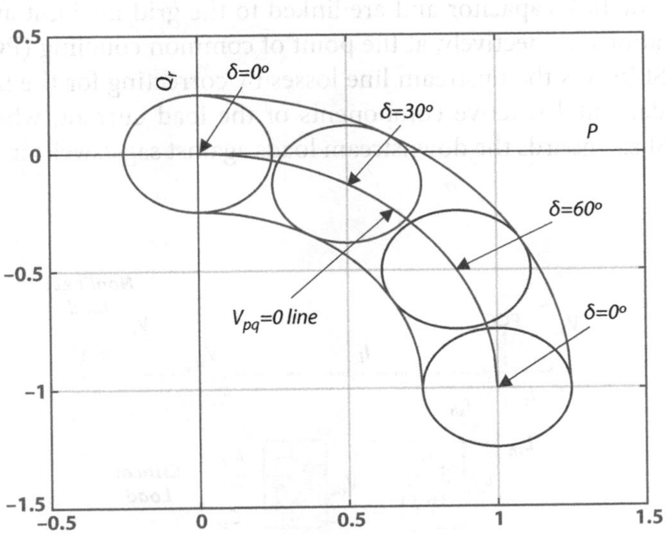

Figure 12.8 Control area P and Q (at the receiving end), with a UPQC-controlled transmission line at δ = 0°, 30°, 60°, and 90°.

Qr (and Qs) are will be zero. As the Vpq rotates completely for the range ($0 < \rho < 2\pi$) with its highest amplitude V_{pqmax}, the loci of the associated Qr, and P are derived as a circle centered on the origin of the {Qr, P} coordinate system.

All values of P and Qr which are obtained by varying the amplitude Vpq and its angle ρ fall inside the area of this circle. In other words, all Qr and P values that are possible with the UPQC of a particular rating are defined by the circular region in the {Qr, P} coordinate system.

For instance, it can be seen that a UPQC with 0.5 p.u. voltage rating will maintain 0.5 p.u. power flow in both the directions without causing a reactive power requirement on either side of the system (sending and receiving end). Surely, UPQC, can compel the network at one end to provide Q or consume it from the other side. Figure 12.8 illustrates similar control features for P and Qr for phase angle $\delta = 0°, 30°, 60°$, and $90°$.

12.5 Multilevel Converter-Based UPQC

The UPQC has been considered as a universal solution for a number of power quality issues that arise at all levels of utilization in the smart grid due to the ever-increasing inclusion of distributed generation and nonlinear loads. The commonly used power circuit of the UPQC consists of two voltage source inverters (VSI) that, as shown in Figure 12.9, share a common dc link capacitor and are linked to the grid in shunt and series configurations, respectively, at the point of common coupling (PCC). The shunt VSI lowers the upstream line losses by correcting for the harmonic and fundamental reactive components of the load current, whereas the series VSI safeguards the downstream loads against sags/swells in the PCC voltage.

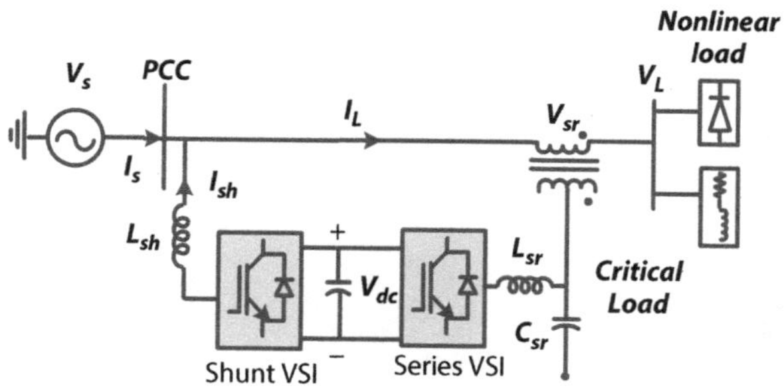

Figure 12.9 Twelve switch converter-based unified power quality conditioner.

In certain microgrid scenarios with significant harmonic distortion in the PCC (point of common coupling) voltage, the series voltage source inverter (VSI) is employed to prevent these distortions from propagating into the load. In parallel, the shunt VSI is responsible for correcting the load voltage, while the series VSI regulates the source current. Remarkably, both VSIs can effectively perform their tasks in a reciprocal manner. Despite the advantages of the standard 12-switch UPQC, which allows separate control of both VSIs and superior disturbance inhibition, it is not extensively utilized due to the underutilization of the series VSI. The series VSI often remains dormant or operates at extremely low modulation indices under typical circumstances, leading to various computational issues.

Modular multilevel converters (MMCs) have attracted a lot of attention over the years because they can lower the voltage rating of switching devices while significantly improving the quality of the output voltage waveform. MMCs were initially created for high-voltage direct current transmission systems, but they have since shown to be useful in a variety of other applications. Researchers investigated the use of neutral-point-clamped converters or hybrid multilevel-converter-based UPQC to improve power quality in medium/high voltage power systems [14]. This strategy provides a practical means of using low-loss and inexpensive devices. A three-phase ac–dc–ac multilevel UPQC based on cascaded three-leg converters was one of the unique topologies suggested by other investigations [15]. Four three-leg converters make up this topology, three of which have single-phase dc links and the fourth of which has a three-phase dc link. A multilevel three-phase UPQC comprising single-phase full-bridge cells with efficient series and parallel connections was used in another method [16]. In order to lower the current and voltage rating of switching devices, the power is distributed evenly across the cells in this configuration. For high-power compensation, a modular UPQC based on pairs of H-bridge submodules was also examined [17]. In this configuration, the series connection of the H-bridge module in the shunt component is made through a multi-winding transformer, while the series connection of the H-bridge is made directly by inserting it into the distribution line, doing away with the requirement for a large series injection transformer. Two different types of power cells were used to create a parallel multicell-based UPQC, which improved harmonic cancellation performance [18]. While the other adjusts for harmonic currents, the first kind corrects disturbances at the fundamental frequency. A cascaded multilayer inverter-based UPQC was also suggested [19]. In this configuration, the series converter and shunt converter are separate devices that each include several full-bridge submodules. A UPQC based on a modular multilevel converter (MMC-UPQC) was introduced in the context of medium/high-voltage grid applications [20]. In this setup, the series converter and shunt converter are made up of 2N half-bridge submodules that provide (2N + 1)-level output voltages. The modular and redundancy architecture makes it simple to replace defective modules, increasing the dependability of the MMC-UPQC. They are connected by a shared DC link. Most recent UPQC configuration research has concentrated on the back-to-back inverter. To avoid clumsy multi-winding transformers and typical DC buses, the modular UPQC in combination with direct ac–ac converters offers substitute choices for comprehensive power quality enhancement.

To enhance the topological structure of UPQC and reduce the system's cost, the concept of reduced switch multilevel converter topologies is integrated. This effort results in an improved setup, as depicted in Figure 12.10,

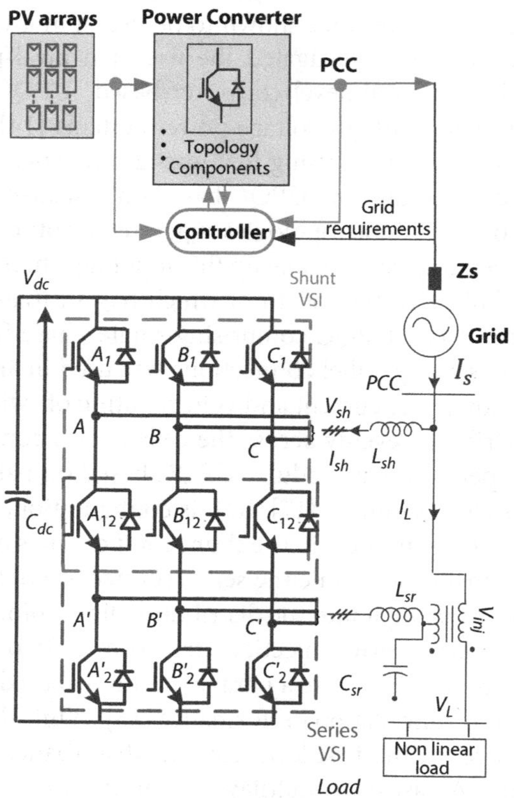

Figure 12.10 Nine switch converter-based UPQC with solar PV integration.

which enhances UPQC performance while employing fewer devices to execute all its capabilities effectively. The configurations of power converters using nine switches and ten switches are especially practical and advantageous for UPQC applications. These setups offer a convenient solution to enhance microgrid performance while effectively managing harmonic distortions and voltage regulation at the PCC.

The nine-switch converter-based UPQC has specific output terminal connections per phase, allowing them to be linked exclusively to either the positive or negative DC rail. The upper terminal connects to the upper positive DC-link, while the lower terminal links to the lower negative DC-link. This is a notable difference compared to conventional back-to-back power converters, where each three-phase full bridge side is independently controllable. Importantly, the top and lower terminals of the 9S-UPQC cannot be interconnected to the negative or positive DC-link, respectively.

12.6 Conclusion

In particular, P and the Q requirement at the receiving end (Qr), at any specified transmission angle δ, can be readily regulated by the UPQC inside the circle (boundary) derived in the {Qr, P} coordinates by rotating Vpq with its highest amplitude, a complete rotation. The reactive part of the line current and the related Q can really be regulated with regard to the voltage chosen at any arbitrary point along the line, despite the fact that the explanation above concentrates on the Q at the receiving end (Qr). Figure 12.8 clearly shows that the UPQC offers a strong, previously unreachable new device for transmission system management thanks to its unique ability to separately regulate the P and Q flow at any transmission angle δ. The receiving end Q demand is often a major issue because it affects the steady-state system losses, the over-voltage during rejection of load, and the fluctuation of the system-voltage with change in load. Naturally, a similar comparison could be conducted at the sending end or at any other locations along the line, but the results would be rather consistent for every realistic δ. The reduced switch counts multilevel inverter such as nine-switch and 10-switch converters give much flexibility in the application of UPQC.

References

1. R. A. Walling, R. Saint, R. C. Dugan, J. Burke, and L. A. Kojovic, "Summary of distributed resources impact on power delivery systems," *IEEE Trans. Power Del.*, vol. 23, no. 3, pp. 1636–1644, Jul. 2008.
2. M. Singh, V. Khadkikar, A. Chandra, and R. K. Varma, "Grid interconnection of renewable energy sources at the distribution level with power-quality improvement features," *IEEE Trans. Power Del.*, vol. 26, pp. 307-315, 2011.
3. J.A. Martinez and J.M. Arnedo, "Voltage sag studies in distribution networks- part I: System modeling," *IEEE Trans. Power Del.*, vol. 21,no. 3, pp. 338–345, Jul. 2006.
4. M. Sullivan, T. Vardell, and M. Johnson, "Power interruption costs to industrial and commercial consumers of electricity, *IEEE Trans. Ind. App.*, vol. 33, no. 6, pp. 1448–1458, Nov. 1997.
5. J. G. Nielsen and F. Blaabjerg, "Control strategies for dynamic voltage restorer compensating voltage sags with phase jump," in *Proc. Annu. Appl. Power Electron. Conf. Exposition*, 2001, no. 2, pp. 1267–1273.
6. Y. W. Li, P. C. Loh, F. Blaabjerg, and D. M. Vilathgamuwa, "Investigation and improvement of transient response of DVR at medium voltage level," in *Proc. Annu. Appl. Power Electron. Conf. Exposition*, 2006, pp. 1074–1080.
7. "*Voltage Characteristics of Electricity Supplied by Public Distribution Systems*," ver. BS EN 50160:2007, 2000.
8. S. B. Karanki, N. Geddada, M. K. Mishra, B. K. Kumar., "A DSTATCOM topology ith reduced DC Link voltage rating for load compensation with nonstiff source," *IEEE Trans. Power Electron.*, vol.27, no.3, pp.1201-1211, March 2012.
9. B. Singh, S.R. Arya, "Adaptive theory-based improved linear sinusoidal tracer control algorithm for DSTATCOM," *IEEE Trans. Power Electron.*, vol.28, no.8, pp. 3768-3778, Aug. 2013.
10. A. Y. Goharrizi, S. H. Hosseini, M. Sabahi, G. B. Gharehpetian, "Three-phase HFL-DVR with independently controlled phases," *IEEE Trans. Power Electron.*, vol.27, no.4, pp.1706-1718, April 2012.
11. C. Wessels, F. Gebhardt, F.W. Fuchs, "Fault ride-through of a DFIG wind turbine using a dynamic voltage restorer during symmetrical and asymmetrical grid faults," *IEEE Trans. Power Electron.*, vol.26, no.3, pp.807-815, March 2011.
12. Y. W. Li, D. MahindaVilathgamuwa, F. Blaabjerg and P. C. Loh, "A robust control scheme for medium-voltage-level DVR implementation," *IEEE Trans. Ind. Electron.*, vol. 54, pp. 2249-2261, 2007.
13. C. Benachaiba and B. Ferdi, "Voltage quality improvement using DVR," *Elect. Power Qual. Util. J.*, vol. XIV, no. 1, 2008.
14. I. A. Rubilar, J. R. Espinoza, J. A. Munoz, and L. A. Moran, "DC link voltage unbalance control in three-phase UPQCs based on NPC topologies," in *Proc.*

42nd IAS Annu. Meet. Conf. Rec. IEEE Ind. Appl. Conf., Nov. 5–8, 2007, pp. 597–602.
15. M. Ayslan, C. N. Jacobina, C. B. Carlos, and A. A. Gregory, "A new three-phase AC-DC-AC multilevel converter based on cascaded threeleg converters," in *Proc. IEEE Energy Convers. Congr. Expo.*, 2015, pp. 4685–4692.
16. J. A. Munoz, J. R. Espinoza, L. A. Moran, and C. R. Baier, "Design of a modular UPQC configuration integrating a components economical analysis," *IEEE Trans. Power Del.*, vol. 24, no. 4, pp. 1763–1772, Oct. 2009.
17. B. Han, B. Bae, S. Baek, and G. Jang, "New configuration of UPQC for medium-voltage application," *IEEE Trans. Power Del.*, vol. 21, no. 3, pp. 1438–1444, Jul. 2006.
18. J. A. Munoz, J. R. Espinoza, I. A. Rubilar, L. A. Moran, and P. E. Melin, "A modular approach for integrating harmonic cancellation in a multi-cell based UPQC," in *Proc. 34th Annu. Conf. IEEE Ind. Electron. Soc.*, 2008, pp. 3069–3074.
19. N. Farokhnia, S. H. Fathi, and H. R. Toodeji, "Voltage sag and unbalance mitigation in distribution systems using multi-level UPQC," in *Proc. 1st Power Quality Conf.*, 2010, pp. 1–5.
20. X. N. Xiao, J. J. Lu, Y. Chang, and Y. C. Yang, "A 10kV 4MVA unified power quality conditioner based on modular multilevel inverter," in *Proc. IEEE Int. Elect. Mach. Drives Conf.*, 2013, pp. 1352–1357.

13

Efficiency Evaluation and Harmonic Investigation of a High-Efficiency FrSPWM-Controlled Infinite-Level Inverter

Aishwarya V.[1,2]

[1]*Department of Electrical and Electronics Engineering, APJ Abdul Kalam Technological University, Trivandrum, India*
[2]*Department of Electrical and Electronics Engineering, Toc H Institute of Science and Technology, Arakunnam, Ernakulam District, Kerala, India*

Abstract

This chapter demonstrates the power loss computation, efficiency assessment, and harmonic study of a three-phase infinite-level inverter (TILI). The bi-level voltage fed inverter (VFI)-based electric drives have the drawbacks of poor efficiency and reliability and high switching power losses. Multi-level inverters (MLI) promise greater utilization of the DC-link and offer reduced output distortion and switching stresses at the cost of a complex control circuit, increased component count and cost. A TILI secures infinite voltage levels, with lesser components. It offers an excellent DC-link voltage utilization factor of 1.23, reduced input voltage requirement, minimized switching power losses, decreased voltage stress, increased working efficiency, and enhanced output power quality. A fully rectified sinusoidal PWM (FrSPWM) is deployed for the inverter control. The power loss and efficiency of TILI are theoretically estimated and compared with that obtained from the simulation and experimental results. The TILI achieved 93% efficiency and a power loss of only 7%. The harmonic investigation of the experimental results confirms reduced voltage and current distortions of 2.22% and 1.71%, respectively, which are within the permissible limits of IEEE 519-2014 Standard.

***Keywords*:** Efficiency, fully-rectified sinusoidal pulse width modulation (FrSPWM), harmonic analysis, power loss, three-phase infinite-level inverter

---

Email: aishwarya@tistcochin.edu in; Orcid ID: 0000-0002-4567-1380

Salman Ahmad, Farhad Ilahi Bakhsh, and P. Sanjeevikumar (eds.) Multilevel Converters, (253–280)

Nomenclature

f_s	switching frequency
f_o	line frequency
V_{DC}	DC-link voltage
V_{ot}	Buck converter output voltage
d	duty cycle of Buck converter
M	modulation index
V_{ll}	output line voltage of TILI
V_{dcref}	reference DC voltage
V_{oBuck}	output voltage of Buck converter
$V_{l,RMS}$	RMS value of output line voltage of TILI
V_{ac}	output AC voltage of TILI
P_{con}	conduction loss
P_{sw}	switching loss
P_{ototal}	total three-phase power output of TILI
V_{ph}	phase voltage output of TILI
I_{ph}	phase current output of TILI
$\cos\phi$	power factor
$P_{o/ph}$	per-phase output power of TILI
V_o	output voltage amplitude
I_o	load current
$P_{sw,HB}$	switching losses in single-phase H-bridge inverter
V_{ds}	drain–source voltage of power MOSFET
I_d	drain current of power MOSFET
t_r	rise time of power MOSFET
t_f	fall-time of power MOSFET
$P_{con,HB}$	conduction losses in single-phase H-bridge inverter
R_{dsON}	drain–source ON-state resistance of power MOSFET
$P_{HBlosses}$	total losses in single-phase H-bridge inverter
$P_{in,HB}$	input power to single-phase H-bridge inverter
P_{in}	total input power of TILI
I_{in}	input current of Buck converter
η	efficiency of Buck converter
P_{oBuck}	output power of Buck converter
$V_{o,avg}$	average output voltage of Buck converter
I_{oBuck}	output current of Buck converter
I_1	current in MOSFET at the start of turn-on period
I_2	current in MOSFET at the culmination of turn-on period
ΔI_L	peak-to-peak ripple current of Buck inductor

$I_{RMS,Buck}$	RMS current through the Buck converter switch
P_L	power loss in inductor
R_{DCR}	DC resistance of inductor
$P_{sw,Buck}$	switching losses in Buck converter
$P_{con,Buck}$	conduction losses in Buck converter
P_{Diode}	power loss in diode
V_f	forward voltage of diode
Q_{rr}	reverse recovery charge of power MOSFET
t_{rr}	reverse recovery time of power MOSFET
$P_{Buck,losses}$	total power loss in Buck converter
$P_{TotalLosses/ph}$	total per-phase power loss in TILI
$P_{TotalLosses}$	total three-phase power loss in TILI

13.1 Introduction

High-power applications demand the need for purely sinusoidal waveforms. The voltage waveform of the traditional two-level voltage-fed inverter (VFI) is non-sinusoidal and is rich in harmonic content. The VFI switches are controlled through pulse width modulation (PWM) scheme which curbs the power quality and input power factor. High switching power losses are associated with the VFI switches due to the high-frequency PWM operation. Conventional two-level VFI suffers from the drawbacks of high switching power losses, generation of common-mode voltages, presence of high-frequency harmonic components in the output, reduced working efficiency, and decreased DC-link voltage utilization (Aishwarya & Gnana Sheela, 2021a, 2022; Aishwarya & Sheela, 2021b; Chaturvedi *et al.*, 2008; Jagabar Sathik *et al.*, 2020; Lai, 2009; Rodríguez *et al.*, 2002; Tarisciotti *et al.*, 2020; Vahedi & Trabelsi, 2019). Multilevel inverters have gained popularity and have outperformed conventional VFIs in various applications in the past decade. Multilevel inverters are favored in many applications requiring high-performance power conversion due to their benefits, which include improved power quality, diminished operational frequencies, ameliorated common-mode noise and dV/dt on switches, and boosted electromagnetic compatibility (Aishwarya & Gnana Sheela, 2021b, 2022; Rajalakshmi & Rangarajan, 2019). The foundation of MLI is the idea of creating an AC voltage output from various DC bus voltage levels (Sarwar & Asghar, 2010). As the DC-side level count rises, the generated voltage becomes progressively equivalent to a sinusoidal wave (Rodríguez *et al.*, 2002; Sarwar & Asghar, 2010). Standard MLIs are grouped as diode-clamped (DCMLI), floating capacitor (FCMLI), and cascaded H-bridge (CHBMLI) (Chen & He, 2006; Kala & Arora, 2017; Omer *et al.*, 2020; Prabaharan & Palanisamy, 2017).

Joy *et al.* (2017) and Aishwarya and Sheela (2021b) proposed a novel infinite-level inverter with an improved utilization of the DC-link. It comprises a high-frequency front-end Buck converter and a fundamental-frequency H-bridge inverter (Aishwarya & Sheela, 2021b). The switching power losses get ameliorated through the line-frequency operation of the H-bridge inverter (Aishwarya & Sheela, 2021b). It achieves higher output voltage levels with the same component count by raising only the operational frequency of the Buck switch. It solves issues with shoot-through and dead-time requirements in PWM inverters that cause less energy to be transferred to the load. The switching power loss at high-frequency operation and the device constraints are significantly minimized in the TILI due to lesser input requirement. TILI is a simple and versatile structure that achieves a DC-link voltage utilization factor of 1.23, minimized DC voltage requirement, decreased switching power losses and switching stresses, and improved power quality (Aishwarya & Sheela, 2021b).

The fundamental switching components used in inverter circuits are diodes, IGBTs, and MOSFETs. When the inverter is operated, power losses occur in these switching components. Hence, the calculation of power losses is very critical in achieving the design goals of reduced size, minimizing the power losses, decreasing the cost, or a combination of any of these goals (Kim *et al.*, 2001). The different power losses in a switching device are categorized as conduction power loss, off-state loss, and switching power loss (Zhou *et al.*, 2006). The off-state loss in a switching component is usually neglected, as the leakage current in the device is infinitesimally small. Thus, the complete power loss in a switching component is composed of conduction power loss and switching power loss (Aghdam *et al.*, 2005; Aishwarya & Sheela, 2023; Kim *et al.*, 2001). Several methods for the analysis of switching power loss and conduction loss are presented in Aghdam *et al.* (2005), Aishwarya and Sheela (2023), Alamri *et al.* (2015), Alamri and Darwish (2014), De Caro *et al.* (2016), Karasani *et al.* (2017), Kim *et al.* (2001), Salem *et al.* (2021), and Wang *et al.* (2016).

Mecke (2015) compared the efficiency of bi-level VFI and a five-level diode-clamped multilevel inverter. It was inferred that the use of wide-bandgap semiconductors with less standing voltage will decrease the power losses, cooling effort, and filter sizes (Mecke, 2015). The efficiency of a three-level active neutral-point clamped MLI for different PWM schemes was investigated by Floricau *et al.* (2009). a loss comparison of bi-level and tri-level inverters was carried out by Sharkh *et al.* (2014). Wang *et al.* (2016)

Table 13.1 Comparative summary of different multilevel inverter configurations.

Multilevel inverter configuration	No. of levels	No. of DC sources	No. of active switches	No. of gate drivers	Entire component count (ECC)
Salem *et al.* (2021)	4	3	18	18	39
Jagabar Sathik *et al.* (2020)	7	1	8	8	22
Tarisciotti *et al.* (2020)	15	1	36	36	84
Rajalakshmi and Rangarajan (2019)	9	4	8	8	23
Prabaharan and Palanisamy (2017)	13	4	14	14	32
Karasani *et al.* (2017)	9	7	24	24	63
De Caro *et al.* (2016)	25	5	30	30	35
Alamri *et al.* (2015)	7	3	12	12	27
Sarwar and Asghar (2010)	4	1	6	6	15
Aghdam *et al.* (2005)	9	12	48	48	108
Sharkh *et al.* (2014)	3	1	12	12	34
Proposed TILI	$\frac{f_s}{f_o}$	3	15	15	33

estimated the efficiency of 99.27% for a four-level hybrid-clamped MLI employing SiC MOSFETS. A reduced switching loss PWM scheme for a three-level DCMLI was developed by Chaturvedi *et al.* (2011). A comparison of power losses in AC/DC/AC PWM converters was conducted by Nguyen and Lee (2015). Monmasson and Cirstea (2007) reviewed the design process for FPGAs used in industrial control systems. Due to the parallelism present in its topologies, an FPGA's control algorithm executes in a very short amount of time. The control performance is significantly improved while keeping high flexibility using a low-cost FPGA-based controller (Aishwarya & Sheela, 2021b).

Table 13.1 showcases the comparative summary of various MLI configurations. The nine-level inverter demonstrated by Aghdam *et al.* (2005) possesses the highest component count of 108. The inverter proposed by De Caro *et al.* (2016) generates a 25-level AC voltage with only 35 components. From Table 13.1, it is noted that the proposed TILI needs just 33 components for developing an AC voltage having extendable voltage levels that vary with the operational frequency of the Buck converter (f_s) and the line frequency of the H-bridge (f_o). The number of levels generated by a TILI is equal to $\frac{f_s}{f_o}$.

This chapter deals with the power loss computation, efficiency assessment, and harmonic study of a three-phase infinite-level inverter (TILI). The efficiency and power loss of TILI are computed theoretically using the design parameters, verified with the simulation results, and finally validated using the experimental outcomes. Fully rectified sinusoidal PWM control for the TILI is produced using an FPGA (Aishwarya & Sheela, 2021b). A harmonic investigation of the experimental results is conducted using a Fluke 434 Power Quality analyzer.

The chapter is divided into five sections. Section 13.2 introduces the concept of a TILI. The power loss evaluation and efficiency assessment of a TILI is presented in Section 13.3. Simulation results are presented in Section 13.4. Hardware implementation and results are presented in Section 13.5. Section 13.6 elaborates the results and discussion. Lastly, the chapter is summarized and concluded in Section 13.7.

13.2 Three-Phase Infinite-Level Inverter (TILI)

TILI is a specialized inverter that combines a Buck stage with an H-bridge stage. The fundamental structure of the TILI configuration consists of a Buck unit connected in cascade with an H-bridge unit (Aishwarya & Sheela, 2021b; Joy *et al.*, 2017). This circuit uses three infinite-level inverter structures, each of which operates autonomously with a phase-shift of 120 electrical degrees from the others. This phase-shift ensures balanced operation and effective utilization of the power sources. The TILI circuit includes three individual Buck converters, each functioning independently and controlled separately. These Buck converters produce three completely rectified voltages, each one having the same magnitude as the DC input voltage. Importantly, the outputs of the three Buck converters have a phase-shift of 120 degrees from the others. These rectified voltages generated by the Buck converters serve as inputs to the three H-bridge units. These inverters, operating in the supply frequency, convert the rectified inputs into three sinusoidal voltages, each of which has a phase-shift of 120 electrical degrees. The three sinusoidal voltages produced by the H-bridge units are then applied to the phases of a balanced three-phase load. By maintaining the 120 electrical degree phase-shift between the voltages, the TILI circuit ensures balanced and synchronized power delivery to the three-phase load (Aishwarya & Sheela, 2021a; Joy *et al.*, 2017).

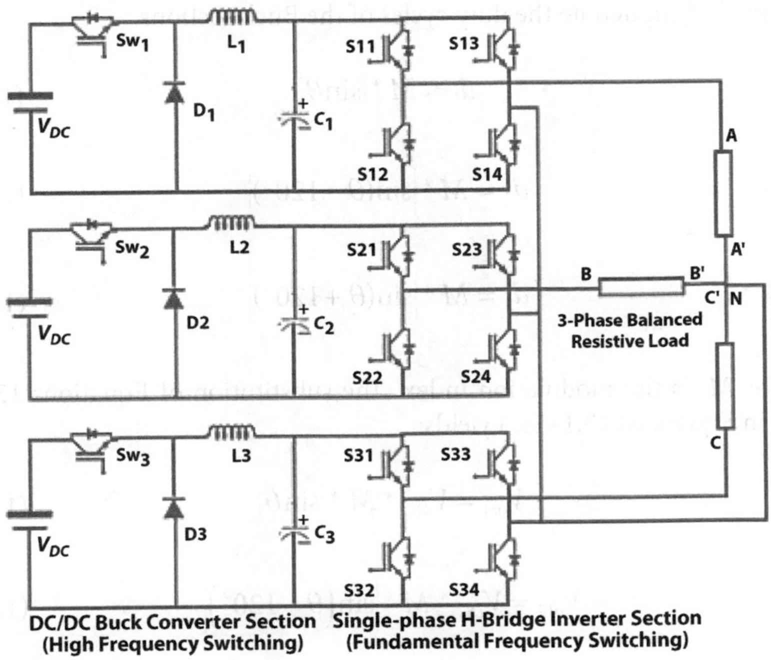

Figure 13.1 Circuit of a TILI.

TILI offers a very small total harmonic distortion (THD) with sinusoidal output. Her only the Buck switch operates at a greater frequency. The switches of the H-bridge toggle at the supply frequency, resulting in minimized switching losses. The key component of the TILI is the Buck converter which acts as a wave-shaping circuit. This wave is unfolded gradually by the H-bridge inverter at the line frequency. Figure 13.1 represents the circuit of TILI.

13.2.1 Mathematical Modeling of TILI

Let V_{DC} be the input voltage to the three Buck converters of a TILI, and $V_{ot1}, V_{ot2}, V_{ot3}$ be their corresponding output voltages.

$$V_{ot1} = d_1 * V_{DC} \tag{13.1}$$

$$V_{ot2} = d_2 * V_{DC} \tag{13.2}$$

$$V_{ot3} = d_3 * V_{DC} \tag{13.3}$$

where d_1, d_2, d_3 denote the duty cycles of the Buck sections.

$$d_1 = M * |\sin\theta| \tag{13.4}$$

$$d_2 = M * |\sin(\theta - 120^\circ)| \tag{13.5}$$

$$d_3 = M * |\sin(\theta + 120^\circ)| \tag{13.6}$$

where M is the modulation index. The substitution of Equations 13.4–13.6 in Equations 13.1–13.3 yields:

$$V_{ot1} = V_{DC} * M * |\sin\theta| \tag{13.7}$$

$$V_{ot2} = V_{DC} * M * \left|\sin\left(\theta - 120^\circ\right)\right| \tag{13.8}$$

$$V_{ot3} = V_{DC} * M * \left|\sin(\theta + 120^\circ)\right| \tag{13.9}$$

The line voltages $V_{ll1}, V_{ll2}, V_{ll3}$ are derived from Equations 13.7–13.9 and are expressed
by Equations 13.10–13.12.

$$V_{ll1} = \sqrt{3} * V_{DC} * M * \left|\sin(\theta + 30^\circ)\right| \tag{13.10}$$

$$V_{ll2} = \sqrt{3} * V_{DC} * M * \left|\sin(\theta - 90^\circ)\right| \tag{13.11}$$

$$V_{ll3} = \sqrt{3} * V_{DC} * M * \left|\sin(\theta - 210^\circ)\right| \tag{13.12}$$

Using the assumption that M=1 and the RMS value of line voltage, $V_{l,RMS} = 400V$ the necessary
DC-link voltage may be computed from Equation 13.13:

$$V_{DC} = \frac{\sqrt{3} * V_{l,RMS}}{\sqrt{2} * M} = 325V \tag{13.13}$$

From Equation 13.13, it can be concluded that, for a TILI:

$$V_{l,RMS} = 1.23 * V_{DC} \tag{13.14}$$

13.2.2 Hardware Setup

The duty cycles for the three Buck units of the TILI are developed using the fully rectified sinusoidal PWM (FrSPWM) method (Aishwarya & Sheela, 2021b; Joy *et al.*, 2017). In this method, a completely rectified reference signal is compared with a high-frequency carrier signal. The amplitude of the reference signal generated at the supply frequency is identical to the voltage input (V_{DC}). The derived PWM pulse-train is then applied to the Buck switches. By autonomously controlling the three Buck units, their output voltages have a magnitude identical to the voltage input (V_{DC}) and assume a rectified-sine waveform having a 120-degree phase-shift. The H-bridge switches are also controlled autonomously at the supply frequency that is significantly lower than the operating frequency of the Buck stages. Hence, the outputs of the H-bridge units manifest as three sinusoidal voltages at the supply frequency having an identical magnitude as that of the voltage input (V_{DC}) with a 120-degree phase-shift. The implementation of this control strategy leads to ameliorated switching loss in the TILI. By carefully

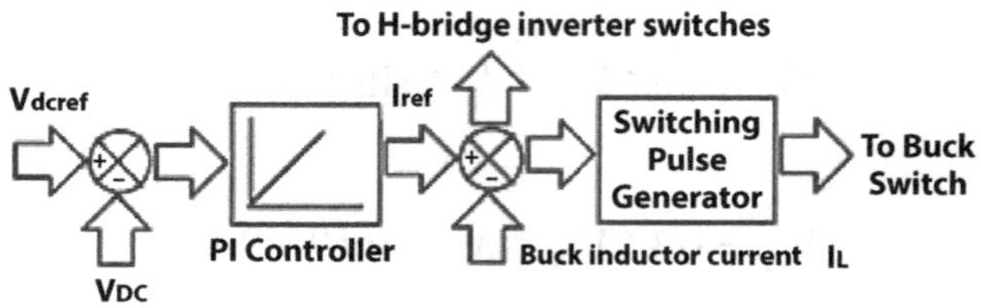

Figure 13.2 Control circuit of TILI.

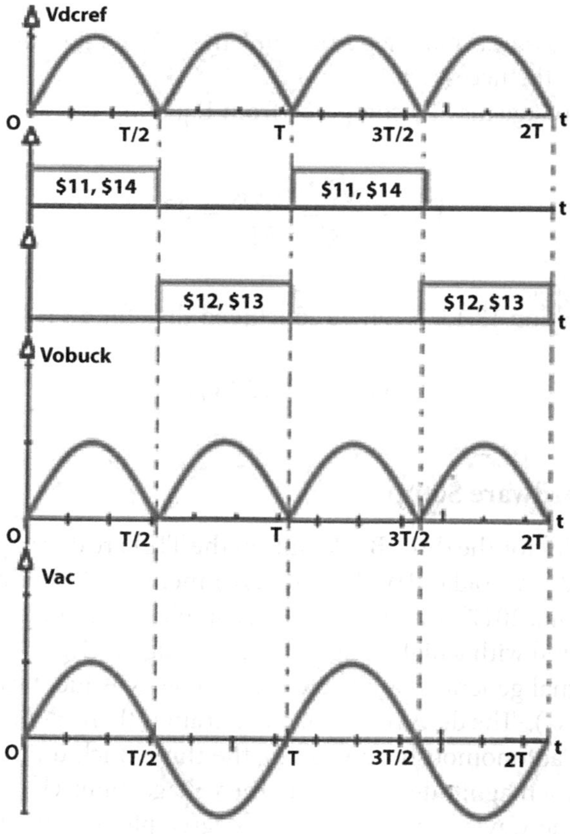

Figure 13.3 Waveforms of TILI.

managing the duty cycles and phase shifts, the TILI can achieve efficient power conversion with improved overall performance.

The control circuit of a TILI is depicted in Figure 13.2, illustrating the arrangement and interconnection of the Buck converters, H-bridge inverters, and associated control elements. Figure 13.3 depicts the waveforms of a TILI where $V_{dcref}, V_{oBuck}(=V_{ot}), V_{ac}$ denote the DC reference voltage, the Buck stage output, and AC voltage of the H-bridge section respectively.

13.3 Power Loss Evaluation and Efficiency Assessment of TILI

The switching loss (P_{sw}) and conduction loss (P_{con}) of each individual switch/diode are evaluated through a thorough analytical assessment of the power losses (Viswanathan & Jeevananthan, 2017). Due to its simplicity, sinusoidal PWM is presumed for the calculation of analytical losses. The complete power loss in the TILI is the sum total of the losses in the Buck stage and the H-bridge stage. The power losses in TILI are modeled as follows:

The total three-phase power output of TILI, P_{ototal}, is obtained from Equation 13.15.

$$P_{ototal} = 3 * V_{ph} * I_{ph} * \cos\phi = 150W \tag{13.15}$$

The output power in one phase of TILI is computed from Equation 13.16.

$$P_o/ph = \frac{P_{ototal}}{3} = \frac{150}{3} = 50W \tag{13.16}$$

Assuming an output voltage (V_o) of 30V at UPF, load current I_o is obtained from Equation 13.17.

$$I_o = \frac{P_o}{3 * V_o * \cos\phi} = \frac{50}{3 * \frac{30}{\sqrt{2}} * 1} = 0.786A \tag{13.17}$$

3. Calculation of switching losses in the MOSFET of single-phase H-bridge inverter: Switching losses in the H-bridge unit, $P_{sw,HB}$ is computed from Equation 13.18.

$$P_{sw,HB} = \frac{1}{2} * V_{ds} * I_d * (t_r + t_f) * f_o = 0.0267mW \quad (13.18)$$

where $V_{ds} = V_o, I_d = I_o, f_o = 50Hz$, rise-time $t_r = 35ns$, and fall-time $t_f = 29ns$. The practical MOSFET IXFK230N20T is used as the H-bridge switch and is operated at the supply frequency of 50 Hz.

Calculation of conduction losses in the MOSFET of H-bridge inverter: Conduction losses in the single-phase H-bridge inverter, $P_{con,HB}$ is obtained using Equation 13.19.

$$P_{con,HB} = I_d^2 R_{dsON} = (0.786)^2 * 0.0075 = 4.635mW \quad (13.19)$$

Calculation of total losses in the H-bridge inverter: Total power loss in the four MOSFETs of the

H-bridge unit, $P_{HBlosses}$ is computed

as:

$$P_{HBlosses} = 4 * (P_{sw,HB} + P_{con,HB}) = 4 * (0.0267 + 4.633)mW \quad (13.20)$$

$$P_{HBlosses} = 4 * 4.66mW = 18.64mW \quad (13.21)$$

Calculation of power input to H-bridge inverter: Thus, the power input to the H-bridge inverter is calculated from Equation 13.22.

$$P_{in,HB} = P_{o,Buck} = P_{o/ph} + P_{HBlosses} = 50.0186W \quad (13.22)$$

Calculation of RMS current through the MOSFET of Buck stage: Presuming a theoretical efficiency of 97% for the Buck converter, input current of the Buck converter I_{in} is calculated as:

$$I_{in} = \frac{P_{in}}{\eta * V_{DC}} = \frac{50.0186}{0.97 * 30} = 1.719A \quad (13.23)$$

The output current of the Buck converter I_{oBuck} is calculated from Equation 13.24.

$$I_{oBuck} = \frac{P_{oBuck}}{V_{o,avg}} = \frac{50.0186}{19.1} = 2.76A \tag{13.24}$$

Calculation of RMS value of current through MOSFET of Buck stage: At the start of the turn on period, the MOSFET current of Buck stage is calculated from Equation 13.25 (Viswanathan & Jeevananthan, 2017).

$$I_1 = \left[\frac{P_{oBuck}}{\eta * V_{DC}} - \frac{\Delta I_L}{2}\right] + \left[\frac{P_{oBuck}}{V_o} - \frac{\Delta I_L}{2}\right] = 4.296A \tag{13.25}$$

At the termination of the turn on period, the MOSFET current of Buck converter is estimated from Equation 13.26 (Viswanathan & Jeevananthan, 2017).

$$I_2 = \left[\frac{P_{oBuck}}{\eta * V_{DC}} + \frac{\Delta I_L}{2}\right] + \left[\frac{P_{oBuck}}{V_o} + \frac{\Delta I_L}{2}\right] = 4.378A \tag{13.26}$$

The RMS current through Buck converter switch $I_{RMS,Buck}$ is calculated from Equation 13.27.

$$I_{RMS,Buck} = \sqrt{\frac{1}{3}[I_1^2 + I_1 I_2 + I_2^2] * d} = 3.461A \tag{13.27}$$

Calculation of conduction loss in the Buck switch: The practical MOSFET IXFK230N20T is used as the Buck converter switch. The conduction loss in the Buck switch $P_{con,Buck}$ is calculated from Equation 13.28.

$$P_{con,Buck} = I_d^2 R_{dsON} = I_{RMS,Buck}^2 R_{dsON} = 0.089W \tag{13.28}$$

Calculation of switching loss in the Buck switch: Switching loss in the Buck switch $P_{sw,Buck}$ is calculated from Equation 13.29.

$$P_{sw,Buck} = \frac{1}{2} * f_s * V_o * [(I_1 * t_r) + (I_2 * t_f)] = 0.0397W \quad (13.29)$$

Calculation of inductor loss in the Buck converter: Losses in the inductor of the Buck converter P_L are modeled using Equation 13.30.

$$P_L = I_{RMS}^2 * R_{DCR} = I_{O,Buck}^2 * R_{DCR} = 0.0156W \quad (13.30)$$

where R_{DCR} is the DC resistance of the inductor and $R_{DCR} = 0.0022\Omega$.

Calculation of diode loss in the Buck stage: Losses in the diode of the Buck stage P_{Diode} are modeled using Equation 13.31.

$$P_{Diode} = V_f * I_{o,Buck} = 1.35 * 2.67 = 3.6045W \quad (13.31)$$

Calculation of total losses in the Buck stage: The total power loss in the Buck stage $P_{Buck,losses}$ is calculated from Equation 13.32.

$$P_{Buck,losses} = P_{sw,Buck} + P_{con,Buck} + P_L + P_{Diode} = 3.749W \quad (13.32)$$

Thus, the complete losses in one phase of TILI $P_{TotalLosses/ph}$ is obtained from Equation 13.33 as:

$$P_{TotalLosses/ph} = P_{HB,losses} + P_{Buck,losses} = 3.767W \quad (13.33)$$

The complete power loss in TILI $P_{TotalLosses}$ is obtained from Equation 13.34.

$$P_{TotalLosses} = 3 * P_{TotalLosses/ph} = 11.301W \quad (13.34)$$

The total input power of TILI P_{in} is obtained using Equation 13.35.

$$P_{in} = P_{o,total} + P_{TotalLosses} = 161.301W \quad (13.35)$$

Thus, the efficiency of the proposed TILI in percentage $\%\eta$ is estimated from Equation 13.36.

$$\%\eta = \left(\frac{P_{o,total}}{P_{in}}\right) * 100\% = \left(\frac{150}{161.301}\right) * 100\% = 93\% \qquad (13.36)$$

It is observed that the TILI has the lowest losses in the Buck stage and inverter section compared to its conventional counterparts. Moreover, the losses in the diode bridge rectifier section are absent in TILI. Thus, the proposed TILI has the least loss percentage of 11.3% and the highest efficiency

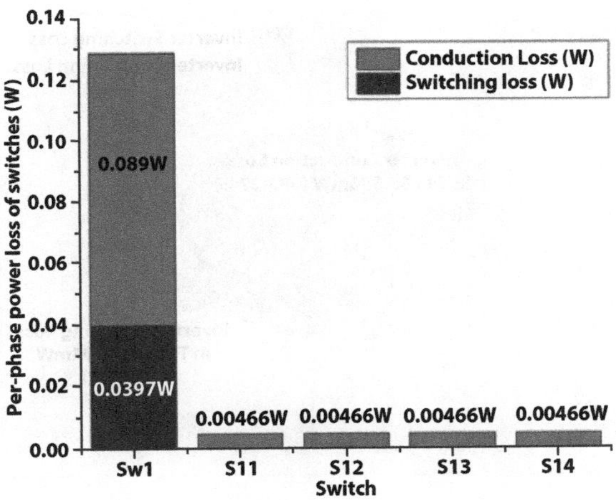

Figure 13.4 Per-phase power loss distribution in TILI switches.

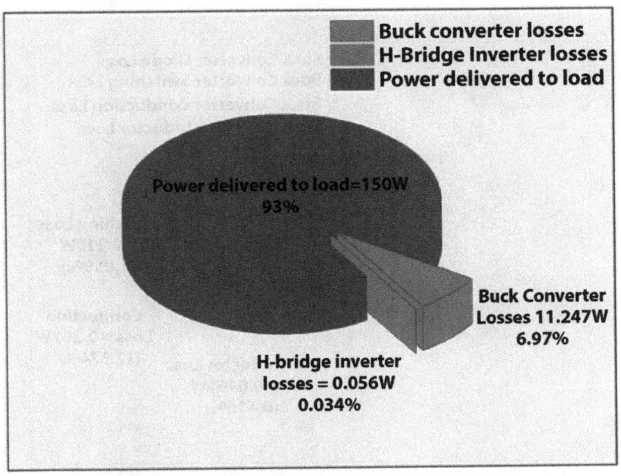

Figure 13.5 Power loss distribution in TILI.

of 93% among all the other inverters. Figure 13.4 portrays how power loss is spread among the switches of one of the phases of the TILI. Figure 13.5 depicts the distribution of complete losses in various parts of the TILI. Figures 13.6 and 13.7 depict the power losses in the H-bridge inverter section and Buck section of the TILI, respectively.

Figure 13.8 showcases how the efficiency of TILI and switching power losses change with the variation in the frequency of operation of the Buck unit. The switching power loss in the Buck switch increases from 7.8 mW

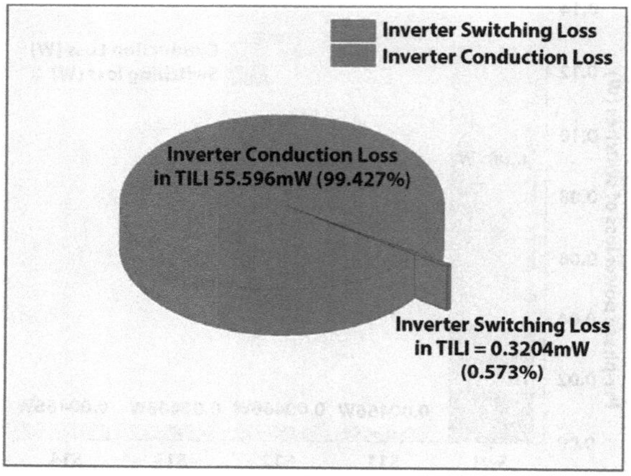

Figure 13.6 Power loss distribution in the inverter section of TILI.

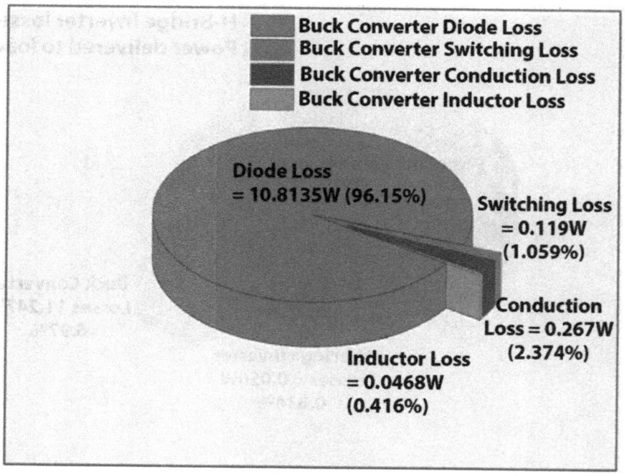

Figure 13.7 Power loss distribution in the Buck converter stage of TILI.

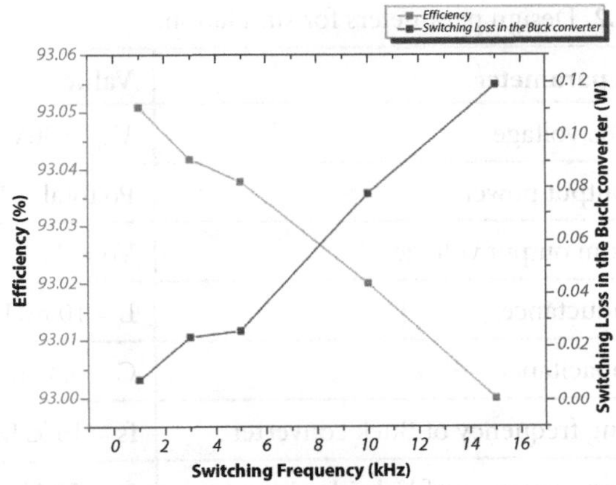

Figure 13.8 Variation of efficiency and switching power losses in TILI with switching frequency.

to 0.1191 W as the switching frequency is increased from 1 kHz, and the efficiency of TILI fluctuates proportionally from 93.05% to 93%. Thus, it can be concluded that a sharp rise in the operating frequency of the Buck converter switch has no effect on the switching power losses. According to the chosen control approach, the TILI can be operated securely at high operating frequencies with a respectable high efficiency and highest DC-link utilization.

13.4 Simulation Results

The MATLAB/Simulink SimPowersystems toolbox was deployed for simulation. When the reference DC voltage changes in a TILI, the Buck stage regulates the inverter's DC voltage. Through voltage mode control, the DC-link voltage is controlled. The design parameters used for simulation are listed in Table 13.2. The three-phase output voltage, per-phase voltage, and current of a TILI are each depicted in Figures 13.9–13.11, respectively. Figures 13.12 and 13.13 display the FFT analysis of a TILI's output phase voltage and output current waveforms. The phase voltage and current waveforms are found to have extremely low THD of 2.20% and 1.69%, respectively, which are both substantially within the IEEE-519 standards (Committee *et al.*, 2014). Table 13.2 provides the design parameters used for simulation.

Table 13.2 Design parameters for simulation.

Design parameter	Value
Input DC voltage	$V_{DC} = 30V$
Rated output power	Po,total = 150 W
Maximum output voltage	Vo = 30 V
Buck inductance	L = 10 mH
Buck capacitance	C = 0.33 μ F
Switching frequency of Buck converter	fs = 15 kHz
Switching frequency of H-bridge inverter	fo = 50 Hz
Resistive load	R = 100 Ω

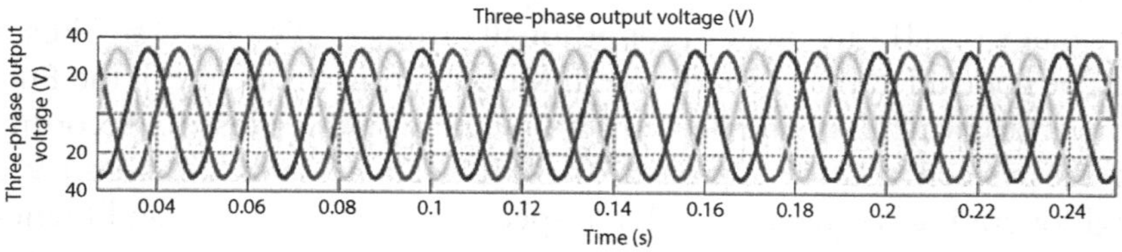

Figure 13.9 Three-phase output voltage of TILI.

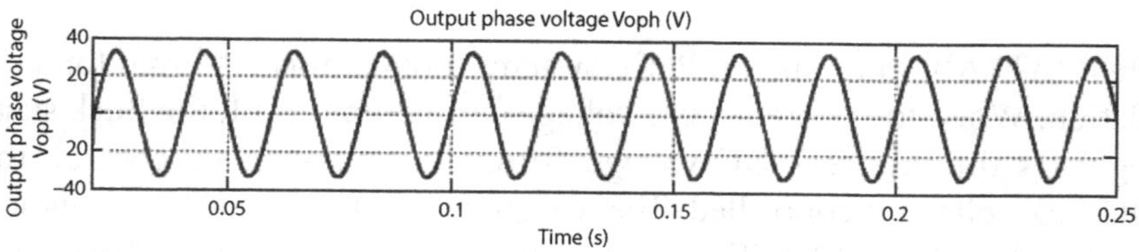

Figure 13.10 Per-phase output voltage of TILI.

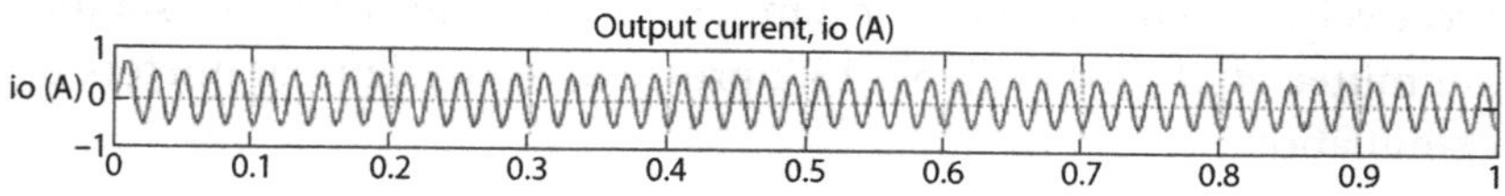

Figure 13.11 Output current of TILI.

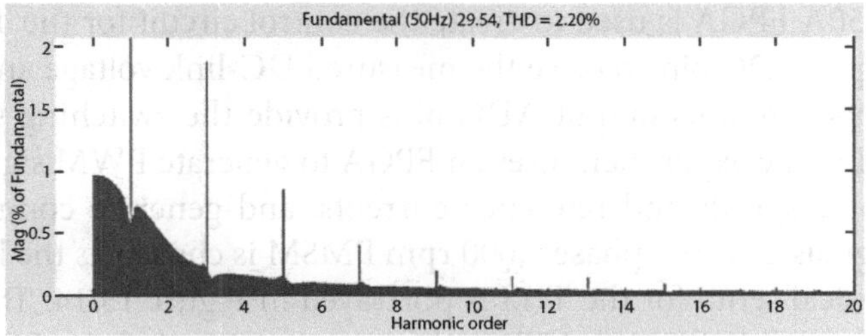

Figure 13.12 Harmonic spectrum of output voltage of TILI.

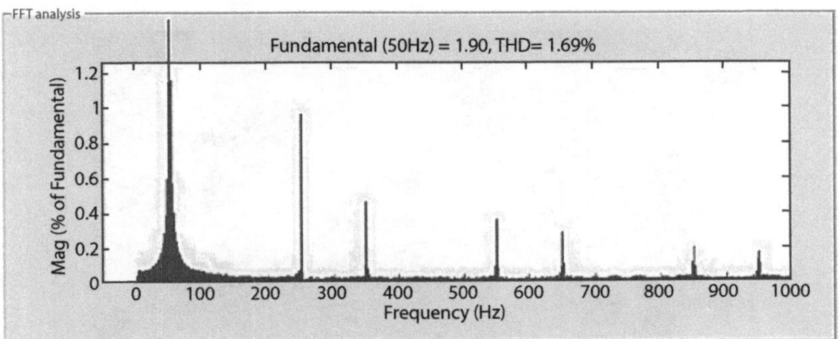

Figure 13.13 Harmonic spectrum of output current of TILI.

13.5 Hardware Development and Results

13.5.1 Experimental Setup

The physical setup of TILI is constructed to validate the theoretical analysis and simulation results. The entire setup of the TILI circuit is assembled on an acrylic board elevated on a platform. The DC/DC Buck converter section, inverter section, a triple Buck driver circuit, an ADC circuit, and a Hall-signal detection circuit are designed in separate PCBs and are mounted on the board. Three different Buck converters, each with its own distinct driver circuit, are incorporated on a single PCB to constitute the DC/DC Buck converter section. Three H-bridge inverter circuits are constructed on separate PCBs together with their corresponding driver circuits to constitute the inverter part. The inductor currents are sensed using three current sensors. The measured currents and voltages are converted into digital signals by an ADC circuit and fed into the FPGA control board. One 30-V battery serves as the source of power for the TILI. Xilinx Spartan

3A, XC3S50A FPGA is used to create the control circuit for the TILI. The FPGA's input ADC pins receive the measured DC-link voltage and inductor currents, while its output ADC pins provide the switching sequence for the TILI. The experiment uses an FPGA to generate PWM signals, calculate motor speeds and reference currents, and generate commutation control signals. A three-phase 3,000 rpm PMSM is chosen as the load. The entire physical setup for the TILI is portrayed in Figure 13.14. The details of the MOSFET IXFK230N20T used as the active switch in TILI is presented in Table 13.3.

Figure 13.14 Complete experimental setup of TILI.

Table 13.3 Specifications of MOSFET IXFK230N20T.

Parameter	Value
Continuous drain current	Id = 230 A
Drain–source breakdown voltage	Vds = 200 V
On-state resistance	RdsON = 7.5 mΩ
Rise time	tr = 35 ns
Fall time	tf = 29 ns
Diode forward voltage	Vf = 1.3 V
Reverse recovery charge	Qrr = 0.74 μC
Reverse recovery time	trr = 200 ns

13.5.2 FPGA-Based PWM Signal Generation

The PWM sequences from the FPGA controller are generated using a program code written in VHDL to activate the inverter switches. The VHDL program code makes it simple to control the PWM pulse's parameters, such as pulse count, operating frequency, duty cycle, etc.

PWM generation: FPGA is used to produce fully rectified sine PWM (FrSPWM) pulses since it is affordable and has a rapid circuit response owing to its concurrent execution process. PWM pulses are produced using fully rectified sine PWM (FrSPWM) for the inverter's active switches. The FrSPWM approach compares a high-frequency triangle carrier signal with a fully rectified sinusoidal control signal. The switches are toggled up whenever the control signal is higher than the carrier signal and thrown out whenever it is lower.

VHDL code generation: For the open-loop control of the drive system, a VHDL code is built. Xilinx design environment, the ISE Design Suite, is used to write, verify, and simulate the code. The ModelSim.SE.v6.2c Simulator software is used to observe the validated generated code.

13.5.3 Experimental Outcomes and Verification

The output AC voltage and harmonic spectra of the TILI are displayed in Figures 13.15 and 13.16, respectively. The output phase voltage is fully sinusoidal, with a 30-V amplitude and a harmonic distortion of 2.22%. The harmonic profile of the output current is displayed in Figure 13.17. The phase current waveform has a relatively small THD of 1.71% and is

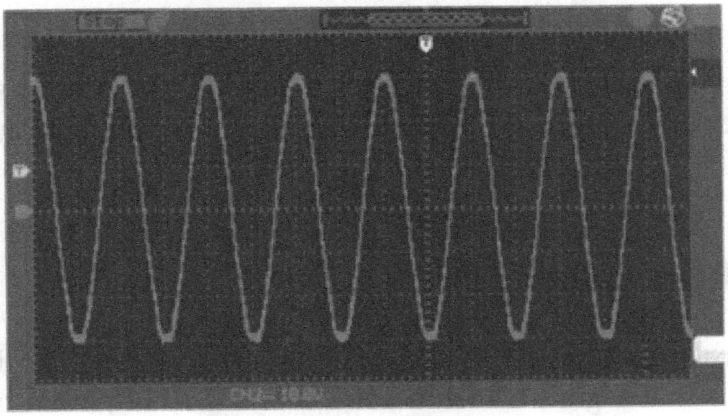

Figure 13.15 Output AC voltage of TILI.

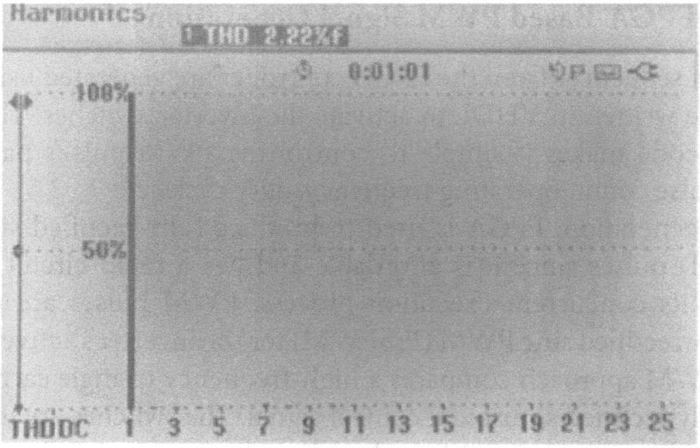

Figure 13.16 Harmonic profile of output voltage.

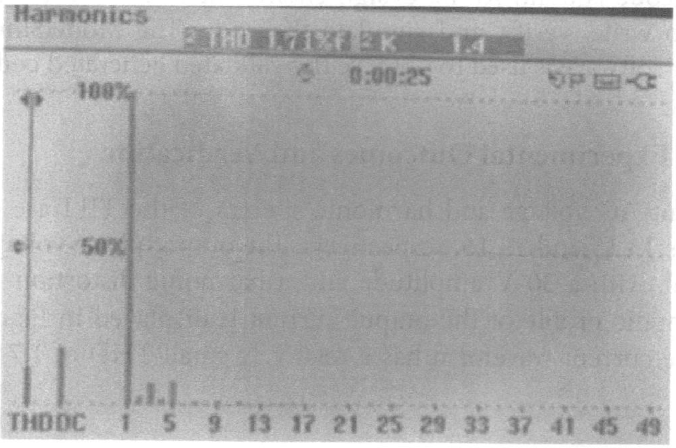

Figure 13.17 Harmonic profile of output current.

sinusoidal. Thus, it is established that the THD of the TILI phase voltage and phase current are under 5% and falls perfectly inside the range of IEEE 519.

13.6 Results and Inference

The proposed TILI has a 93% efficiency, which is quite high, and only 7% of power losses. Figure 13.4 shows the distribution of power loss in the switches of one phase of the TILI. It is discovered that while power losses in Buk switches are fairly large, they are far less significant in H-bridge switches. Only 0.034% of the overall power losses in the TILI are accounted for the power losses in the H-bridge part, compared to 6.97% in the Buck converter part. Figure 13.5 establishes this fact. Since the inverters are switched at the supply frequency, the inverter switching loss component is very small (see Figure 13.6), and 99.427% of the H-bridge inverter loss

Table 13.4 Comparison of power losses in theoretical computations, simulation, and experimental results.

	Device	Theoretical results	Simulation results	Experimental results
Power loss (W)	Sw1	0.128	0.15	0.145
	S11	0.0047	0.0053	0.0051
	S12	0.0047	0.0053	0.0051
	S13	0.0047	0.0053	0.0051
	S14	0.0047	0.0053	0.0051
	Sw2	0.128	0.15	0.14
	S21	0.0047	0.0053	0.0051
	S22	0.0047	0.0053	0.0051
	S23	0.0047	0.0053	0.0051
	S24	0.0047	0.0053	0.0051
	Sw3	0.128	0.15	0.14
	S31	0.0047	0.0053	0.0051
	S32	0.0047	0.0053	0.0051
	S33	0.0047	0.0053	0.0051
	S34	0.0047	0.0053	0.0051
	L	0.046	0.05	0.065
	D	10.814	11.515	11.707
Total power loss		**11.301 W**	**12.079 W**	**12.258 W**
Efficiency (%)		**@ 93%**	**92.55%**	**92.445%**

is attributed to conduction losses in the H-bridge switches, with switching losses in the inverter switches contributing only 0.573% of the inverter loss.

Conduction losses in the diodes, inductor losses, switching losses in the Buck switches, and conduction losses in the Buck switches are all responsible for the power losses in the Buck stage. Figure 13.7 shows that 96.15% of all Buck converter losses come from diode conduction losses. The switching losses in the Buck switch amount to 2.374%, while the conduction losses account for 1.059%. Only 0.416% of the overall Buck converter losses come from inductor losses.

Based on theoretical calculations, simulations, and experimental findings, Table 13.4 compares power loss and efficiency. The findings of the simulation produce 12.079 W of power loss and an efficiency of 92.55%, in contrast to the theoretical results' 11.296 W of power loss and 93% efficiency. Thus, it can be concluded that both theoretical and simulation results are congruent with the power loss computation and efficiency evaluation. The output phase voltages of the TILI are pure sinusoidal, with a tiny voltage distortion of 2.22% and an output current distortion of 1.71% as can be observed in Figure 13.16. The TILI power quality measures are well under the permissible bounds of IEEE 519.

13.7 Conclusion

This chapter discusses the power loss estimation, efficiency assessment, and harmonic study of an infinite-level inverter (TILI). With a very high DC-link voltage utilization factor of 1.23, the proposed TILI also exhibits excellent operating efficiency, lower switching power losses, and excellent output power quality. The inverter switches are controlled using fully rectified SPWM (FrSPWM). This inverter topology reduced power losses by 7% while achieving 93% efficiency at an operating frequency of 15 kHz. The power losses in the H-bridge stage represent only 0.034% of total power losses in the TILI compared to 6.97% in the Buck stage. The conduction losses in the H-bridge switches are responsible for 99.427% of the H-bridge inverter loss. The switching losses in the inverter switches only contribute 0.573%. The inverter switching loss component is very low because the inverters are switched at the supply frequency. It was found that the switching losses of the Buck stage change very little as the switching frequency of TILI is raised. In the Buck switch, the conduction losses and switching losses sum up to 1.059% and 2.374%, respectively. Only 0.416% of the overall Buck converter losses come from inductor losses. It also includes a comparison of power loss and efficiency on the basis of simulation, experiment, and theoretical estimation. The harmonic voltage and current distortions of TILI are extremely low, coming in at 2.22% and 1.7%, respectively, which is substantially below the IEEE 519-2014 standard's permissible threshold. The power loss and efficiency, as computed from theoretical calculations, are observed to be 11.296 W and 93%, respectively. The power loss and efficiency calculated from the simulation results are 12.079 W and 92.55%, and those computed from the experimental results are 12.258 W and 92.445%, respectively. The harmonic voltage and current distortions of the TILI were reduced to 2.22% and 1.71%, respectively, which are under the permitted ranges of the IEEE 519-2014 standard. This resulted in improved power quality.

The proposed inverter makes a great grid-tied inverter and is simple to integrate with renewable energy sources like solar cells, fuel cells, etc. The development of appropriate control algorithms for integrating TILI with renewable energy sources is part of the future scope. This will allow for closed-loop operation with increased output voltage and enhanced power quality.

References

1. Aghdam, M. G. H., Fathi, S. H., & Ghasemi, A. (2005). The analysis of conduction and switching losses in three-phase OHSW multilevel inverter using switching functions. *2005 International Conference on Power Electronics and Drives Systems, 1*, 209–218. https://doi.org/10.1109/peds.2005.1619687
2. Aishwarya, V., & Gnana Sheela, K. (2021a). Comparative Analysis of a Gradationally Controlled Voltage Inverter. In *Lecture Notes in Electrical Engineering* (Vol. 616). https://doi.org/10.1007/978-981-16-1978-6_6
3. Aishwarya, V., & Gnana Sheela, K. (2021b). Review of reduced-switch multilevel inverters for electric vehicle applications. *Int. J. Circuit Theory Appl.*, *49*(9). https://doi.org/10.1002/cta.3087
4. Aishwarya, V., & Gnana Sheela, K. (2022). Topologies and Comparative Analysis of Reduced Switch Multilevel Inverters for Renewable Energy Applications. In M. S. Bhaskar, N. Gupta, S. Padmanaban, J. B. Holm-Nielsen, & U. Subramaniam (Eds.), *Power Electronics for Green Energy Conversion* (pp. 221–264). John Wiley & Sons, Ltd. https://doi.org/https://doi.org/10.1002/9781119786511.ch7
5. Aishwarya, V., & Sheela, K. G. (2021a). FPGA-Based Reduced Switch Novel Infinite-Level Inverter. *J. Circuits Syst.*, https://doi.org/10.1142/S0218126621503023

6. Aishwarya, V., & Sheela, K. G. (2021b). FPGA-Based Reduced Switch Novel Infinite-Level Inverter. *J. Circuits Syst.*, https://doi.org/10.1142/S0218126621503023
7. Aishwarya, V., & Sheela, K. G. (2023). Energy efficiency evaluation of a three-phase extendable-level inverter. *Eng. Res. Express*, *5*(1), 015060. https://doi.org/10.1088/2631-8695/ACC157
8. Alamri, B., Alshahrani, S., & Darwish, M. (2015). Losses investigation in SPWM-controlled cascaded H-bridge multilevel inverters. *2015 50th International Universities Power Engineering Conference (UPEC)*, 1–5. https://doi.org/10.1109/UPEC.2015.7339805
9. Alamri, B., & Darwish, M. (2014, October 22). Precise modelling of switching and conduction losses in cascaded h-bridge multilevel inverters. *Proceedings of the Universities Power Engineering Conference.* https://doi.org/10.1109/UPEC.2014.6934613
10. Chaturvedi, P. K., Jain, S., & Agarwal, P. (2011). Reduced switching loss pulse width modulation technique for three-level diode clamped inverter. *IET Power Electron.*, *4*(4), 393–399. https://doi.org/10.1049/iet-pel.2010.0311
11. Chaturvedi, P. K., Jain, S., Agrawal, P., Nema, R. K., & Sao, K. K. (2008). Switching losses and harmonic investigations in multilevel inverters. *IETE J. Res.*, *54*(4), 297–307. https://doi.org/10.4103/0377-2063.44233
12. Chen, A., & He, X. (2006). Research on hybrid-clamped multilevel-inverter topologies. *IEEE Trans. Ind. Electron.*, *53*(6), 1898–1907. https://doi.org/10.1109/TIE.2006.885154
13. Committee, D., Power, I., & Society, E. (2014). IEEE Std 519-2014 (Revision of IEEE Std 519-1992). In *IEEE Std 519-2014 (Revision of IEEE Std 519-1992)* (Vol. 2014, pp. 1–29). https://doi.org/10.1109/IEEESTD.2014.6826459
14. De Caro, S., Foti, S., Scimone, T., Testa, A., Cacciato, M., Scarcella, G., & Scelba, G. (2016, September). THD and efficiency improvement in multi-level inverters through an open end winding configuration. *ECCE 2016 - IEEE Energy Conversion Congress and Exposition, Proceedings.* https://doi.org/10.1109/ECCE.2016.7855496
15. Floricau, D., Gateau, G., Leredde, A., & Teodorescu, R. (2009). The efficiency of three-level active NPC converter for different PWM strategies. *2009 13th European Conference on Power Electronics and Applications*, EPE '09.
16. Jagabar Sathik, M., Sandeep, N., Almakhles, D., & Blaabjerg, F. (2020). Improved "K" type seven-level switched capacitor inverter topology with Self-voltage balancing. *Int. J. Circuit Theory Appl.*, *48*(10), 1800–1819. https://doi.org/10.1002/cta.2811
17. Joy, M. C., Chaithanya, V., & Jayanand, B. (2017, February 13). Three-phase Infinite Level Inverter. *1st IEEE International Conference on Power Electronics, Intelligent Control and Energy Systems, ICPEICES 2016.* https://doi.org/10.1109/ICPEICES.2016.7853364
18. Kala, P., & Arora, S. (2017). A comprehensive study of classical and hybrid multilevel inverter topologies for renewable energy applications.

Renewable Sustainable Energy Rev., *76*, 905–931. https://doi.org/10.1016/j.rser.2017.02.008

19. Karasani, R. R., Borghate, V. B., Meshram, P. M., Suryawanshi, H. M., & Sabyasachi, S. (2017). A three-phase hybrid cascaded modular multilevel inverter for renewable energy environment. *IEEE Trans. Power Electron.*, *32*(2), 1070–1087. https://doi.org/10.1109/TPEL.2016.2542519
20. Kim, T. J., Kang, D. W., Lee, Y. H., & Hyun, D. S. (2001). The analysis of conduction and switching losses in multi-level inverter system. *2001 IEEE 32nd Annual Power Electronics Specialists Conference (IEEE Cat. No.01CH37230)*, *3*, 1363–1368. https://doi.org/10.1109/pesc.2001.954310
21. Lai, J. S. (2009). Power conditioning circuit topologies. *IEEE Ind. Electron. Mag.*, *3*(2), 24–34. https://doi.org/10.1109/MIE.2009.932580
22. Mecke, R. (2015, October 27). Energy efficiency of two-level and multilevel inverters - A drive system comparison. *2015 17th European Conference on Power Electronics and Applications, EPE-ECCE Europe 2015*. https://doi.org/10.1109/EPE.2015.7309068
23. Monmasson, E., & Cirstea, M. N. (2007). FPGA design methodology for industrial control systems - A review. *IEEE Trans. Ind. Electron.*, *54*(4), 1824–1842. https://doi.org/10.1109/TIE.2007.898281
24. Nguyen, H. V., & Lee, D. C. (2015). Comparison of power losses in single-phase to three-phase AC/DC/AC PWM converters. *9th International Conference on Power Electronics - ECCE Asia: "Green World with Power Electronics", ICPE 2015-ECCE Asia*, 940–945. https://doi.org/10.1109/ICPE.2015.7167894
25. Omer, P., Kumar, J., & Surjan, B. S. (2020). A Review on Reduced Switch Count Multilevel Inverter Topologies. *IEEE Access*, *8*, 22281–22302. https://doi.org/10.1109/ACCESS.2020.2969551
26. Prabaharan, N., & Palanisamy, K. (2017). Analysis of cascaded H-bridge multilevel inverter configuration with double level circuit. *IET Power Electron.*, *10*(9), 1023–1033. https://doi.org/10.1049/iet-pel.2016.0506
27. Rajalakshmi, S., & Rangarajan, D. P. (2019). Investigation of modified multilevel inverter topology for PV system. *Microprocess. Microsyst.*, *71*, 102870. https://doi.org/10.1016/j.micpro.2019.102870
28. Rodríguez, J., Lai, J. S., & Peng, F. Z. (2002). Multilevel inverters: A survey of topologies, controls, and applications. *IEEE Trans. Ind. Electron.*, *49*(4), 724–738. https://doi.org/10.1109/TIE.2002.801052
29. Salem, A., Van Khang, H., Robbersmyr, K. G., Norambuena, M., & Rodriguez, J. (2021). Voltage Source Multilevel Inverters with Reduced Device Count: Topological Review and Novel Comparative Factors. *IEEE Trans. Power Electron.*, *36*(3), 2720–2747. https://doi.org/10.1109/TPEL.2020.3011908
30. Sarwar, A., & Asghar, M. S. J. (2010). Multilevel converter topology for solar PV based grid-tie inverters. *2010 IEEE International Energy Conference and Exhibition, EnergyCon 2010*, 501–506. https://doi.org/10.1109/ENERGYCON.2010.5771733

31. Sharkh, S. M., Abusara, M. A., Orfanoudakis, G. I., & Hussain, B. (2014). Loss Comparison of Two- and Three-Level Inverter Topologies. In *Power Electronic Converters for Microgrids* (pp. 51–72). John Wiley & Sons, Singapore Pte. Ltd. https://doi.org/10.1002/9780470824054.ch4
32. Tarisciotti, L., Costabeber, A., Diaz, M., & Cardenas, R. (2020). Improved modular multilevel converter topology for low voltage variable speed drives. *Proceedings of the IEEE International Conference on Industrial Technology, 2020-February*, 837–842. https://doi.org/10.1109/ICIT45562.2020.9067144
33. Vahedi, H., & Trabelsi, M. (2019). *Single-DC-Source Multilevel Inverters* (pp. 11–18). Springer, Cham. https://doi.org/10.1007/978-3-030-15253-6_3
34. Viswanathan, V., & Jeevananthan, S. (2017). Hybrid converter topology for reducing torque ripple of BLDC motor. *IET Power Electron.*, *10*(12), 1572–1587. https://doi.org/10.1049/IET-PEL.2015.0905/CITE/REFWORKS
35. Wang, B., Localzo, G., Murr, G. El, Wang, J., Griffo, A., Gerada, C., & Cox, T. (2016). Overall assessments of dual inverter open winding drives. *2015 IEEE International Electric Machines & Drives Conference (IEMDC)*, 1029–1035. https://doi.org/10.1109/IEMDC.2015.7409188
36. Zhou, Z., Khanniche, M. S., Igic, P., Jankovic, S. N., Batcup, S. G., & Mawby, P. A. (2006). Dynamic thermal simulation of power devices operating with PWM signals. *2006 25th International Conference on Microelectronics, MIEL 2006 - Proceedings*, 183–186. https://doi.org/10.1109/ICMEL.2006.1650925

14

Modeling and Analysis of Direct Torque Control Space-Vector Modulation of DFIG

Vishal Rathore[1]* and Dhananjay Kumar[2]

[1]*Department of Electrical and Electronics Engineering, Bhopal Institute of Technology and Science, Bhopal, India*

[2]*Department of Electrical & Electronics Engineering, Government Engineering College, Bihar, India*

Abstract

This chapter presents modelling, analysis and control of a doubly fed induction generator (DFIG) using direct torque control space vector pulse with modulation (DTC-SVPWM). A DFIG is primarily utilized in systems for converting wind energy, It employs only a partially rated converter and offers a direct grid connection. However, the traditional control techniques used to manage the DFIG are either cumbersome to use or ineffective. Some provide distorted results, while others call for more intricate adjustments of proportional-integral controllers. This chapter mainly focuses on the modelling and evaluation of DTC-SVPWM of the DFIG. Despite the fact that DFIG can be controlled through different approaches, due to its strong suitability and simplicity, DTC-SVPWM is thoroughly analyzed in this chapter in terms of quality of stator current, dynamic response and the steady-state response.

Keywords: DTC, SVPWM, DFIG, steady-state response

14.1 Introduction

There is presently a world energy crisis. Over the world, fossil fuels have long been the main source of energy used to produce power. Due to the early surplus, this worked nicely. Unfortunately, this fuel source is under

**Corresponding author*: vishalrathore01@gmail.com

Salman Ahmad, Farhad Ilahi Bakhsh, and P. Sanjeevikumar (eds.) Multilevel Converters, (281–296)

pressure as a result of population growth, urbanization, and the formation of numerous industries. The depletion of our ozone layer has also been dramatically accelerated by the hazardous emissions created by burning fossil fuels. This has led to a worsening of the detrimental effects of global warming on Earth. Over time, there has been an increase in demand for reliable, safe energy sources, and it is predicted that, in a few years, only these sources will be used to provide electrical energy.

Additionally, the production of electrical energy *via* coal releases toxic byproducts that destroy the ozone layer, which is an extremely valuable resource. These explanations explain why there has been a global trend towards producing electrical energy from renewable resources [1–3]. The sun is one such source that can be exploited to generate tiny amounts of electricity [4]. Although using solar panels requires little upkeep, it is expensive because batteries are needed. Moreover, just a few hundred watts of electricity are generated by each solar panel [5]. As a result, it is implied that a solar farm will have a substantial physical footprint but limited electrical output.

Tidal energy is another sort of renewable energy. A generator can be made to rotate using the ocean's natural current [4, 6], yet there are considerable drawbacks to this as well. Due to location limitations, harmful effects on marine life, and prohibitively expensive plant development, tidal energy suffers tremendously [6]. The process of obtaining heat from beneath the earth's surface is known as geothermal energy [4]. Geothermal energy is pricey, is location-specific, and contributes to land instability. Biomass is an additional renewable energy source. Biomass energy is defined as energy produced by burning plant material [4]. Although it is far more expensive and could cause deforestation, this alternative to coal is nonetheless quite effective.

Wind energy is the last sustainable energy source [4]. Wind energy is captured *via* wind energy conversion systems using wind turbines. The generator's shaft is turned by the wind that was caught. Wind energy conversion systems (WECS) utilize a tiny surface to generate significant power. However, their initial cost is substantial [7]. Moreover, the intermittent nature of wind energy may be problematic. High-powered electronic switches are used by WECS to control the system's output [8]. WECS employ both synchronous and asynchronous generators [9]. WECS-derived electrical energy is quickly gaining popularity and seems to be a good replacement for coal-based power generation [10], yet the varying characteristics of the wind must be taken into consideration because the electrical energy generated must be consistent. The effective and efficient production of electrical power from WECS is essential for easing the present global energy crisis.

14.2 Modeling of DFIG

14.2.1 DFIG Analysis in Steady State

Figure 14.1 displays the DFIG's simplified steady-state equivalent model [11, 12].

The steady-state currents are expressed as [11, 12]:

$$\begin{aligned}\bar{I}_S = &\frac{1}{D}\left[D_A\left(cX_m\sin(\alpha)+\bar{V}_S\frac{R_r}{s}\right)+D_B\left\{\bar{V}_S(X_r+X_m)-X_m\frac{\bar{V}_r}{s}\cos(\alpha)\right\}\right]\\ &+j\frac{1}{D}\left[D_A\left\{\bar{V}_S(X_r+X_m)-X_m\frac{\bar{V}_r}{s}\sin(\alpha)\right\}-D_B\left(X_m\frac{\bar{V}_r}{s}\sin(\alpha)+\bar{V}_S\frac{R_r}{s}\right)\right]\end{aligned} \tag{14.1}$$

$$\begin{aligned}\bar{I}_r = &\frac{1}{D}\left[D_A\frac{\bar{V}_r}{s}\left(R_r\cos(\alpha)-(X_r+X_m)\sin(\alpha)\right)+D_B\left\{\frac{\bar{V}_r}{s}\left(\cos(\alpha)(X_r+X_m)+R_s\sin(\alpha)\right)-\bar{V}_SX_m\right\}\right]\\ &+j\frac{1}{D}\left[D_A\left(\frac{\bar{V}_r}{s}\left((X_r+X_m)+R_s\sin(\alpha)-\bar{V}_SX_m\right)\right)-\left(D_B\frac{V_r}{s}\left\{R_s\cos(\alpha)-(X_r+X_m)\sin(\alpha)\right\}\right)\right]\end{aligned} \tag{14.2}$$

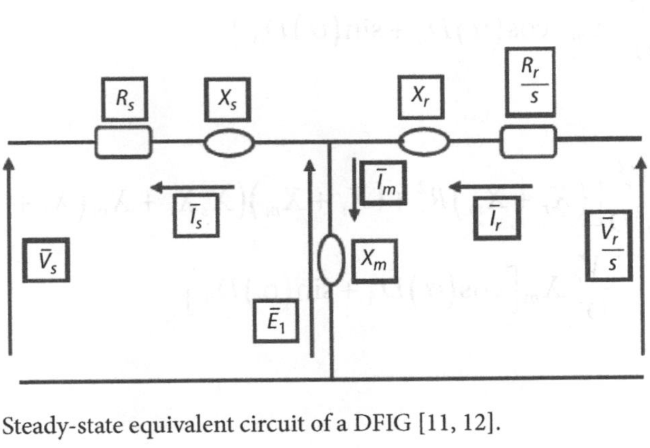

Figure 14.1 Steady-state equivalent circuit of a DFIG [11, 12].

The following are the stator's reactive and active powers during a steady-state operation [13, 14]:

$$P_s = \text{Re}(\overline{V}_s \overline{I}_s^*) \tag{14.3}$$

$$P_s = \frac{V_s^2}{D}\left[R_s \frac{R_r^2}{s^2} + R_s\left(X_r + X_m\right)^2 + X_m^2 \frac{R_r}{s}\right] - \frac{V_s V_r}{sD} X_m\left[\cos(\alpha) D_B - \sin(\alpha) D_A\right] \tag{14.4}$$

$$Q_s = \text{Im}(\overline{V}_s \overline{I}_s^*) \tag{14.5}$$

$$\begin{aligned} Q_s = {} & \frac{V_s V_r}{sD} X_m\left[\cos(\alpha) D_A - \sin(\alpha) D_B\right] \\ & + \frac{V_s^2}{D}\left[\left(X_m + X_r\right)\left(X_m X_s + \left(X_s + X_m\right) X_m\right)\right] + \frac{R_r^2}{s^2}\left(X_m + X_s\right) \end{aligned} \tag{14.6}$$

In a steady-state operation, the rotor's active and reactive powers are given as [11, 12]:

$$\begin{aligned} P_r = {} & \frac{V_r^2}{s^2 D}\left[R_s \frac{R_s R_r}{s} + \frac{R_r}{s}\left(X_m + X_r\right) X_s + X_m\left(\frac{R_r}{s} X_s + \left(R_s + \frac{R_r}{s}\right) X_m\right)\right] \\ & + \frac{V_s V_r}{sD} X_m\left[\cos(\alpha) D_B + \sin(\alpha) D_A\right] \end{aligned} \tag{14.7}$$

$$\begin{aligned} Q_r = {} & \frac{V_r^2}{s^2 D}\left[\left(X_r + X_m\right) R_s^2 + \left(X_s + X_m\right)\left(X_s X_r + X_m\left(X_s + X_r\right)\right)\right] \\ & + \frac{V_s V_r}{sD} X_m\left[\cos(\alpha) D_A + \sin(\alpha) D_B\right] \end{aligned} \tag{14.8}$$

Finally, [11, 12] are used to calculate the DFIG's steady-state electromagnetic torque.

$$T_{em} = -\frac{3V_sV_r}{w_s sD} X_m \left[\cos(\alpha)\left(\frac{R_r}{s}(X_m + X_s) - (X_m + X_r)R_s \right) + \sin(\alpha)\left(\frac{R_sR_r}{s} + (X_r + X_s)X_m + X_rX_s \right) \right] \tag{14.9}$$

14.2.2 DFIG α–β Reference Frame Dynamic Modeling

The following are the DFIG's voltage equations in the α–β reference frame [15]:

$$V_{\alpha s} = R_s I_{\alpha s} + \frac{d\phi_{\alpha s}}{dt} \tag{14.10}$$

$$V_{\beta s} = R_s I_{\beta s} + \frac{d\phi_{\beta s}}{dt} \tag{14.11}$$

$$V_{\alpha r} = R_s I_{\alpha r} + \frac{d\phi_{\alpha r}}{dt} + w_m \phi_{\beta r} \tag{14.12}$$

$$V_{\beta r} = R_s I_{\beta r} + \frac{d\phi_{\beta r}}{dt} + w_m \phi_{\alpha r} \tag{14.13}$$

The equations of flux is given as [15, 6]:

$$\phi_{\alpha s} = L_m I_{\alpha r} + L_s I_{\alpha s} \tag{14.14}$$

$$\phi_{\beta s} = L_m I_{\beta r} + L_s I_{\beta s} \tag{14.15}$$

$$\phi_{\alpha r} = L_m I_{\alpha s} + L_r I_{\alpha r} \tag{14.16}$$

$$\phi_{\beta r} = L_m I_{\beta s} + L_r I_{\beta r} \tag{14.17}$$

The electromagnetic torque is given as [15]:

$$T_{em} = \frac{3}{2}\frac{pL_m}{L_s}\left(\phi_{\alpha s} I_{\beta r} - \phi_{\beta s} I_{\alpha r}\right) \tag{14.18}$$

The reactive and active powers of the stator and rotor are expressed as [15]:

$$P_s = \frac{3}{2}\left(V_{\alpha s} I_{\alpha s} + V_{\beta s} I_{\beta s}\right) \tag{14.19}$$

$$Q_s = \frac{3}{2}\left(-V_{\alpha s} I_{\beta s} + V_{\beta s} I_{\alpha s}\right) \tag{14.20}$$

$$P_r = \frac{3}{2}\left(V_{\beta r} I_{\beta r} + V_{\alpha r} I_{\alpha r}\right) \tag{14.21}$$

$$Q_r = \frac{3}{2}\left(V_{\beta r} I_{\alpha r} - V_{\alpha r} I_{\beta r}\right) \tag{14.22}$$

14.2.3 DFIG q–d Reference Frame Dynamic Modeling

The q–d reference frame DFIG voltage equations are expressed as [17–20]:

$$V_{ds} = \mathrm{R}_s\, I_{ds} + \frac{d\phi_{ds}}{dt} - w_s \phi_{qs} \tag{14.23}$$

$$V_{qs} = \mathrm{R}_s\, I_{qs} + \frac{d\phi_{qs}}{dt} + w_s \phi_{ds} \tag{14.24}$$

$$V_{dr} = \mathrm{R}_r\, I_{dr} + \frac{d\phi_{dr}}{dt} - w_r \phi_{qr} \tag{14.25}$$

$$V_{qr} = \mathrm{R}_r\, I_{qr} + \frac{d\phi_{qr}}{dt} + w_r \phi_{dr} \tag{14.26}$$

The flux equations are as follows [21, 22]:

$$\phi_{ds} = L_s I_{ds} + L_m I_{dr} \tag{14.27}$$

$$\phi_{qs} = L_s I_{qs} + L_m I_{qr} \tag{14.28}$$

$$\phi_{dr} = L_s I_{dr} + L_m I_{ds} \tag{14.29}$$

$$\phi_{qr} = L_s I_{qr} + L_m I_{qs} \tag{14.30}$$

The electromagnetic torque is [15]:

$$T_{em} = \frac{3}{2}\frac{pL_m}{L_s}\left(\phi_{ds} I_{qr} - \phi_{qs} I_{dr}\right) \tag{14.31}$$

The reactive and active powers of the stator and rotor are expressed as [23]:

$$P_s = \frac{3}{2}\left(V_{qs} I_{qs} + V_{ds} I_{ds}\right) \tag{14.32}$$

$$Q_s = \frac{3}{2}\left(I_{ds} V_{qs} - I_{qs} V_{ds}\right) \tag{14.33}$$

$$P_r = -sP_s \tag{14.34}$$

$$Q_r = sQ_s \tag{14.35}$$

Figures 14.2 and 14.3 show the DFIG circuit in both α–β and q–d reference frames. In Figure 14.2, the stator DC voltage, which includes the stator DC loss and the change in stator DC flux, is reduced by multiplying the synchronous revolutions and the quadrature stator flux. The rotor DC voltage is formed by combining the change in DC flux and DC voltage drop of the rotor, and the product of the quadrature rotor flux and electrical slip velocity is then subtracted from it.

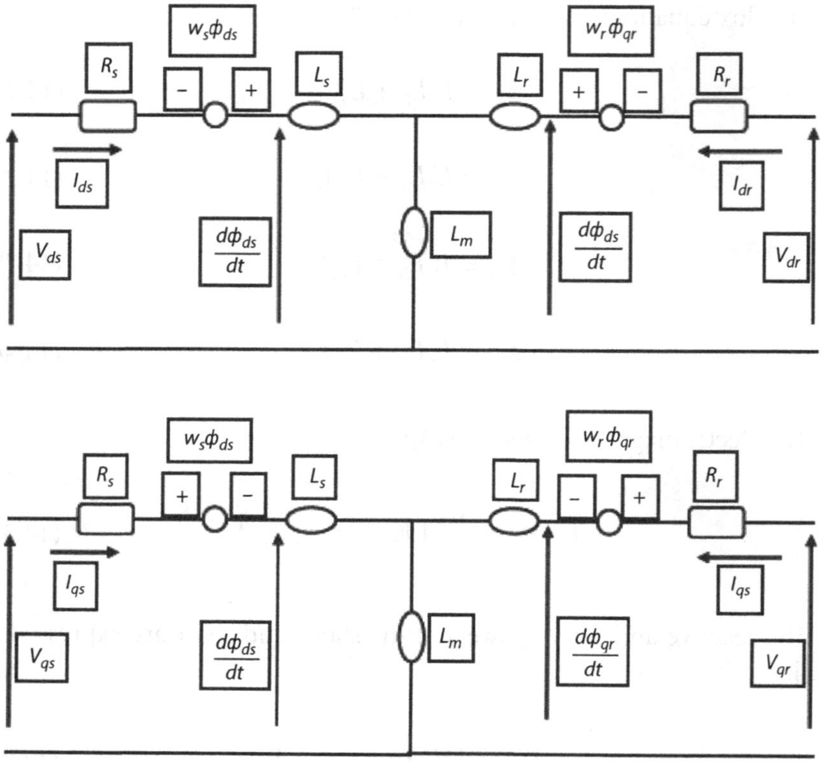

Figure 14.2 q–d reference frame equivalent circuit.

The multiplication of the stator direct flux and synchronous speed, along with the sum of any changes in the stator quadrature flux and the quadrature stator voltage drop, makes up the stator quadrature voltage. Equations can demonstrate this. The rotor quadrature voltage is the result of the slip electrical speed and the multiplication of the direct flux and the quadrature voltage drop added together of the rotor. Equations can demonstrate this. The magnetizing inductance and the appropriate leakage inductances were taken into account with regard to the change in flux.

According to Figure 14.3, the stator voltage is made up of the stator voltage drop and the stator flux change. The rotor's voltage, on the other hand, is made up of the rotor's voltage drop, the rotor's flux fluctuation, and the sum of the rotor's electrical speed and flux. Additionally, the stator's voltage is determined by adding the stator's voltage drop and stator's flux variation.

The change in rotor flux and the decrease in rotor voltage must be added to determine the rotor voltage. After that, multiply the outcome by the total

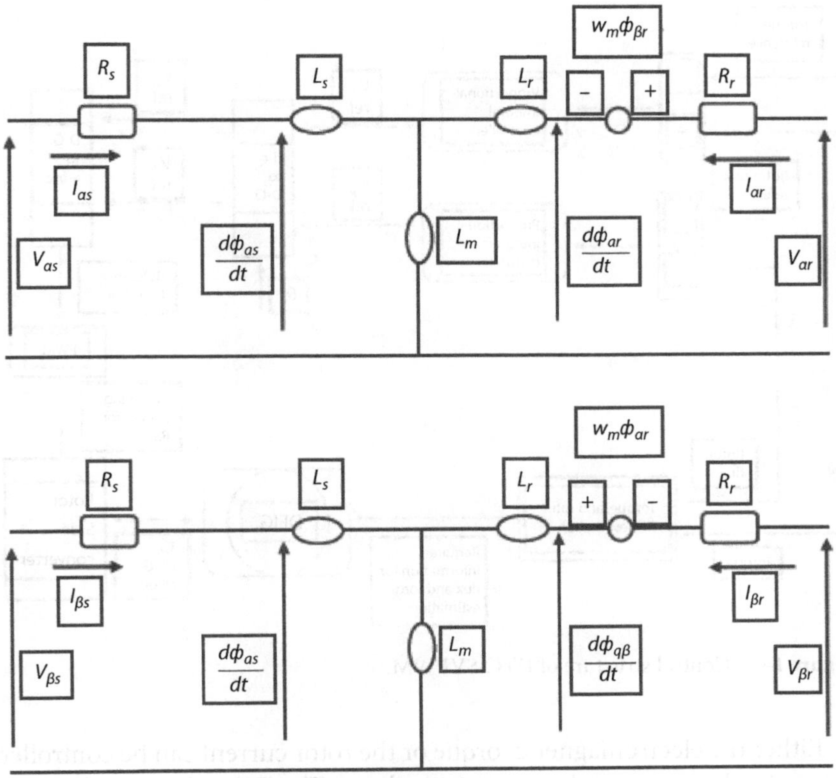

Figure 14.3 α–β reference frame equivalent circuit [15].

of the electrical speed and rotor flux. Regarding the flux change, the necessary leakage inductances and the magnetizing inductance were considered (both stator and rotor).

14.3 DTC Using SVPWM

Direct control is used to manage the electromagnetic torque and rotor flux. Two PI controllers are used in the control approach: one controls the electromagnetic torque and the other the rotor flux. To deliver the necessary voltage to the rotor of the DFIG, the proportional–integral controllers produce a rotor reference voltage. In order to create switching signals to power the converter, the slip angle is used to transformed the voltage and then sent *via* modulator. This process ensures that grid synchronization is maintained [21].

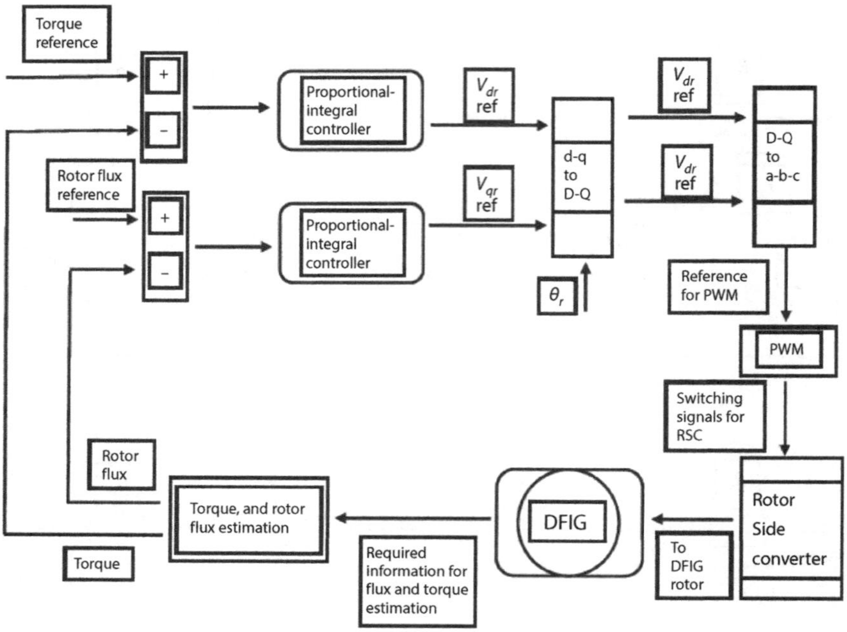

Figure 14.4 Control structure of DTC-SVPWM.

Either the electromagnetic torque or the rotor current can be controlled using the direct or quadrature rotor voltage. The DC voltage of the rotor regulates the flux, and the electromagnetic torque is regulated by the quadratic rotor voltage when the stator angle is used to calculate the slip angle. The inverse is true if the slip angle is extracted using the stator voltage angle. Before estimating the differences between the two electromagnetic torque values, they are negated. Figure 14.4 depicts the DTC-control SVPWM's arrangement.

14.4 Results and Analysis

The DTC-SVPWM control system directly manages the rotor flux and electromagnetic torque. Nevertheless, it uses proportional–integral controllers and a modulator rather than look-up table and hysteresis controllers. The approach is similar to FOC, but it directly manages the electromagnetic torque and rotor flux. It is vital to know the stator and rotor resistance values. However, using proportional–integral controllers increases complexity and makes the generator's control strategy distinct from other systems.

Additionally, proper rotor angle knowledge is necessary due to the slip angle requirement.

The control method was assessed by implementing it to a 3-MW DFIG and simulated it by employing MATLAB/Simulink. The machine was only tested for RSC control. Hence, it only operated at sub-synchronous speeds. To evaluate the control system's performance at varied rotor speeds, two sub-synchronous rates have been chosen. At 169 rad/s, one was closer to the synchronous speed, while another moved more quickly at 185 rad/s. To ensure that the control system had enough time to reach a steady state, we simulated the speed of each rotor for 6 s. The rotor speed of 169 rad/s was employed for the first 6 s to gauge the effectiveness of the control strategy. After increasing the speed from 169 to 185 rad/s, the performance of the control method was assessed for the next 6 s at the increased speed.

The frequency of switching was set to 100 kHz, and the time interval for sampling was 10 s. The stator voltage could be oriented towards the direct axis to determine the slip angle. The quadrature voltage of the rotor controls the rotor flux, while the direct voltage of the rotor manages the electromagnetic torque. The rotor flux and the electromagnetic torque would have been controlled by the rotor direct voltage and quadrature rotor voltage, respectively, if the stator flux had been aligned with the direct axis.

For the purpose of extracting the most active power possible, the optimal torque approach was applied, and the calculation of the reference value employed an optimal coefficient of $k_{opt} = 0.296$. It denotes that the rotor's speed was used to determine the reference electromagnetic torque. To provide the optimal reaction, the PI-controllers were fine-tuned. The

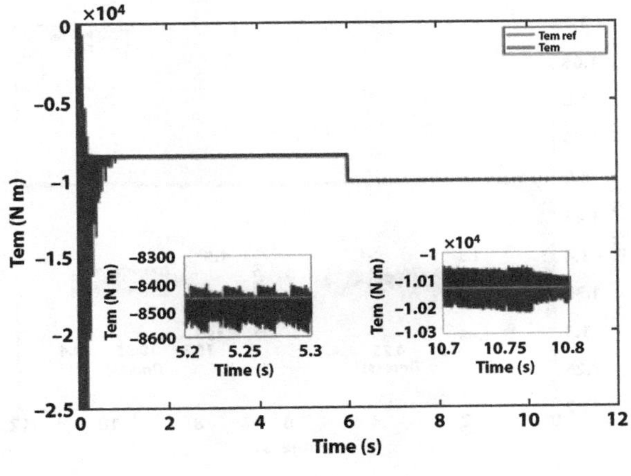

Figure 14.5 Electromagnetic torque waveform.

reference stator's reactive power was set to 0 Var. In Figures 14.5 to 14.10, the DTC-performance SVPWM's is depicted.

In Figure 14.5, the electromagnetic torque ripple is displayed to be 184 N m and 172 N m at minimum and maximum speeds, respectively. Similarly, in Figure 14.6, the ripple in stator active power follows the expected pattern with values of 34.68 and 32.42 kW at minimum and maximum speeds, respectively. Moreover, 1.13 ms pass before the stator active power reaches the subsequent reference value.

When factoring in the flux ripple, the rotor flux shown in Figure 14.7 results in 8.5 mWb at the minimum speed and 8.1 mWb at the maximum

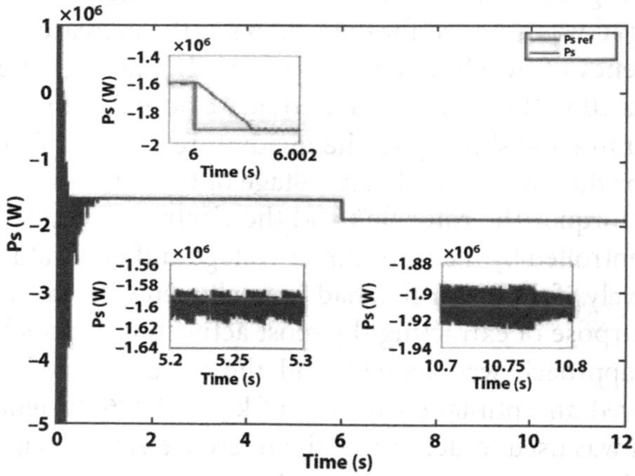

Figure 14.6 Active power of the stator.

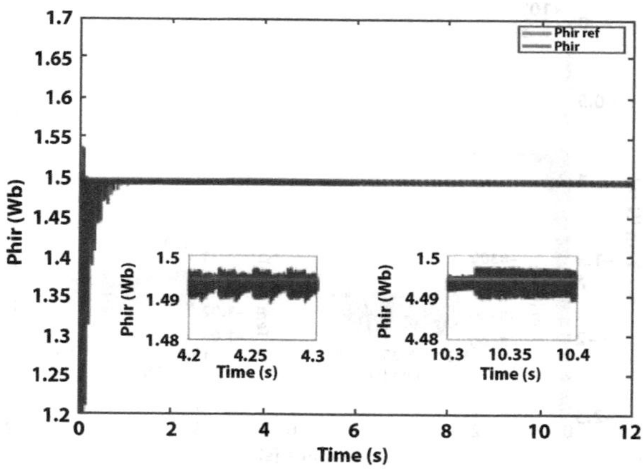

Figure 14.7 Rotor flux waveform.

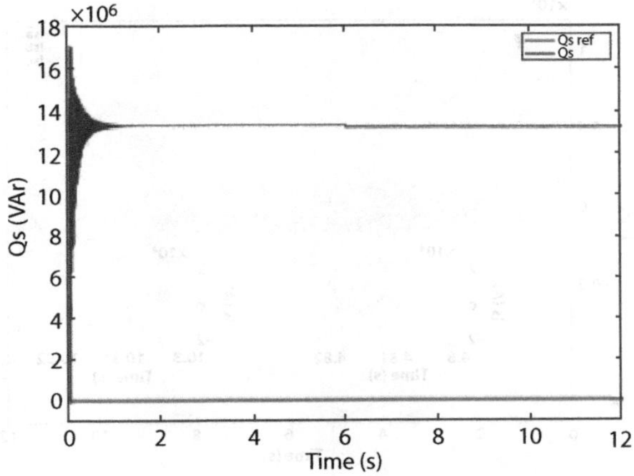

Figure 14.8 Reactive power of the stator.

speed. The reactive power of the stator, as displayed in Figure 14.8, deviates significantly from the target value. This is due again to the absence of direct stator reactive power control.

The three-phase rotor current in Figure 14.9 travels along the anticipated path. Additionally, there is little, if any, distortion and the waveforms are practically perfectly balanced. Both speeds result in three-phase stator current as depicted in Figure 14.10, which is of an acceptable value. A tiny distortion and waveform unbalance are present at both speeds.

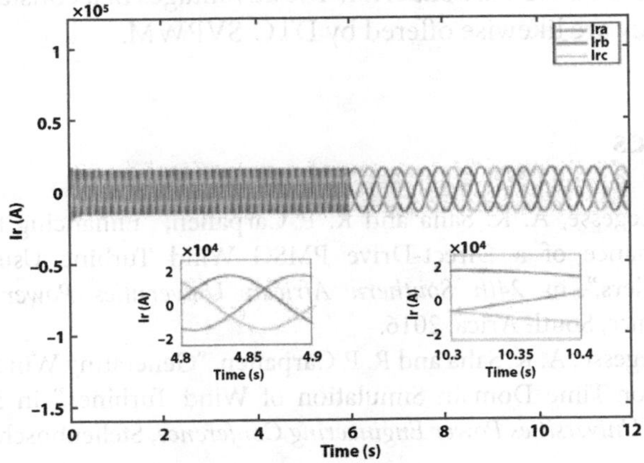

Figure 14.9 Three-phase rotor current.

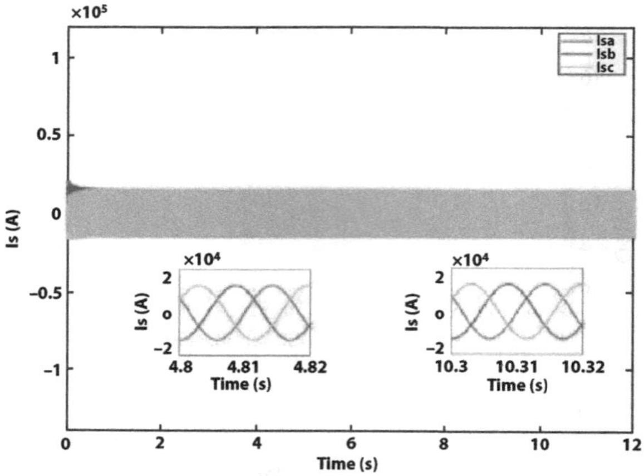

Figure 14.10 Three-phase stator current.

14.5 Conclusion

The study presents an analysis of a traditional control technique used on the DFIG. At both speed values, the proposed approach yields the lowest ripples, according to the data. At both speeds, results in best response were in terms of stator current distortion. Due to control technique requirements, the rotor and stator resistances had to be understood for DTC-SVPWM. Reactive control is not offered by DTC-SVPWM. The proposed technique resulted in a lesser distortion when the stator current was taken into account. At both speeds, the DTC-SVPWM method provides a smoother and more stable ripple. Additionally, the stator current waveforms generated by DTC-SVPWM are superior. The advantages of a constantly switching frequency are likewise offered by DTC-SVPWM.

References

1. A. N. Legesse, A. K. Saha and R. P. Carpanen, "Enhancing the Dynamic Performance of a Direct-Drive PMSG Wind Turbine Using Damping Controllers," in *24th Southern African Universities Power Engineering Conference*, South Arica, 2016.
2. A. N. Legesse, A. K. Saha and R. P. Carpanen, "Generating Wind Speed Time Series for Time Domain Simulation of Wind Turbines," in *5th Southern African Universities Power Engineering Conference*, Stellenbosch, 2017.

3. A. N. Legesse, A. K. Saha and R. P. Carpanen, "Characterisation of wind speed series and power in Durban," *J. Energy South. Afr.*, vol. 28, no. 3, pp. 66-78, 2017
4. Anwar KM. Environmental & Financial Benefits of 360 kW Photo Voltaic Solar System (On-Grid) in University of Wah. *J. Sustain. Perspect.* 2022 Aug 1;2.
5. Zientara B. How much electricity does a solar panel produce?. *Solar Power Rocks.* 2019 Feb;11.
6. Shetty C, Priyam A. A review on tidal energy technologies. *Mater. Today Proc.* 2022 Jan 1;56:2774-9.
7. Valentine SV. Understanding the variability of wind power costs. *Renew. Sustain. Energy Rev.* 2011 Oct 1;15(8):3632-9.
8. Kumar Y, Ringenberg J, Depuru SS, Devabhaktuni VK, Lee JW, Nikolaidis E, Andersen B, Afjeh A. Wind energy: Trends and enabling technologies. *Renew. Sustain. Energy Rev.* 2016 Jan 1;53:209-24.
9. Legesse AN, Saha AK, Carpanen RP. Damping local oscillations of a direct-drive PMSG wind turbine. *Int. J. Eng. Technol.* 2017;9(1):158-68.
10. Saidur R, Islam MR, Rahim NA, Solangi KH. A review on global wind energy policy. *Renew. Sustain. Energy Rev.* 2010 Sep 1;14(7):1744-62.
11. Vicatos MS, Tegopoulos JA. Steady state analysis of a doubly-fed induction generator under synchronous operation. *IEEE Trans. Energy Convers..* 1989 Sep;4(3):495-501.
12. Islam MR, Guo Y, Zhu JG. Steady state characteristic simulation of DFIG for wind power system. In *International Conference on Electrical & Computer Engineering (ICECE 2010)* 2010 Dec 18 (pp. 151-154). IEEE.
13. Shukla RD, Singh A, Singh SP. Generators for variable speed wind energy conversion systems: A comparative study. *IJESCC.* 2012 Jul;3(2):103-17.
14. Duong MQ, Sava GN. Coordinated reactive power control of DFIG to improve LVRT characteristics of FSIG in wind turbine generation. In *2017 International Conference on Electromechanical and Power Systems (SIELMEN)* 2017 Oct 11 (pp. 256-260). IEEE.
15. Abad G, Lopez J, Rodriguez M, Marroyo L, Iwanski G. *Doubly fed induction machine: modeling and control for wind energy generation.* John Wiley & Sons; 2011 Sep 28.
16. Wiam A, Ali H. Direct torque control-based power factor control of a DFIG. *Energy Procedia.* 2019 Apr 1;162:296-305.
17. Bedoud K, Ali-Rachedi M, Bahi T, Lakel R, Grid A. Robust control of doubly fed induction generator for wind turbine under sub-synchronous operation mode. *Energy Procedia.* 2015 Aug 1;74:886-99.
18. Byeon GS, Park IK, Jang GS. Modeling and control of a doubly-fed induction generator (DFIG) wind power generation system for real-time simulations. *J. Electr. Eng. Technol.* 2010;5(1):61-9.
19. Chafouk H, Ouyessaad H. Fault detection and isolation in DFIG driven by a wind turbine. *IFAC-PapersOnLine.* 2015 Jan 1;48(21):251-6.

20. Reddy K, Saha AK. A comparative study on direct torque control algorithms applied to a doubly fed induction generator. In *2021 Southern African Universities Power Engineering Conference/Robotics and Mechatronics/ Pattern Recognition Association of South Africa (SAUPEC/RobMech/PRASA)* 2021 Jan 27 (pp. 1-6). IEEE.
21. Vishal Rathore, Krishna B. Yadav "Mathematical Modelling and Numerical Analysis of SPIM Drive Using Modified SVPWM Technique" *ECTI Transactions on Electrical Engineering*, vol.20, no.2, pp. 152-162.
22. Kumeshan R, Saha AK. A review of swarm-based metaheuristic optimization techniques and their application to doubly fed induction generator. *Heliyon*. 2022 Oct 7:e10956.
23. Vishal Rathore and K.B.Yadav (2021) "Direct Torque Control of Asymmetrical Multiphase (6-Phase) Induction Motor Using Modified Space Vector Modulation". In: Kumar, J., Jena, P. (eds) Recent Advances in Power Electronics and Drives. *Lecture Notes in Electrical Engineering*, vol 707. Springer, Singapore. https://doi.org/10.1007/978-981-15-8586-9_45.

15

Observer-Based Sliding Mode Control of Static Var Compensator: A Voltage Control Application in a Hybrid Power System

Zahid Afzal Thoker*,† and Shameem Ahmad Lone†

Department of Electrical Engineering, National Institute of Technology, Srinagar, India

Abstract

The complementary operation of wind power and diesel generators offers a reliable source of electrical power to the off-grid consumers. Due to abrupt load changes and fluctuating nature of wind power, voltage instability is an issue that occurs due to a mismatch in the reactive power of the system. Therefore, a novel robust sliding mode control strategy is proposed for the static var compensator (SVC) for reactive power support to reduce the mismatch in the system reactive power. To improve the performance of a sliding mode controller (SMC) and eliminate the chattering problem, a sliding mode observer (SMO) is constructed to estimate the SVC parameters effectively. Based on the estimated parameters and the design of a suitable sliding surface, the sliding mode controller generates a proper control law with which the operation and power output of SVC is controlled through converter control. Lyapunov stability analysis is conducted to guarantee the stability of the controller. Simulink/MATLAB is used to validate the efficacy of the proposed controller through simulation studies.

***Keywords*:** Wind–diesel power system, static var compensator, sliding mode controller, sliding mode observer

**Corresponding author*: thokerzahid_10phd1@nitsri.net
†Equally contributing authors

Salman Ahmad, Farhad Ilahi Bakhsh, and P. Sanjeevikumar (eds.) Multilevel Converters, (297–316)

15.1 Introduction

Globally, humankind is facing the challenge of climate change due to the emission of greenhouse gases. In this aspect, electrical power generation using conventional sources is having a significant contribution. Therefore, the need of electrical power generation out of renewable energy sources is urgent. Wind energy offers a promising source of renewable energy. However, the wind power output is fluctuating in nature; therefore, with diesel power supply acting as a back supply, wind–diesel power system in the complementary form can be used to meet electrical energy demands [1–3]. Voltage instability is an issue with this system due to random load, fluctuating wind power, and other disturbances like parameter uncertainty and system disturbance. Considering the importance of voltage stability, it becomes a critical problem to deal with reactive power compensation of isolated wind–diesel hybrid power system [4, 5].

To deal with reactive power management problems, some FACTS devices can be used in these systems. The incorporation of static var compensator (SVC) can compensate the reactive power dynamics of the system to a large extent to keep voltage deviations to a minimum [7, 8]. In previous works, SVC with additional control strategies to enhance the system voltage stability and sub-synchronous oscillation damping is designed in [9]. In [10], SVC with conventional PI control technique is developed for an isolated wind–diesel system to deal with reactive power management. In [11], with mine blast algorithm, optimized PID controller-based FACTS is used for simultaneous frequency and voltage control. However, in case disturbances occur and operating point changes, it lacks robustness. With SVC as VAR compensating device, a model predictive control for voltage stability in an autonomous wind diesel power system is reported in [13, 12], but its calculation burden demands precise mathematical modeling of SVC. Moreover, the work reported does not give validation in the events of wind speed changes. Intelligent control techniques like genetic algorithms and artificial neural networks are employed to enhance the voltage stability of the wind–diesel power system through proper optimization of PI-controller parameters of SVC [14, 15], but these algorithms are too complicated to be practically applicable. In addition to that, wind power disturbance study has not been covered.

Sliding mode controller is a well-established controller having its wide range of applications in power systems. It possesses the feature of robustness to parameter variations and external disturbances [16, 17]. In [18], a variable structure controller is designed for SVC with application in transmission line model. An adaptive neural network-based sliding mode controller is designed for SVC in [19], but wind energy is not considered. In [20], load frequency control in a multi-area power system is reported using prediction-based SMC. In [21], voltage stability improvement using a sliding mode controller in the transmission system is implemented in the two-area power system, but wind power generation is not considered. In sliding mode controller, chattering issue occurs on account of external disturbances and parameter variations [16]. With sliding mode observer (SMO), sliding mode controller offers a robust and simple control technique to design a reactive power compensation controller, as it can estimate the state variables of an uncertain model effectively [22, 23].

Considering the abovementioned related literature, the contribution of the paper is summarized as follows:

(i) A wind–diesel hybrid system is established with interconnection of static var compensator to carry out the reactive power control.
(ii) With the help of a hyper-surface design, model-based sliding mode control strategy is developed over SVC to bring the reactive power balance and hence the voltage control.
(iii) To avoid the chattering phenomenon in the SMC, a sliding mode observer is designed to estimate the uncertainty in the SVC parameters and aid the SMC to generate the effective control signal.
(iv) The proposed study is validated using MATLAB/Simulations in the events of sudden load changes and wind power disturbances.

The rest of the chapter is organized as follows: mathematical modeling of the system components is given in Section 15.2. In Section 15.3, the sliding mode control strategy over SVC along with stability analysis is discussed. Simulation results are presented in Section 15.4. Finally, the conclusions about the chapter are given in Section 15.5.

15.2 Mathematical Modeling of the System

To carry out the mathematical modeling, a block diagram of an isolated wind–diesel power system shown in Figure 15.1 is considered, where we have the synchronous generator (SG) equipped with IEEE-type1 excitation system, wind energy conversion system connected with an induction generator (IG), and load and SVC all connected to a common bus-bar. SVC is supposed to supply the additional reactive power demanded by the system apart from the reactive power supplied by SG. Any mismatch in reactive power supply and demand is reflected in the voltage deviation of the bus-bar. Under a steady-state condition, the reactive power balance equation is as follows:

$$\Delta Q_{SG} + \Delta Q_{SVC} = \Delta Q_{IG} + \Delta Q_{L} \tag{15.1}$$

where ΔQ_{SG}, ΔQ_{IG} are the deviations in reactive power of synchronous generator and wind energy induction generator systems, respectively, ΔQ_L is deviation in reactive power demand of load, and ΔQ_{SVC} is deviation in reactive power supplied by static var compensator. The system voltage deviates as a result of reactive power load disturbance, and it may cause severity in system safety and performance. The voltage deviation as a function of reactive power mismatch is given as

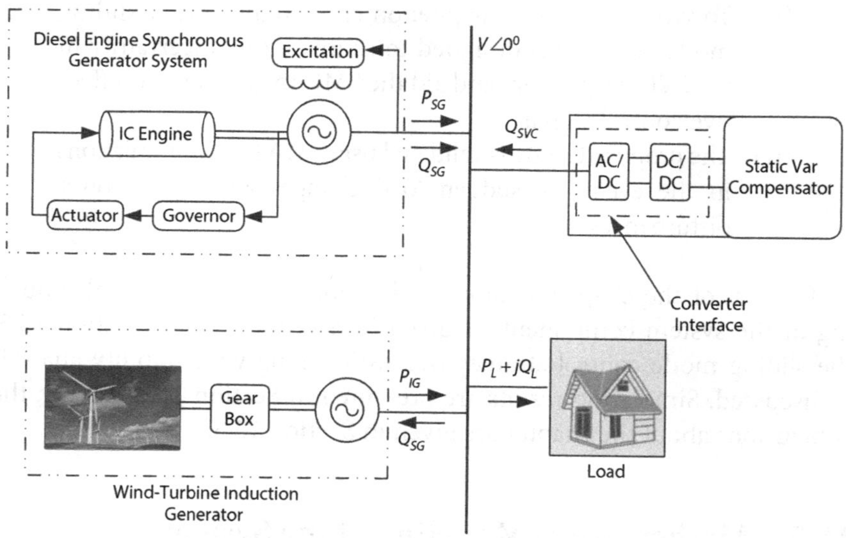

Figure 15.1 Block diagram representation of a wind–diesel hybrid system.

$$\Delta V(s) = \frac{K_{vg}}{1 + sT_{vg}} [\Delta Q_{SG} + \Delta Q_{SVC} - \Delta Q_L - \Delta Q_{IG}] \tag{15.2}$$

where K_{vg}, T_{vg} is the gain and time constant for the reactive power flow, respectively.

15.2.1 Modeling of Synchronous Generator and Excitation System

A synchronous generator acts as a primary source of reactive power to the system. The reactive power deviation in SG [24] is given as:

$$\Delta Q_{SG}(s) = K_1 \Delta E'_q(s) + K_2 \Delta V(s) \tag{15.3}$$

$$K_1 = \frac{V cos\delta}{X'_d}, K_2 = \frac{E'_q cos\delta - 2V}{X'_d} \tag{15.4}$$

where X'_d and X_d are the direct axis transient and steady-state reactances, respectively. E'_q is the transient voltage at armature terminals, and δ is the power angle. The equation corresponding to flux linkage in the synchronous generator is defined as:

$$\Delta E'_q(s) = \frac{1}{1 + sT_G} (K_3 \Delta E_{fd}(s) + K_4 \Delta V(s)) \tag{15.5}$$

$$K_3 = \frac{X'_d}{X_d}, K_4 = \frac{(X_d - X'_d) cos\delta}{X_d} \tag{15.6}$$

$$T_G = \frac{X'_d T'_{d0}}{X_d} \tag{15.7}$$

where ΔE_{fd} defines the voltage deviation in an excitation system.

The excitation system equipped to a synchronous generator is modeled as IEEE-type1 [25].

15.2.2 Modeling of Induction Generator

An induction generator supplies active power to the system and absorbs the reactive power from the system. Considering the induction generator modeling [25], the reactive power deviation of IG is given as:

$$\Delta Q_{IG}(s) = K_5 \Delta V(s) + K_6 \Delta P_{WT}(s) \tag{15.8}$$

$$K_5 = \frac{2VX_e}{R_v^2 + X_e^2}, K_6 = \frac{2R_v X_e}{2R_v R_p - (R_v^2 + X_e^2)} \tag{15.9}$$

$$R_p = \frac{R_2'(1-s_1)}{s_1}, R_v = R_p - R_e \tag{15.10}$$

R_e, X_e represents the equivalent resistance and reactance of IG, R'_2 is rotor resistance from primary side, and s_1 is slip of IG.

15.2.3 Modeling of Static Var Compensator

The static var compensator is modeled based on [26]; the reactive power deviation in SVC is given as:

$$\Delta Qsvc = K_7 \Delta V(s) + K_8 \Delta B_{svc}(s) \tag{15.11}$$

$$K_7 = 2VB_{svc}, < K_8 = V^2 \tag{15.12}$$

The control mechanism of SVC based on sliding mode controller and observer design is shown in Figure 15.2, where ΔV represents the change in bus-bar voltage, ΔV_{ref} is change in reference voltage, K_p is the proportional gain constant, K_t is the integral gain constant, α_0 is the firing angle of thyristor as nominal value, $\Delta\alpha$ is the firing angle deviation of thyristor, $\Delta\alpha_1$

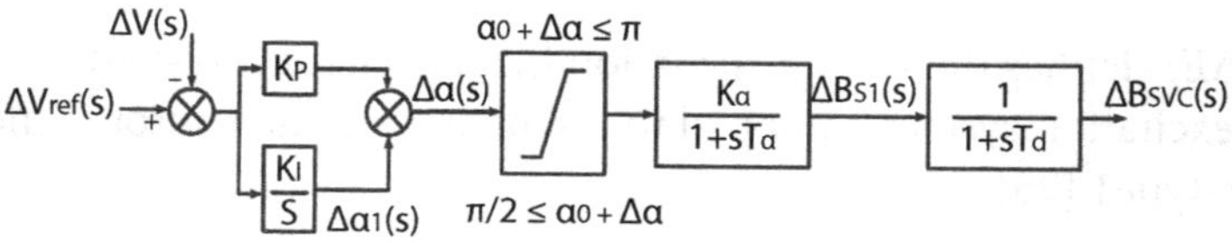

Figure 15.2 Block diagram of the SVC model.

is the integral of deviation in firing angle, K_α is the gain constant of firing delay, T_α is the delay time, T_d is dead time, and ΔB_{s1} and ΔB_{svc} represents the change in susceptance before and after the phase delay, respectively.

15.3 Sliding Mode Control Strategy for SVC

A novel robust sliding mode control structure is developed over SVC to maintain the reactive power balance of an isolated wind–diesel system so as to ensure the voltage stability. For improving the accuracy and reduce the chattering phenomenon of sliding mode controller (SMC), a sliding mode observer (SMO) is designed which estimates the SVC state variables effectively even in the presence of disturbance. Considering Figures 15.2 and 15.3, the state equation of the SVC system model can be given as:

$$\dot{x}(t) = Ax(t) + Bu(t) + Ny(t)$$

$$y(t) = Cx(t) \tag{15.13}$$

where A is system matrix, 6 is input matrix, C is output matrix, and N is disturbance matrix.

$$A = \begin{bmatrix} -\frac{1}{T_d} & \frac{1}{T_d} & 0 \\ 0 & -\frac{1}{T_\alpha} & \frac{K_\alpha}{T_\alpha} \\ 0 & 0 & 0 \end{bmatrix}, B = \begin{bmatrix} 0 & \frac{K_\alpha K_P}{T_\alpha} & K_I \end{bmatrix}^T$$

$$C = \begin{bmatrix} 1 & 0 & 0 \end{bmatrix}, N = \begin{bmatrix} 0 & \frac{K_\alpha K_P}{T_\alpha} & K_I \end{bmatrix}^T$$

Whenever the changes in load and wind power occur, the power system parameters deviate; therefore, with uncertainties the state equations of the system can be represented as:

$$\dot{x}(t) = (A + \Delta 4) + Bu(t) + (N + \Delta V)y(t) \tag{15.14}$$

where ΔA and ΔG are the parameter uncertainties. For the convenience of design of sliding mode controller, uncertainties in the lumped form can be given as:

$$q(t,x) = \Delta Ax(t) + (N + \Delta N)y(t) \tag{15.15}$$

So Equation 15.14 can be written as:

$$\dot{x}(t) = Ax(t) + Bu(t) + q(t, x) \tag{15.16}$$

The following assumptions are made while designing the sliding mode observer and controller.

Assumption 1 [27]: The matrix pairs (A, B) and (A, C) are fully controllable and fully observable, respectively.

Assumption 2: System uncertainty $q(t, x)$ is assumed to satisfy the Lipschitz condition as:

$$||q(t,x) - q(t,\hat{x})|| \leq K\,||x - \hat{x}|| \tag{15.17}$$

where K represents the Lipschitz constant.

Assumption 3: Assume that a positive constant $\bar{\mu}$ exits with condition $||q(t,x)|| \leq \bar{\mu}$, where $||\ *\ ||$ represents the Euclidean norm.

15.3.1 Sliding Mode Observer Design

To enhance the controller performance, sliding mode observer is designed to estimate the unknown parameters of the SVC system with external disturbance [28, 29]. Since effective estimation of system states is not possible with traditional Luenbergers observer (TLO) due to wind power fluctuations, abrupt load change, and unknown system parameters, sliding mode observer (SMO) can therefore be constructed in the following form:

$$\begin{aligned} \dot{\hat{x}}(t) &= A\hat{x}(t) + Bu(t) + q(t,\hat{x}) - L(C\hat{x}(t) - y(t)) + Fv_0 \\ \hat{y}(t) &= C\hat{x}(t) \end{aligned} \tag{15.18}$$

where $\hat{x}(t)$ represents the vector of estimated states, L is the observer gain matrix to be designed, v_0 is the compensation input, and F is a constant matrix satisfying the conditions as CF is non-singular and (A, F) is full-rank.

Defining the estimated state vector as $e(t) = x(t) - \hat{x}(t)$, then the system error dynamics can be obtained from Equations 15.16 and 15.18 as:

$$\begin{aligned} \dot{e}(t) &= A_{LC}e(t) + \phi(t,\hat{x},x) - Fv_0 \\ e_y(t) &= Ce(t) \end{aligned} \tag{15.19}$$

where $A_{LC} = (A - LC)$, $\phi(t,\hat{x},x) = q(t,x) - q(t,\hat{x})$; the matrix L is chosen based on assumption 1 to make $(A - LC)$ stable.

Lemma 1: There exist matrices X and Y with appropriate dimensions, satisfying the following inequality:

$$X^T Y + XY^T \leq \epsilon_1 X^T X + \epsilon_1^{-1} Y^T Y \tag{15.20}$$

Theorem 1: There exists a positive definite matrix R satisfying the following inequality:

$$A_{LC}^T R + RA_{LC} + \zeta_1 K^T K + \zeta_1^{-1} R^2 < 0 \tag{15.21}$$

and the SMO satisfies:

$$v_0 = m\frac{Ce}{||Ce||} \tag{15.22}$$

$$F = R_1 C^T m^{-1} \tag{15.23}$$

where R_1 is the inverse of R and is obtained through linear matrix inversion (LMI) solver [30].

15.3.2 Sliding Mode Controller Design

Based on the estimated parameters of SVC through SM observer design, the sliding mode controller for SVC is designed to compensate for the additional reactive power demand in the system. The design procedure

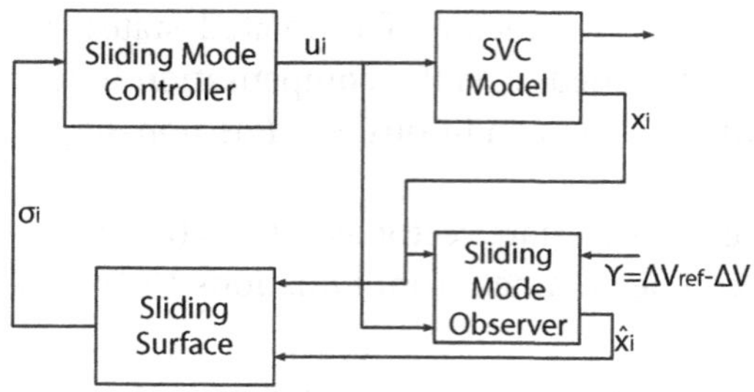

Figure 15.3 Block diagram of the control strategy for SVC.

involves two steps: the first step is the sliding surface design and the second step is the design of control law [31, 32].

15.3.2.1 Sliding Surface Design

Based on the hybrid power system model (Equation 15.16), the sliding surface designed is given as [6]:

$$\sigma = Hx \tag{15.24}$$

where H is a matrix with constant entries and proper dimensions and HB matrix is non-singular [17]. When the trajectories of the system reach the designed sliding surface, the conditions $\sigma(t) = 0$ and $\dot{\sigma}(t)$ are satisfied by the sliding mode equation. Therefore, the controller in equivalent form $u_{eq}(t)$ can be obtained from Equations 15.16 and 15.24 given as:

$$u_{eq}(t) = -(HB)^{-1}\,[HAx(t) + Eq(t,x)] \tag{15.25}$$

Using Equations 15.16 and 15.25, the sliding mode equation obtained is given as:

$$\dot{x}(t) = (A - B(HB)^{-1}HA)x(t) + (I - B(HB)^{-1}H)q(t,x) \tag{15.26}$$

In generalized form, Equation 15.26 can be rewritten as follows

$$\dot{x}(t) = A_1x(t) + B_1q(t, x) \tag{15.27}$$

where $A_1 = (A - B(HB)^{-1}\,HA)$ and $B_1 = (I - B(HB)^{-1}\,H)$.

Theorem 2: Based on the assumption (1) and (3), the system performance is stable in sliding mode when $||x(t)|| > \dfrac{2\bar{\mu}\,||R||||B_1||}{\lambda_{min}(S)}$ at any time [28].

15.3.2.2 SM Control Law Design

On the basis of the designed sliding surface, the reaching law [33] is used to generate the control law given as:

$$\dot{\sigma} = -\hat{\mu}sgn(\sigma) - k\sigma \tag{15.28}$$

where $\bar{\mu} > 0$, $k > 0$ and *sgn*() denotes the sign function and is defined as:

$$sgn(\sigma) = \begin{cases} +1 \text{ if } \sigma > 0 \\ 0 \text{ if } \sigma = 0 \\ -1 \text{ if } \sigma < 0 \end{cases}$$

For the controller to possess robustness

$$\hat{\mu} = n|\sigma| \tag{15.29}$$

where n is some positive constant.

Using Equations 15.16, 15.24, and 15.8, the constructed sliding mode controller is given as:

$$u(t) = -(HB)^{-1}(HA\hat{x}(t) + H\bar{\mu} + \hat{\mu}sg(\sigma) + k\sigma) \tag{15.30}$$

Proof: Taking the Lyapunov function

$$V_c(t) = \frac{1}{2}\sigma^2 + \frac{1}{2n}\tilde{\mu}^2 \tag{15.31}$$

and defining the estimated error variable $\tilde{\mu} = \hat{\mu} - \bar{\mu}$, then the derivative of Equation 15.31 is given by:

$$\dot{V}_c(t) = \sigma\dot{\sigma} + \frac{1}{n}\tilde{\mu}\dot{\tilde{\mu}} \tag{15.32}$$

As system 15.16 satisfies the assumptions, and using Equations 15.24, 15.28, and 15.29, we have

$$\begin{aligned}\dot{\sigma} &= H\dot{x} \\ &= HA(x-\hat{x}) + H(q(t,x) - \bar{\mu}) - \hat{\mu}sgn(\sigma) - k\sigma\end{aligned} \tag{15.33}$$

Using assumptions 15.3, 15.29, and 15.33, Equation 15.32 can be rewritten as:

$$\begin{aligned}&\dot{V}_c(t) = \sigma\dot{\sigma} + \frac{1}{n}\tilde{\mu}\dot{\tilde{\mu}} = \sigma HA(x-\hat{x}) + \sigma H(q(t,x) - \bar{\mu}) - \\ &\hat{\mu}|\sigma| - k\sigma^2 + \frac{1}{n}(\hat{\mu} - \bar{\mu})\dot{\hat{\mu}} \\ &\leq |\sigma|\,\|HA\|\,\|x-\hat{x}\| + |\sigma|\,\|H\|\,\|q(t,x) - \bar{\mu}\| - k\sigma^2 - \\ &\frac{1}{n}\bar{\mu}\dot{\hat{\mu}} + \hat{\mu}(\frac{1}{n}\dot{\hat{\mu}} - |\sigma|) \\ &\leq |\sigma|\,\|H\|\,\|q(t,x) - \bar{\mu}\| - k\sigma^2 - \bar{\mu}|\sigma| \\ &\leq 0\end{aligned} \tag{15.34}$$

Hence, the system is driven to reach the designed sliding surface (15.24) *via* control law (15.30), and motion in sliding phase is ensured to remain stable with locally bounded.

15.4 Simulation Results

In order to carry out the dynamic voltage stability analysis by controlling the reactive power in the wind–diesel hybrid system through the proposed sliding mode observer-based sliding mode control of SVC. The system with its parameters given in Tables 15.1–15.4 is subjected to two different kinds of disturbances that is the step reactive load changes and wind turbine output power variations generated from the time varying wind gusts.

Table 15.1 Numerical value comparison.

Numerical value-based comparison			
Voltage deviation	**PI**	**SMC**	**SMO-SMC**
Undershoot (p.u)	1.67×10^{-3}	1.41×10^{-3}	1.001×10^{-3}
Overshoot (p.u)	5.69×10^{-4}	1.80×10^{-5}	2.02×10^{-5}
Settling time (s)	2.28×10^{-2}	2.00×10^{-2}	1.40×10^{-2}

Table 15.2 Wind–diesel hybrid system data.

System capacity/load	Value
Base power	250 KVA
Diesel power	150 KW
Wind power	150 KW
Load power	250 KW

Table 15.3 IG, SG, and load parameters.

SG	IG	Load
$P_{SG} = 0.4$ p.u	$P_{IG} = 0.6$ p.u	$P_L = 1.0$ p.u
$Q_{SG} = 0.2$ p.u	$Q_{la} = 0.19$ p.u	$Q_L = 0.75$ p.u
$E_q = 1.136$ p.u	$P_{in} = 0.75$ p.u	p_f(lag) = 0.8
$\delta = 21.05°$	$\eta_{IG} = 80\%$	
$E'_q = 0.9603$ p.u	$R_1 = R'_2 = 0.19$ p.u.	
$V = 1.0$ p.u	$X_1 = X_2 = 0.56$ p.u.	
$X_d = 1.0$ p.u	$s = -0.041$	
$X'_d = 0.15$ p.u	$Pf = 0.9$	
$T'_{d0} = 5.0$ s		

Table 15.4 Excitation system and SVC parameters.

Excitation system	SVC
$K_A = 40$	$Q_{SVC} = 0.74$ p.u
$T_A = 0.05$ s	$Q_C = 0.85$ p.u
$K_F = 0.5$	$\alpha = 2.45$ rad
$K_E = 1.0$	$K_\alpha = 1$
$S_E = 0.0$	$T_\alpha = 0.005$ s
$T_F = 0.715$ s	$T_d = 0.01667$ s
$T_E = 0.55$ s	

Finally, to highlight the improvement in the performance of the system with proposed control-based SVC, a comparison of deviations in system parameters is shown with that of traditional PI and sliding mode control based SVC. The controller parameters are provided in Tables 15.5 and 15.6.

Table 15.5 Controller parameters.

SMC
$H = [15.142\ 9.125\ 2.162]$
$\hat{\mu} = 1,\ n = 0.5,\ k = 0.5$

Table 15.6 Observer parameters.

SMO
$L = [-144\ 746.30\ 7{,}210.21]_T$
$R_1 = \begin{bmatrix} 14.912 & 13.485 & 655.310 \\ 13.485 & 18.736 & 74.674 \\ 65.310 & 74.674 & 340.878 \end{bmatrix}$
$F = [7.456\ 6.743\ 32.655]$
$K = 0.2,\ \zeta_1 = 0.2,\ m = 0.5$

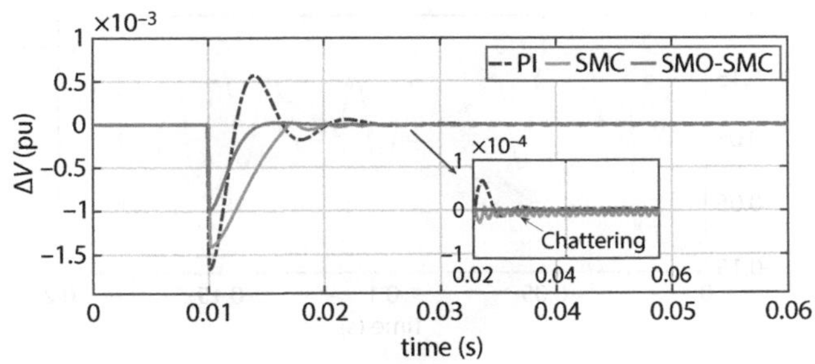

Figure 15.4 Deviation in system voltage.

15.4.1 Case I: Step Disturbance in Load

In this case, the system under consideration is subjected to step disturbance in reactive load. Under step disturbance, the load is incremented by $\Delta Q_L = 0.01\ pu$ at time $t = 0.01$s. The deviation in system voltage and other reactive power parameters is observed and a comparative analysis of proposed control approach, traditional PI, and SM control schemes is shown.

The deviation in system voltage is shown in Figure 15.4. It can be seen that, in comparison to traditional PI control scheme, using SMC these deviations are reduced in terms of undershoot, overshoot, and settling time. However, chattering occurs with it shown in zoom in view (Figure 15.4). Then, with SMO-SMC, apart from chattering elimination, improved performance in terms of undershoot, overshoot, and settling time is observed and is given in Table 15.1.

15.4.2 Case II: Operation Under Varying Wind Power

In this case study, to verify the robustness and efficacy of the proposed controller, time-varying wind turbine output power is generated with the system subjected to continuously time-varying wind gusts shown in Figure 15.5. This happens practically in a wind–diesel hybrid system, under varying weather conditions and seasons, which may lead to frequent changes in the operating points. Thus, continuously varying reactive power deviations occur in the various system parameters, and hence the impact on the system voltage gets reflected.

From Figure 15.6, with the sliding mode observer-based sliding mode control approach, deviation in system voltage can be minimized effectively. On the other hand, these deviations oscillate and require large settling time with the use of traditional PI control strategy and thus can affect the

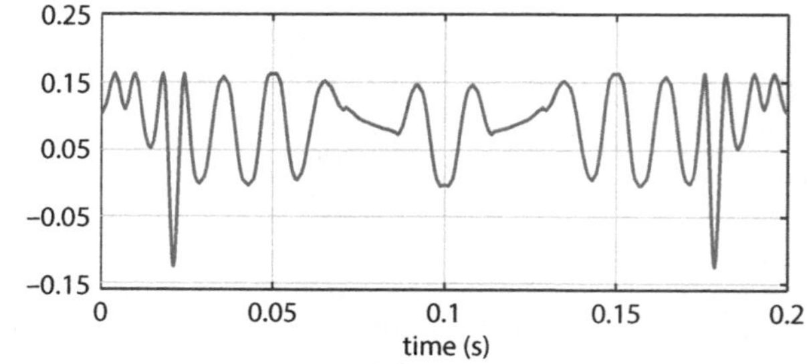

Figure 15.5 Deviation in wind turbine output power in the system subjected to wind gusts.

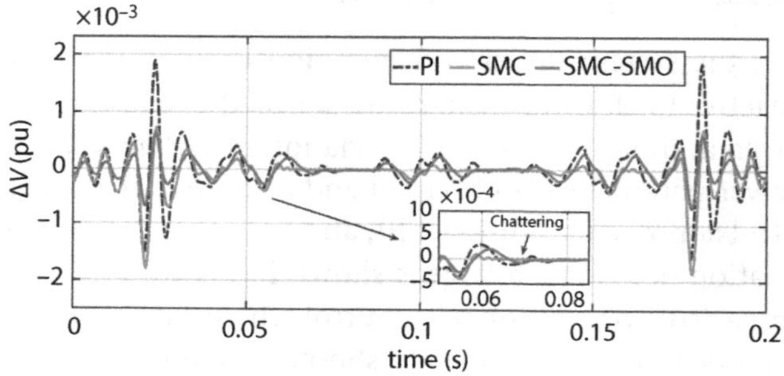

Figure 15.6 Deviation in system voltage.

stability of the system. With SM, control scheme phenomenon of continuous and minute oscillations known as chattering is also observed. Since the system is subjected to disturbance of continuously varying wind gusts, therefore from this case study, the robustness of the proposed control scheme is verified.

15.5 Conclusion

In this chapter, a wind–diesel hybrid system model with static var compensator (SVC) is established. With sliding surface design, a sliding mode controller (SMC) is developed over static var compensator (SVC) which gives improved voltage control. With the help of sliding mode observer (SMO), effective estimation of SVC parameters is guaranteed and a proper control law is generated with which controller robustness and chattering-free response are achieved. The stability of the controller is proven using Lyapunov stability analysis. Simulations studies performed in the events of sudden load changes and wind power disturbances showed the efficacy of the proposed controller.

References

1. M. Amirthalingam, "A novel technology utilizing renewable energies to mitigate air pollution, global warming climate change," in *2009 1st International Conference on the Developments in Renewable Energy Technology (ICDRET)*, pp. 1–3, Dec 2009.
2. A. Mohanty, M. Viswavandya, P. K. Ray, and S. Patra, "Stability analysis and reactive power compensation issue in a microgrid with a dfig based wecs," *Int. J. Electr. Power Energy Syst.*, vol. 62, pp. 753 – 762, 2014.
3. D. Yadav and T. Bhatti, "Voltage control through reactive power support for wecs based hybrid power system," *Int. J. Electr. Power Energy Syst.*, vol. 62, pp. 507 – 518, 2014.
4. L. Wang and D. Truong, "Stability enhancement of dfig-based offshore wind farm fed to a multi-machine system using a statcom," *IEEE Trans. Power Syst.*, vol. 28, pp. 2882–2889, Aug 2013.
5. Nieto, Alejandro. (2016). Power Quality Improvement in Power Grids with the Integration of Energy Storage Systems. *Int. J. Eng. Technol. Res.* 5. 438-443.
6. Veselić, B., Milosavljević, Č., Draženović, B. *et al.* Sliding Manifold Design for Higher-order Sliding Mode Control of Linear Systems. *Int. J. Control Autom. Syst.* 19, 2652–2663 (2021). https://doi.org/10.1007/s12555-020-0493-7
7. Xu, W., Li, Z., Cui, G. *et al.* Fuzzy Adaptive Finite Time Command Filter Backstepping Control of Power System. *Int. J. Control Autom. Syst.* 19, 3812–3822 (2021).
8. Sitthidet Vachirasricirikul, Issarachai Ngamroo, Somyot Kaitwanidvilai, Coordinated SVC and AVR for robust voltage control in a hybrid wind-diesel system, *Energy Convers. Manag.*, Volume 51, Issue 12, 2010, Pages 2383-2393.
9. R. K. Varma and S. Auddy, "Mitigation of subsynchronous oscillations in a series compensated wind farm with static var compensator," in *2006 IEEE Power Engineering Society General Meeting*, pp. 7 pp.–, June 2006.
10. R. C. Bansal, "Automatic reactive-power control of isolated wind–diesel hybrid power systems," *IEEE Trans. Ind. Electron.*, vol. 53, pp. 1116–1126, June 2006.
11. Sudhanshu Ranjan, Abdul Latif, Dulal Chandra Das, Nidul Sinha, S.M. Suhail Hussain, Taha Selim Ustun, Atif Iqbal, Simultaneous analysis of frequency and voltage control of the interconnected hybrid power system in presence of FACTS devices and demand response scheme, *Energy Rep.*, Volume 7, 2021, Pages 7445-7459.
12. Ahmed M. Kassem, Almoataz Y. Abdelaziz. (2015) Optimal Control of an Autonomous Variable-speed Wind Generation System Based on a Bacterial

Foraging Optimization Technique. *Electr. Power Compon. Syst.* 43:8-10, pages 1006-1017.

13. A. M. Kassem and A. Y. Abdelaziz, "Reactive power control for voltage stability of standalone hybrid wind–diesel power system based on functional model predictive control," *IET Renew. Power Gener.*, vol. 8, no.8, pp. 887–899, 2014.
14. R. Bansal, "Modelling and automatic reactive power control of isolated wind-diesel hybrid power systems using ann," *Energy Convers. Manag.*, vol. 49, no. 2, pp. 357 – 364, 2008.
15. A. Mohanty, M. Viswavandya, and P.K. Ray, "An adaptive fuzzy sliding mode controller for reactive power transient stability management," in *2016 IEEE Region 10 Conference (TENCON)*, pp. 3195–3199, Nov 2016.
16. A.A. Ibraheem, A. Bahgat, M.S. Abdel Motelb, FUZZY LOGIC SLIDING MODE CONTROLLER FOR DC DRIVE, Editor(s): Mohamed F. HASSAN, Said M. MEGAHED, *Current Advances in Mechanical Design and Production VII*, Pergamon, 2000, Pages 75-83.
17. U. Itkis, *Control systems of variable structure.* A Halsted Press book, John Wiley & Sons, Incorporated, 1976.
18. P. V. BALASUBRAMANYAM, A. S. R. MURTHY, and S. PARAMESWARAN, "Design of variable structure controller for static var compensator," *Elec. Mach. Power Syst.*, vol. 26, no. 4, pp. 431–450, 1998.
19. M. El-Sadek, G. EI-Saady, and M. Abo-El-saud, "A variable structure adaptive neural network static var controller," *Electr. Power Syst. Res.*, vol. 45, no. 2, pp. 109 – 117, 1998.
20. A. Dev, V. Léchappé, and M. K. Sarkar, Prediction-based supertwisting sliding mode load frequency control for multi area inter-connected power systems with state and input time delays using disturbance observer, *Int J. Control*, (2019), 1–14.
21. F. Pourboghrat, F. Farid, C. J. Hatziadoniu, M. Daneshdoost, F. Mehdian, and M. Lotfalian, "Local sliding control for damping interarea power oscillations," *IEEE Trans. Power Syst.*, vol. 19, pp. 1123–1134, May 2004.
22. H. Yang, Y. Zhang, J. Liang, J. Gao, P. D. Walker, and N. Zhang, "Sliding-mode observer based voltage-sensorless model predictive power control of pwm rectifier under unbalanced grid conditions," *IEEE Trans. Ind. Electron.*, vol. 65, pp. 5550–5560, July 2018.
23. S. K. Routray, N. Nayak, and P. K. Rout, "A robust fuzzy sliding mode control design for current source inverter based statcom application," *Procedia Technol.*, vol. 4, pp. 342 – 349, 2012. 2nd International Conference on Computer, Communication, Control and Information Technology(C3IT-2012) on February 25 - 26, 2012.
24. O. Elgerd, *Electric Energy Systems Theory: An Introduction*. Tata McGraw-Hill, 1983.
25. R. C. BANSAL, T. S. BHATTI, and D. P. KOTHARI, "Automatic reactive power control of isolated wind-diesel hybrid power systems for variable

wind speed/slip," *Electr. Power Compon. Syst.*, vol. 32, no. 9, pp. 901–912, 2004.

26. R. C. Bansal, T. S. Bhatti, and D. P. Kothari, "Automatic reactive power control of wind-diesel-micro-hydro autonomous hybrid power systems using ann tuned static var compensator," in *Large Engineering Systems Conference on Power Engineering, 2003*, pp. 182–188, May 2003.
27. J. Chang, "Robust discrete-time model reference sliding-mode controller design with state and disturbance estimation," *IEEE Trans. Ind. Electron.*, vol. 55, pp. 4065–4074, Nov 2008.
28. J. Fei and Mingang Hua, "A novel design of adaptive sliding mode observer," in *2010 8th World Congress on Intelligent Control and Automation*, pp. 843–848, July 2010.
29. Wen-Hua Chen, D. J. Ballance, P. J. Gawthrop, and J. O'Reilly, "A nonlinear disturbance observer for robotic manipulators," *IEEE Trans. Ind. Electron.*, vol. 47, pp. 932–938, Aug 2000.
30. Boyd S, El Ghaoui L, Feron E, Balakrishnan V. Linear Matrix Inequalities in System and Control Theory. *Society for Industrial and Applied Mathematics*; 1994.
31. Y. Mi, Y. Fu, C. Wang, and P. Wang, "Decentralized sliding mode load frequency control for multi-area power systems," *IEEE Trans. Power Syst.*, vol. 28, pp. 4301–4309, Nov 2013.
32. M. C.Pai,"Adaptive super-twisting terminal sliding mode control for nonlinear systems withmultiple inputs," *Int. J. Dyn. Control*, vol. 8, pp. 1101–1109, Nov 2019.
33. Weibing Gao, Yufu Wang, and A. Homaifa, "Discrete-time variable structure control systems," *IEEE Trans. Ind. Electron.*, vol. 42, pp. 117–122, April 1995.

16

A Review of Modular Multilevel Converters and Its Applications

Dhananjay Kumar[1]*, Kasinath Jena[2], Jitendra Kumar Tandekar[3], Niraj Kumar Dewangan[4] and Vishal Rathore[5]

[1]*Department of Electrical Engineering, Government Engineering College, Siwan, Bihar, India*

[2]*Department of Electrical and Electronic Engineering, Arka Jain University, Jamshedpur, India*

[3]*Department of Electrical Engineering, Samrat Ashok Technological Institute, Vidisha, India*

[4]*Department of Mechatronics Manipal Institute of Technology, Manipal Academy of Higher Education, Manipal, Karnataka, India*

[5]*Department of Electrical and Electronics Engineering, Bhopal Institute of Technology and Science, Bhopal, India*

Abstract

Modular multilevel converters (MMC) have gained a notable attention of researchers from all over the world. There must be hundreds of individual modules connected in series for high-voltage applications. Depending on the sort of sub-module chosen for an application, an MMC can have anywhere from one-to-many switches built into it. In most cases, 2 or 3 years of trouble-free operation of the converter is necessary for MMC-based applications. Any of several types of electrical failure, including open circuit, short circuit, DC-bus short circuit, and single line to ground, can affect an MMC. If the MMC fails to function properly, the voltage in the system could drop. In light of this, it is essential to create a robust fault-tolerant strategy. Every fault-tolerant plan consists of three distinct phases: pre-fault preparation, fault isolation and localization, and post-fault restoration and control. The primary emphasis of this work is on the post-fault control procedures necessary to quickly and reliably restore voltage levels to those that existed before the fault.

**Corresponding author*: dhananjayk00@gmail.com

Salman Ahmad, Farhad Ilahi Bakhsh, and P. Sanjeevikumar (eds.) Multilevel Converters, (317–338)

***Keywords*:** Fault tolerant, half bridge, modular multilevel converter, switch open circuit fault

16.1 Introduction

The field of modular multilevel converters (MMC) has recently attracted significant attention from scientists all around the world. Key features such as modularity, lower harmonic content in output voltage at high levels, lack of need for bulky capacitors and isolated DC sources, easy scalability to any voltage level, and lower voltage stress on switches make it a promising solution for high- and medium-power applications like HVDC and motor drives.

When it comes to high-powered uses, MMCs can be found in both standard and specialty configurations. Motor drive systems, power transmission networks, and power quality improvement are just a few of the many commercial uses for multi-phase synchronous motors (MMC). When it comes to improving power quality, the unified power flow controller is another popular and effective method (UPFC). UPFC provides two levels of correction due to the series and parallel configuration of the power converters. One method of implementing the UPFC is shown in Figure 16.2, which involves establishing an MMC connection between two parallel modules.

The MMC-I is linked to the power-grid via a transformer. It employs series compensation to address the issues of voltage drop and surge on the AC grid. There is no need for a transformer thanks to the MMC-shunt II's correction and direct connection to the AC grid. The transformer in shunt compensation can be dispensed with when using MMC, which is a major advantage. The deformed load's reactive power and harmonic distortion are both corrected by the shunt compensator. The active filtering capacity of the UPFC system stems from its high-quality output waveforms and rapid current management of MMCs.

Key advantages such as scalability, low output voltage THD, and low fault energy distinguish MMC, a multi-cell voltage source converter, as a prime choice for high voltage and medium voltage drives. It expands the range of possible operations that can be supported by a single DC power supply. Figure 16.1 depicts the MMC's setup. DC is supplied *via* a bus that is wired across the MMC leg. There are two arms (upper and lower) and two inductors (which counteract the circulating current) in each leg of a multi-terminal connector. Midway between the top and lower arm is where you will find the three-phase AC output. Each arm consists of a cascaded series of 'N' number of sub-modules (SMs) that work together to provide

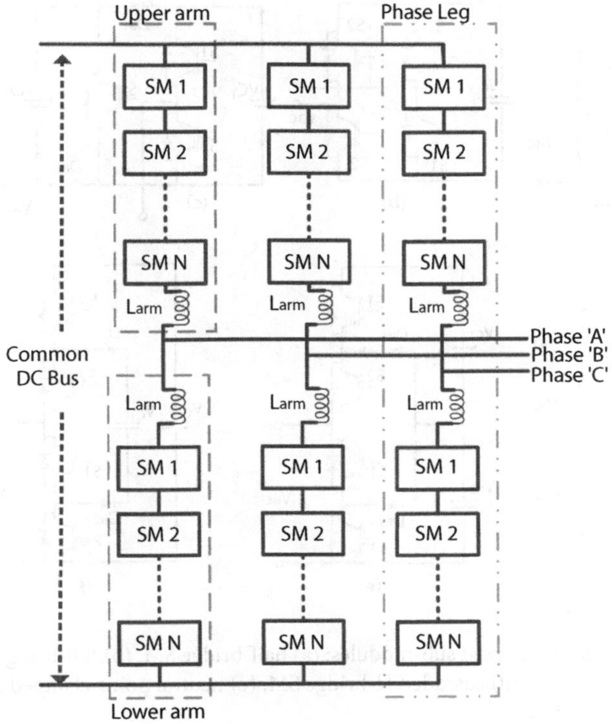

Figure 16.1 Topology of a three-phase MMC with 'N' sub-modules in each arm of a phase leg.

the desired output at any given time. According to the common bus voltage and the voltage rating of semiconductor devices, the number of SMs connected in each arm can be planned.

The function of the module is considered while determining the circuitry and architecture employed in the sub-module. Half-bridge sub-modules, full-bridge sub-modules, flying-capacitor sub-modules, neutral point-clamped sub-modules, cascaded H-bridge sub-modules, and diode-clamped sub-modules are all common examples of sub-modules. Figure 16.2 displays the various circuit configurations of the SMs.

In high-voltage applications, hundreds of individual modules must be linked together in a single circuit. Depending on the sort of sub-module chosen for a given task, each of these units may feature several toggles. In most cases, MMC-based applications need the converter to work reliably for several years. However, the rate of failure in an MMC system increases as the number of switches increases. We face a significant challenge with reliability that will require concerted action in the near future.

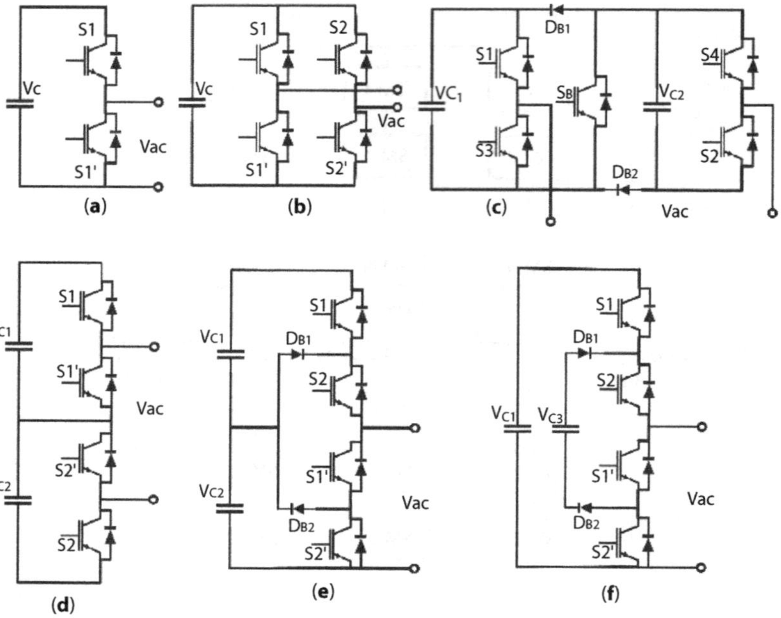

Figure 16.2 Commonly used sub-modules: (a) half-bridge SM, (b) full-bridge SM, (c) clamped-diode SM, (d) cascaded H-bridge SM, (e) neutral point-clamped SM, and (f) flying capacitor SM.

In high-voltage applications, hundreds of individual modules must be linked together in a single circuit. Depending on the sort of sub-module chosen for a given task, each of these units may feature several toggles. In most cases, MMC-based applications need the converter to work reliably for several years. However, the rate of failure in an MMC system increases as the number of switches increases. We face a significant challenge with reliability that will require concerted action in the near future.

16.1.1 Types of Faults in MMC

In general, there are four different kinds of faults that can occur in an MMC, and they are as follows: i.e., DC bus fault, line-to-ground fault on AC side, switch open circuit fault and switch short circuit fault. Detailed explanations of each of these defects can be found, with accompanying graphical representation, in Figure 16.3.

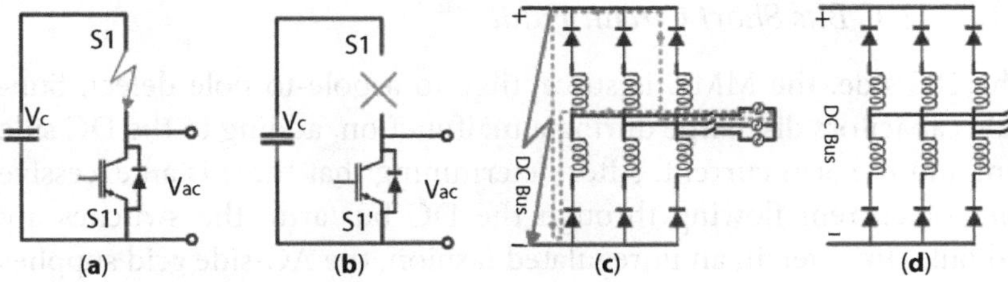

Figure 16.3 Types of faults in MMC: (a) switch short circuit fault, (b) switch open circuit fault, (c) DC bus short circuit, and (d) SLG fault on AC side.

16.1.1.1 Switch Open Circuit Faults

It is critical to first comprehend the origins of open circuit faults before delving into the post fault control measures. Acquiring a deep understanding of these factors is crucial for building fault-tolerant control systems. Thermal or mechanical stress, substantial variations in junction temperature, and improper operation of gate driver circuits are some of the most common causes of open circuit faults in solid-state switching devices. A decrease in the number of levels in the output voltage caused by a faulty switch will have a negative impact on the output voltage and current. Furthermore, since there is no way for the capacitor to discharge, its voltage rises dramatically.

This may cause the capacitor in the sub-module to fail. As a result, researchers have seen a decline in the MMC system's dependability, which requires more investigation.

16.1.1.2 Switch Short Circuit Fault

Switch short circuit problems in MMC can be caused by several factors, including excessive current, voltage, gate driver malfunction, and switch overheating. The normal short circuit withstand time of the switches is less than 10 ms. That is why a quick-response method for detecting switch short circuit failures is essential. In order to identify a SC failure, it is common practice to use the voltage drop across the switch when it is in the on position as a metric. Whenever the switch's on-state voltage drops falls below the OEM-specified value, a SC fault is identified, and the associated SM circuit is isolated.

16.1.1.3 DC Bus Short Circuit Fault

On the DC side, the MMC is susceptible to a pole-to-pole defect. Submodule capacitors discharge during a malfunction, adding to the DC side current and the arm current. After determining that there is an excessive amount of current flowing through the DC bus/arm, the switches are locked out. However, in an unregulated fashion, the AC-side grid supplies the DC bus through the anti-parallel diodes of HB-SM. The converter and the motor loads would be harmed by the fault current that would pass through the MMC in this scenario. Therefore, it is essential to have a reliable fault ride through capability.

16.1.1.4 Single Line to Ground (SLG) Faults

There are several potential causes of SLG problems on the ac side of the MMC system, including transformer insulation failure, transmission conductor cracking, etc. It causes the DC bus to experience double-line frequency oscillations, undervoltage/voltage sags, and extra voltage stress on the switches. Therefore, further research into SLG fault detection, isolation, and post-fault control technique is warranted. It is possible that if the MMC fails to function properly, the voltage in the system will drop as a result of the malfunction. Because of this, it is crucial to work on a solid plan to tolerate failure. Fault isolation, post-fault control, and fault detection and localization are the three components that make up any fault-tolerant method.

This paper primarily focuses on the post fault control strategies that have been reported in the earlier literatures. The rest of the paper has been divided into sub-sections. Section 16.1 focuses on the MMC fault. A discussion on literature review is covered in Section 16.2. Mathematical analysis is discussed in Section 16.3. A brief simulation of results is proposed in Section 16.4. Furthermore, performance analysis is discussed in Section 16.5. Finally, the conclusion and future scope of work are discussed under section 16.6.

16.2 Literature Review

The primary goal of many fault-tolerant techniques is to get the voltage back to where it was before the fault occurred. Internal SM fault bypass techniques for MMC can be broken down into two main groups. To guarantee the symmetrical functioning of the MMC, the first strategy involves bypassing the same number of SMs in each arm as the number of defective sub-modules. The second plan achieves asymmetric mode of operation of MMC by selectively excluding the malfunctioning SMs. One significant advantage of operating in an asymmetrical fashion is the increased likelihood that each arm's submodules will be put to use. When the incorrect phase is converted into an output voltage, however, the results are not symmetrical. The following fault tolerant solutions have been published and debated as a means to alleviate the aforementioned issues.

16.2.1 Adaptive Reference Voltage Method

Capacitor Voltage Increasing (CVI) [1–3] and Zero-Sequence Voltage Injection (ZSVI) [4–6] are two common types of adaptive reference voltage techniques.

16.2.2 Capacitor Voltage Increasing Method

The CVI method is an accurate approach where the voltage of one or more functional SMs in the damaged arm is raised to restore it to its pre-fault state [1–3]. But the number of output voltage levels is minimized, and over-modulation is prevented [7]. Applications with a greater number of submodules benefit most from lower output voltage levels. It's important to remember that the Safe Operating Area (SOA) of SMs fluctuates between 50 and 60 percent of the switch blocking voltage [8]. In this way, the CVI technique is extremely sensitive to the switch voltage stresses and SM capacitor ratings.

16.2.3 Zero-Sequence Voltage Injection Method

By adding a DC component to the phase voltages, it is possible to provide a balanced output voltage. Third-order harmonic voltage injection (THVI) and neutral-shift (NS) are two of the most popular and extensively studied techniques in this area. The THVI approach is appealing because it maximizes the use of the dc-link voltage. However, THVI only works for MMC systems that use sinusoidal PWM modulation, and it also modifies the voltage swings in a sub-module [4, 5]. By adjusting the phase angles of the output voltage, the NS approach achieves line-to-line voltage balance and maximizes line voltage magnitude [6, 9]. The DC component added to the phase voltages accomplishes this. The magnitude of the phase voltages of the MMC must be an integral multiple of the number of the healthy SMs in the respective phases, which in turn determines the converter's output voltage [9]. A fault-tolerant version of the MMC has been built using the NS Method, which is described in [10] and relies on a technique called reference modification. Reference voltages are computed, and the optimal value of the Z-parameter is used to increase the output voltage, as proposed in [11].

16.2.4 Modulation-Based Methods

Altering the MMC modulation technique is another fault-tolerant method that has been reported. Both the space-vector modulation (SVM) approach and over-modulation techniques fall under this broader category of methodologies.

PWM-based methods: The converter can be operated under internal SM fault situations with the use of a modified SVM algorithm, as shown in [12], which allows for the delivery of a balanced three-phase line-to-line voltage.

In [13], SVM technique is proposed that can apply an external DC offset to the problematic MMC phases. However, this method becomes extremely difficult for applications that require a large number of SMs. To determine how many SMs must be added for any given mode of operation, [14] proposes a method that makes use of both a sorting algorithm and a PSC-PWM modulation scheme. In [15], an adaptive PDPWM scheme is implemented using a single carrier and a single reference signal per arm to lessen the computing load and speed up the system's response time.

16.2.5 Over Modulation Methods

In the linear zone, where the AC port voltage is less than the DC port voltage, MMCs are optimized for use in applications like as high-voltage direct current (HVDC). However, by using several other over-modulation techniques, the AC voltage's fundamental component can be boosted even higher. Operating MMC in the over-modulation area, however, would increase the output current's THD [16]. Even with many SMs, the THD would be well within allowable limits [17]. In the over-modulation region, [18] demonstrates a fault tolerant MMC application for STATCOM.

16.2.6 Redundancy-Based Methods

As long as fundamental SMs and their redundant counterparts are protected by MMC, the system's overall performance can be restored after a fault. Redundant SMs are typically run in one of two modes: hot reserve mode and cold reserve mode.

16.2.7 Hot Reserve Mode

In all situations, the redundant SMs work in concert with the primary SMs to distribute the workload. This configuration is commonly used in MV or LV fault tolerant MMC operations [19, 20]. Improved SM use, no charging time required for reserved SMs, and superior transient performance [21, 22] are only a few of the advantages of the hot reserve technique. As a result of these merits, hot reserve techniques have found widespread use in MV and LV settings [23, 24].

16.2.8 Cold Reserve Mode

When an error is identified, this mode comes in and replaces the problematic SMs with backup components. Therefore, there are minimal power losses in the converter when using this technique [25]. However, a supplementary power source is required to charge the initially superfluous submodules [7]. This approach, however, has subpar dynamic behavior because of the constant charging and discharging of the redundant SMs capacitors. A warm redundant SM approach has been disclosed in reference [26] that quickly replaces the damaged submodules with mitigated transients, allowing for more stable charging of the redundant SMs. There is a thorough review of the redundant SM strategy in [25]. Undoubtedly, MMC reliability and symmetrical operation after SM fault are both improved by redundancy strategies. Redundant SMs are advantageous because their spare capacity can be used during times of emergency [27].

Methods for further reducing the computing burden of controllers are provided in [28, 29], such as decentralized control, in which a local controller is assigned to each set of sub-modules of an MMC arm, and the local controllers are interfaced with a central supervisory controller. According to [1], using techniques like the adaptive linear neuron (ADALINE) algorithm and the recursive least squares (RLS) algorithm to estimate SM capacitor voltages rather than actual measurement might significantly reduce the number of voltage sensors required.

16.2.9 Novel Topology Methods

In order to apply the fault-tolerant method in MMC, a number of novel SM and MMC topologies have been published in addition to the standard

HBSM, FBSM, NPCSM, and CDSM. A new SM topology is proposed in [30] that is compatible with standard HBSM. Inherent voltage balancing for the capacitors and high capacity for expansion are two of its distinguishing features. However, due to the fact that each SM requires six switches, switch losses grow dramatically in high-SM-count applications. It has been mentioned in [31] how a director component and wave shaping part can be used in a hybrid MMC with cascaded FBSM to prevent high voltage stress and improve capacitor voltage regulation. Another MMC with lower power losses is presented by the authors in [32], which uses arms made up of HBSM and series-connected triple SM connected in an alternating fashion. Dual-stage construction is proposed in [33], which would allow for fault ride-through and improved AC-side voltage. The first phase is very much like the standard HBSM MMC. Stage II is made up of a thyristor SM group and an FB and is tapped in the sweet spot between the upper and lower am. Several alternative fault-tolerant MMC topologies have been reported as well [34–37].

16.3 Mathematical Modeling

This chapter presents the equivalent circuits and mathematical modeling of modular multilevel converters. The behavior of the system at steady state and dynamic conditions can be better understood by understanding the relation between arm and line quantities. The following are the presumptions made in this evaluation:

a) The sub-module capacitor is large enough to be treated as a reliable DC power source.
b) Since there are more SMs in the MMC leg, any voltage-source-related harmonic components can be disregarded.
c) None of the phase legs are different from one another.
d) Each arm has a pair of identical submodules that work together to instantly equalize the voltage across the capacitor.
e) All three phases of voltage are of the same amplitude, and the loads on all three phases are comparable.

Each arm of a phase leg with N-SMs is considered as a single equivalent controlled voltage source to simplify the analysis. The upper and lower arm currents in phase 'x' (x ϵ a,b,c) of MMC is given by:

$$i_{ux} = \frac{i_x}{2} + i_{diffx}(x\epsilon a,b,c)$$

$$i_{lx} = -\frac{i_x}{2} + i_{diffx}(x\epsilon a,b,c)$$

i_{diffx} in the above equation is the differential current in phase x and is given by:

$$i_{diffx} = -\frac{i_{ux} + i_{lx}}{2} = \frac{i_s}{3} + i_{cirx}(x\epsilon a,b,c)$$

i_{cirx} is the circulating current in the phase leg x.

The dynamic equations of the upper arm voltage is given by:

$$\frac{u_s}{2} + u_{xo} - L_{arm}\frac{di_{ux}}{dt} - R_{arm}i_{ux} - u_{ux} = 0$$

$$\frac{u_s}{2} + u_{xo} - L_{arm}\frac{di_{ux}}{dt} - R_{arm}i_{lx} - u_{lx} = 0$$

The AC output voltage is the difference of the voltages inserted by the lower arm and upper arm sub-modules. Thus, it can be controlled by controlling the number of SMs in the arms. The voltage rating of DC-link capacitor is given by:

$$U_c = \frac{U_s}{N}$$

16.4 Simulation Results

16.4.1 1-Φ 5L -MMC With PS-PWM

Two sinusoidal signals serve as the basis for this modulation system. The sub-modules in the upper arm are based on the negative sinusoidal pulse, while those in the lower arm are based on the positive sinusoidal pulse. It has been proposed that a triangular carrier signal be used to produce gate pulses. If there are n SM in each leg, then n carrier waves are needed. As a result, we have explored four distinct carrier signals. For our system,

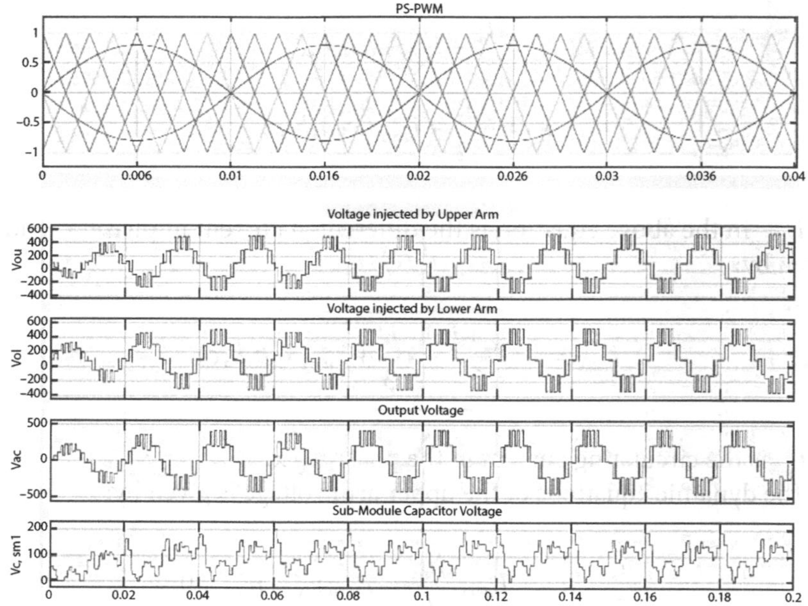

Figure 16.4 Simulation results of 5L-MMC with PS-PWM.

the number of SMs in each arm is 4; therefore, we adjust the phase of the carrier signal by (360°)/N or 90°. For reference, Figure 16.4 depicts the PS-PWM scheme and the resulting waveforms.

16.4.2 1-Φ5L-MMC With POD-PWM

The carrier signals in POD-PWM undergo a level shift in phase with the sinusoidal reference signal, making the system a type of level-shifted PWM. As shown in Figure 16.5, the carrier waves in the lower half of the spectrum are 180° out of phase with respect to those in the upper half of the

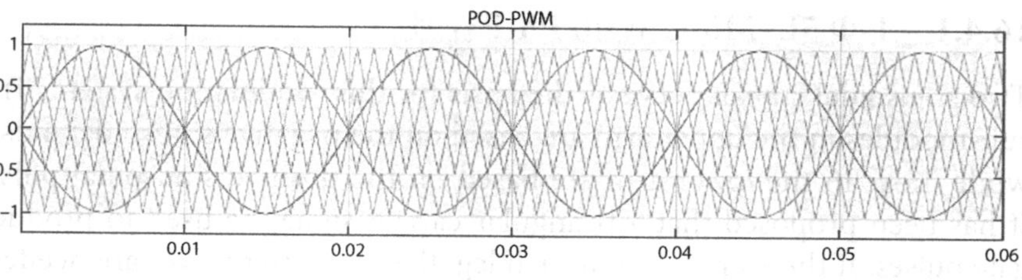

Figure 16.5 POD-PWM scheme for 5L-MMC with PS-PWM.

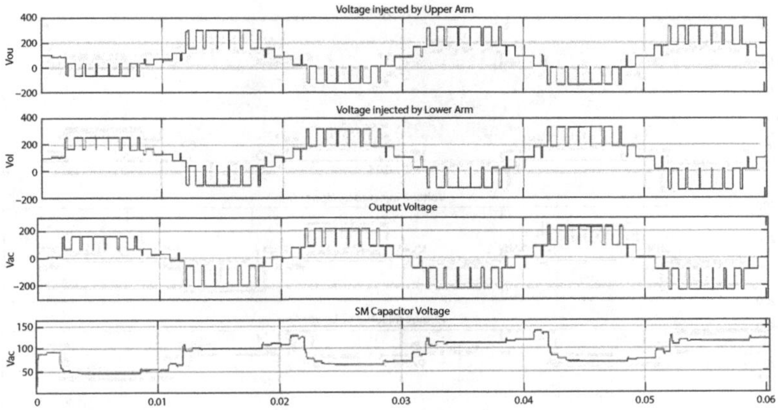

Figure 16.6 Simulation results of 5L-MMC with POD-PWM.

spectrum. Figure 16.6 displays the waveforms of MMC output achieved by POD-PWM.

16.4.3 5L-MMC With APOD-PWM

Level-shifted PWM or APOD-PWM uses four carrier signals to control four SMs along a single arm. In contrast to POD-PWM, the carrier signals in this method are phase-shifted by 180° on alternate cycles. Figure 16.7 depicts the APOD-PWM system and the resulting waveforms.

16.4.4 1-Φ 5L-MMC With Nearest Level Modulation

The number of SMs to be placed in each arm of an NLM is determined so that the voltage produced by an MMC is very close to a sinusoid. It may be shown how many SMs needed to be added as follows:

$$N = round(\frac{u_m}{U_c})$$

where $u_m U_c$ are the modulation waveform amplitude and voltage across the SM capacitor, respectively. Figure 16.8 displays the waveforms obtained by applying POD-PWM to MMC's output.

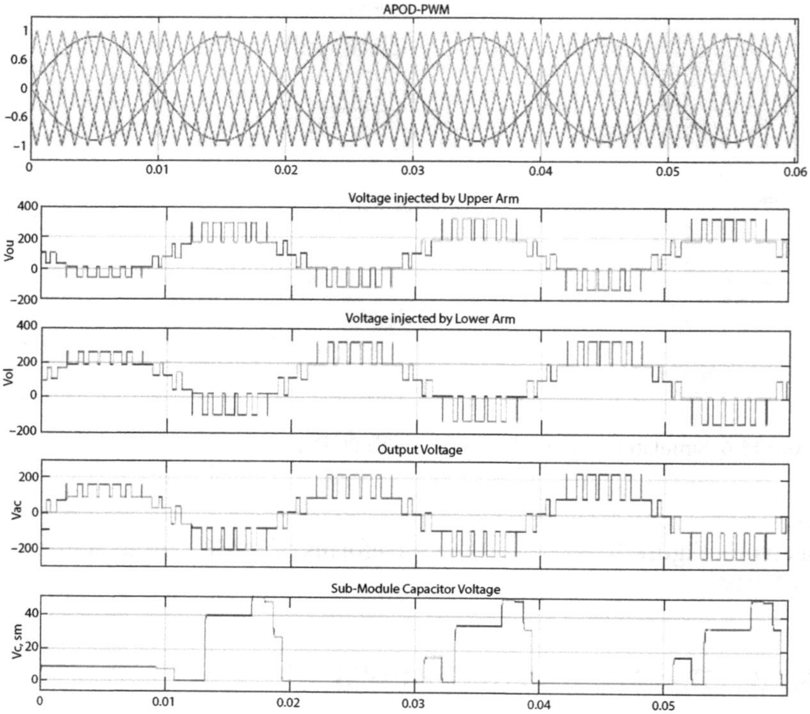

Figure 16.7 Simulation results of 5L-MMC with APOD-PWM.

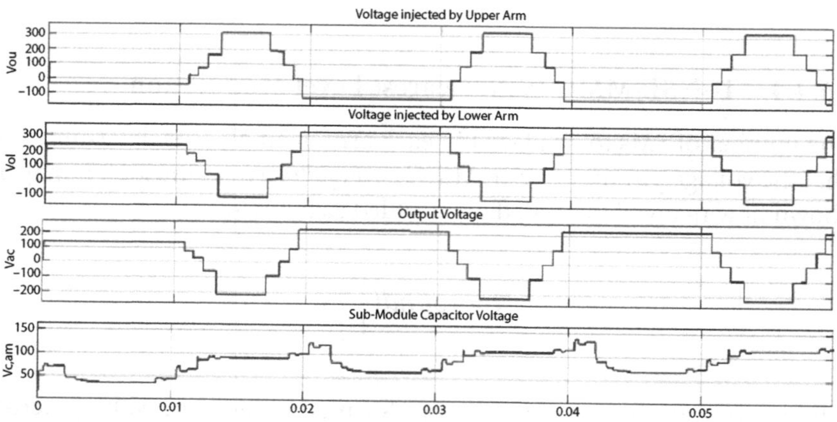

Figure 16.8 Simulation results of 5L-MMC with NLM-PWM.

16.4.5 Three-Phase 9L-MMC With PS-PWM

The Simulink model of three-phase MMC with HBSM is shown in Figure 16.9. Each arm of a phase leg contains four numbers of HBSM cascaded in series to obtain a nine-level output voltage.

The simulation result for three-phase nine-level MMC with PS-PWM is shown in the Figure 16.10. It can be clearly seen that the sub-module capacitor voltage variation is beyond the permissible limit of 10% and thus needs a suitable capacitor voltage balancing algorithm along with a robust fault-tolerant strategy.

16.4.6 Three-Phase 9L-MMC With PD-PWM

The simulation results for three-phase nine-level MMC with PD-PWM are shown in Figure 16.11. The capacitor voltage ripple has been reduced as compared to that of with PS-PWM.

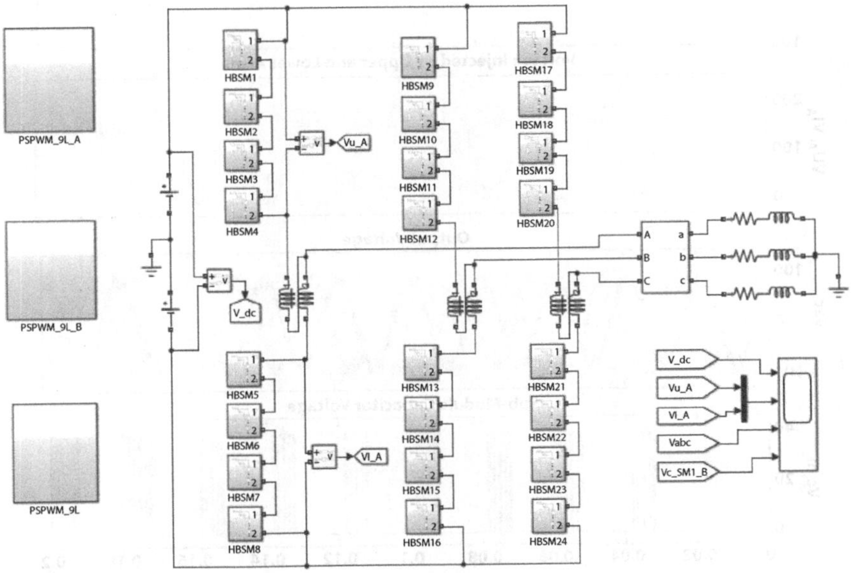

Figure 16.9 Simulink model of three-phase HBSM-based MMC.

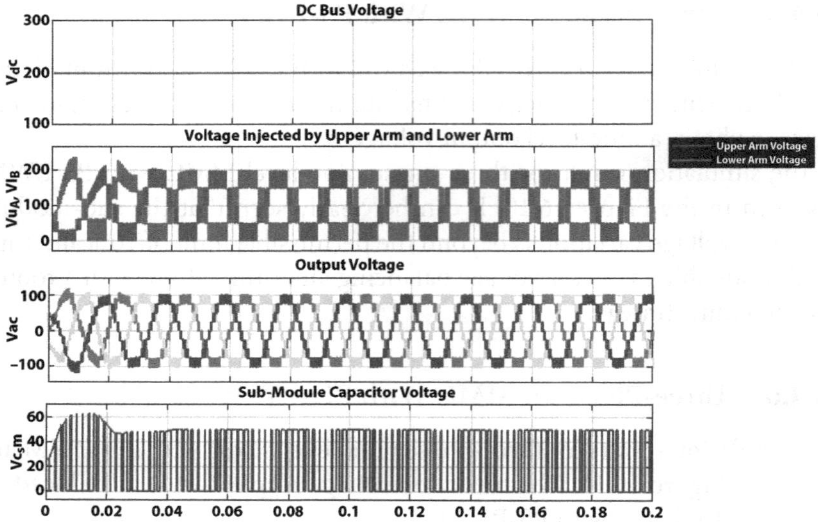

Figure 16.10 Simulation result of nine-level three-phase MMC with PS-PWM.

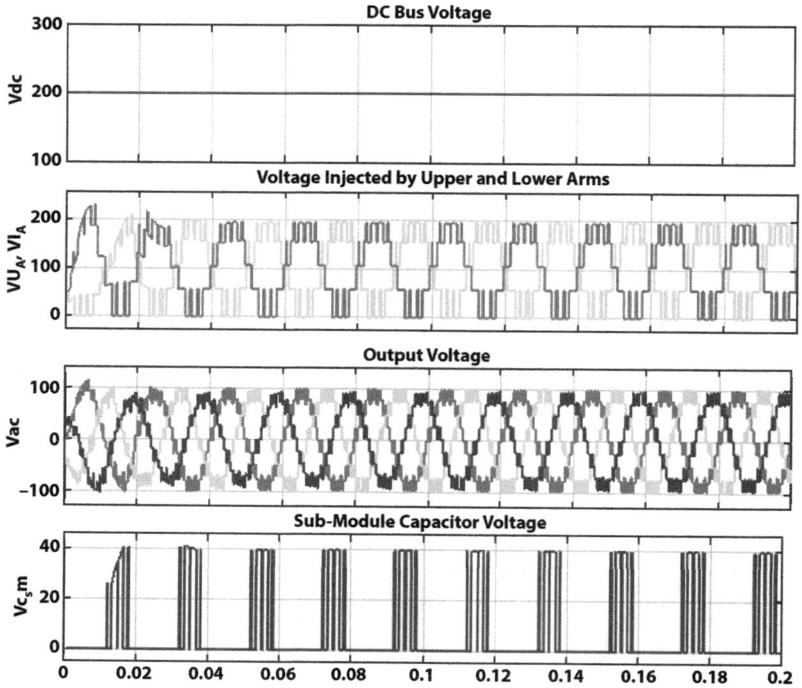

Figure 16.11 Simulation result of nine-level three-phase MMC with PD-PWM.

Table 16.1 The performance of MMC is analyzed while using several different modulation techniques.

Series no.	No. of levels	Modulation scheme	%THD in output voltage
1	5	PS-PWM	37.86%
2	5	POD-PWM	32.35%
3	5	APOD-PWM	41.48%
4	5	NLM-PWM	51.39%
5	9	PS-PWM	16.31%
6	9	PD-PWM	17.44%

16.5 Performance Analysis

The THD in the output voltage using various PWM schemes is presented in Table 16.1.

The THD in output voltage reduces as the number of levels rises, as shown in the table above. Increasing the range of voltages allows for further reduction below the IEEE 519 standard threshold of 5%. POD-PWM was also found to have a lower THD compared to alternative PWM schemes used with 5L MM. The NLM method was shown to have the greatest THD. A lower SM count in NLM has been observed to result in increased THD in the output voltage [21]. So, the NLM method is suggested for use in scenarios with more SMs. Over 5% voltage ripple in capacitors is unacceptable. To keep the MMC in a steady state under all circumstances, it is necessary to employ a control algorithm that regulates the voltage across the sub-module capacitors.

16.6 Conclusion

Reliability is an important parameter for high-power applications. Thus, to enhance the reliability of MMC and widen its scope of industrial application, a robust control algorithm must be deployed with an ability to regain the pre-fault voltage level after the various faults. A comprehensive literature review of various fault-tolerant techniques has been discussed along with the simulation of MMC under varieties of PWM techniques. The performance study graphically shows that as the number of output voltage levels increases, the total harmonic distortion (THD) decreases.

References

1. M. Abdelsalam, M. I. Marei, and S. B. Tennakoon, "An Integrated Control Strategy With Fault Detection and Tolerant Control Capability Based on Capacitor Voltage Estimation for Modular Multilevel Converters," *IEEE Trans. Ind. Appl.*, vol. 53, no. 3, pp. 2840–2851, 2017, doi: 10.1109/TIA.2016.2608940.
2. L. Yang, L. Jia, L. Luo, Y. Zhang, L. Zhao, and Z. Zhang, "A Novel Fault-Tolerant Control Method for Modular Multilevel Converters," in *2019 IEEE 4th International Future Energy Electronics Conference (IFEEC)*, 2019, pp. 1–6. doi: 10.1109/IFEEC47410.2019.9015155.
3. M. Alharbi, S. Isik and S. Bhattacharya, "An Optimized SM Fault-Tolerant Control Method For MMC-based HVDC Applications," *2019 IEEE Energy Conversion Congress and Exposition (ECCE)*, Baltimore, MD, USA, 2019, pp. 1592-1597, doi: 10.1109/ECCE.2019.8912483.
4. G. Guo *et al.*, "Application of Third-Order Harmonic Voltage Injection in a Modular Multilevel Converter," in *IEEE Trans. Ind. Electron.*, vol. 65, no. 7, pp. 5260-5271, July 2018, doi: 10.1109/TIE.2017.2777413.
5. C. Zhao *et al.*, "Energy Storage Requirements Optimization of Full-Bridge MMC With Third-Order Harmonic Voltage Injection," in *IEEE Trans. Power Electron.*, vol. 34, no. 12, pp. 11661-11678, Dec. 2019, doi: 10.1109/TPEL.2019.2906349.
6. Q. Yang, J. Qin and M. Saeedifard, "A Postfault Strategy to Control the Modular Multilevel Converter Under Submodule Failure," in *IEEE Trans. Power Deliv.*, vol. 31, no. 6, pp. 2453-2463, Dec. 2016, doi: 10.1109/TPWRD.2015.2449875.
7. J. V. M. Farias, A. F. Cupertino, H. A. Pereira, S. I. S. Junior and R. Teodorescu, "On the Redundancy Strategies of Modular Multilevel Converters," in *IEEE Trans. Power Deliv.*, vol. 33, no. 2, pp. 851-860, April 2018, doi: 10.1109/TPWRD.2017.2713394.
8. ABB, "Application note 5sya 2051: Voltage ratings of high power semiconductors," 2017.
9. L. Sun, Z. Wu, F. Xiao, X. Cai and S. Wang, "Suppression of Real Power Back Flow of Nonregenerative Cascaded H-Bridge Inverters Operating Under Faulty Conditions," in *IEEE Trans. Power Electron.*, vol. 31, no. 7, pp. 5161-5175, July 2016, doi: 10.1109/TPEL.2015.2477849.
10. A. Das, Lars, Mosgren, and IdunRunde, "Fault tolerant control of modular multilevel converters by reference modification," EP 2 978 092 A1, 2016
11. S. Farzamkia, H. Iman-Eini, A. Khoshkbar-Sadigh and M. Noushak, "A Software-Based Fault-Tolerant Strategy for Modular Multilevel Converter Using DC Bus Voltage Control," in *IEEE J. Emerg. Sel. Top. Power Electron.*, vol. 9, no. 3, pp. 3436-3445, June 2021, doi: 10.1109/JESTPE.2020.3022984.
12. Yi Zang, Xu Wang, Bin Xu and Jihong Liu, "Control Method for Cascaded H-Bridge Multilevel Inverter Failures," *2006 6th World Congress on Intelligent*

Control and Automation, Dalian, 2006, pp. 8462-8466, doi: 10.1109/WCICA.2006.1713629.
13. M. Aleenejad, H. Mahmoudi and R. Ahmadi, "Unbalanced Space Vector Modulation with Fundamental Phase Shift Compensation for Faulty Multilevel Converters," in *IEEE Trans. Power Electron.*, vol. 31, no. 10, pp. 7224-7233, Oct. 2016, doi: 10.1109/TPEL.2015.2509446.
14. J. Wang and Y. Tang, "A Fault-Tolerant Operation Method for Medium Voltage Modular Multilevel Converters With Phase-Shifted Carrier Modulation," in *IEEE Trans. Power Electron.*, vol. 34, no. 10, pp. 9459-9470, Oct. 2019, doi: 10.1109/TPEL.2018.2890536.
15. T. B. Hashfi, S. Mekhilef, M. Mubin, M. Seyedmahmoudian, B. Horan and A. Stojcevski, "Adaptive Carrier-Based PDPWM Control for Modular Multilevel Converter With Fault-Tolerant Capability," in *IEEE Access*, vol. 8, pp. 26739-26748, 2020, doi: 10.1109/ACCESS.2020.2970725.
16. M. López, F. Briz, A. Zapico, D. Diaz-Reigosa and J. M. Guerrero, "Operation of modular multilevel converters under voltage constraints," *2015 IEEE Energy Conversion Congress and Exposition (ECCE)*, Montreal, QC, Canada, 2015, pp. 3550-3556, doi: 10.1109/ECCE.2015.7310162.
17. P. Briff, F. Moreno and J. Chivite-Zabalza, "Extended Controllability of HVDC Converters in the Vector Space," in *IEEE Trans. Power Deliv.*, vol. 32, no. 3, pp. 1505-1515, June 2017, doi: 10.1109/TPWRD.2016.2606576.
18. A. F. Cupertino, H. A. Pereira, S. I. Seleme, and R. Teodorescu, "On Inherent Redundancy of MMC-Based STATCOMs in the Overmodulation Region," *IEEE Trans. Power Deliv.*, vol. 35, no. 3, pp. 1169–1179, 2020, doi: 10.1109/TPWRD.2019.2936784.
19. Y. Chen, X. Yu, and Y. Li, "A Failure Mechanism Cumulative Model for Reliability Evaluation of a k-Out-of-n System With Load Sharing Effect," *IEEE Access*, vol. 7, pp. 2210–2222, 2019, doi: 10.1109/ACCESS.2018.2852730.
20. P. Tu, S. Yang, and P. Wang, "Reliability- and Cost-Based Redundancy Design for Modular Multilevel Converter," *IEEE Trans. Ind. Electron.*, vol. 66, no. 3, pp. 2333–2342, 2019, doi: 10.1109/TIE.2018.2793263.
21. Q. Xiao *et al.*, "A Novel Fault-Tolerant Control Method for Modular Multilevel Converter with an Improved Phase Disposition Level-Shifted PWM," in *IECON 2019 - 45th Annual Conference of the IEEE Industrial Electronics Society*, 2019, vol. 1, pp. 3839–3844. doi: 10.1109/IECON.2019.8927316.
22. S. Yang, Y. Tang, and P. Wang, "Seamless Fault-Tolerant Operation of a Modular Multilevel Converter With Switch Open-Circuit Fault Diagnosis in a Distributed Control Architecture," *IEEE Trans. Power Electron.*, vol. 33, no. 8, pp. 7058–7070, 2018, doi: 10.1109/TPEL.2017.2756849.
23. B. Li, Z. Xu, J. Ding, and D. Xu, "Fault-Tolerant Control of Medium-Voltage Modular Multilevel Converters With Minimum Performance Degradation Under Submodule Failures," *IEEE Access*, vol. 6, pp. 11772–11781, 2018, doi: 10.1109/ACCESS.2018.2811904.

24. K. Li, L. Yuan, Z. Zhao, S. Lu, and Y. Zhang, "Fault-Tolerant Control of MMC With Hot Reserved Submodules Based on Carrier Phase Shift Modulation," *IEEE Trans. Power Electron.*, vol. 32, no. 9, pp. 6778–6791, 2017, doi: 10.1109/TPEL.2016.2628762.
25. J. Guo, X. Wang, J. Liang, H. Pang, and J. Gonçalves, "Reliability Modeling and Evaluation of MMCs Under Different Redundancy Schemes," *IEEE Trans. Power Deliv.*, vol. 33, no. 5, pp. 2087–2096, 2018, doi: 10.1109/TPWRD.2017.2715664.
26. L. Yang, L. Jia, L. Luo, Y. Zhang, L. Zhao, and Z. Zhang, "A Novel Fault-Tolerant Control Method for Modular Multilevel Converters," in *2019 IEEE 4th International Future Energy Electronics Conference (IFEEC)*, 2019, pp. 1–6. doi: 10.1109/IFEEC47410.2019.9015155.
27. G. S. Konstantinou, M. Ciobotaru, and V. G. Agelidis, "Effect of redundant sub-module utilization on modular multilevel converters," in *2012 IEEE International Conference on Industrial Technology*, 2012, pp. 815–820. doi: 10.1109/ICIT.2012.6210039.
28. M. Alharbi, S. Isik, and S. Bhattacharya, "A Novel Submodule Level Fault-Tolerant Approach for MMC With Integrated Scale-Up Architecture," *IEEE J. Emerg. Sel. Top. Ind. Electron.*, vol. 2, no. 3, pp. 343–352, 2021, doi: 10.1109/JESTIE.2021.3061954.
29. S. Yang, Y. Tang, and P. Wang, "Seamless Fault-Tolerant Operation of a Modular Multilevel Converter With Switch Open-Circuit Fault Diagnosis in a Distributed Control Architecture," *IEEE Trans. Power Electron.*, vol. 33, no. 8, pp. 7058–7070, 2018, doi: 10.1109/TPEL.2017.2756849.
30. A. Ghazanfari and Y. A. I. Mohamed, "New Submodule Improving Fault-Tolerant Capability of Modular Multilevel Converters," *IEEE Trans. Energy Convers.*, vol. 35, no. 2, pp. 662–671, 2020, doi: 10.1109/TEC.2020.2971910.
31. M. B. Ghat, A. Shukla, and R. Mishra, "Variable Slope Trapezoidal Reference Signal Based Control for a DC Fault Tolerant Hybrid Modular Multilevel Converter With Cascaded Full Bridges," *IEEE Trans. Ind. Appl.*, vol. 53, no. 4, pp. 3770–3781, 2017, doi: 10.1109/TIA.2017.2682170.
32. T. H. Nguyen, K. A. Hosani, and M. S. E. Moursi, "Alternating Submodule Configuration Based MMCs With Carrier-Phase-Shift Modulation in HVdc Systems for DC-Fault Ride-Through Capability," *IEEE Trans. Ind. Inf.*, vol. 15, no. 9, pp. 5214–5224, 2019, doi: 10.1109/TII.2019.2896990.
33. S. Wang, A. M. Massoud, and B. W. Williams, "A T-Type Modular Multilevel Converter," *IEEE J. Emerg. Sel. Top. Power Electron.*, vol. 9, no. 1, pp. 843–857, 2021, doi: 10.1109/JESTPE.2019.2953007.
34. E. C. Mathew, M. B. Ghat, and A. Shukla, "A Generalized Cross-Connected Submodule Structure for Hybrid Multilevel Converters," *IEEE Trans. Ind. Appl.*, vol. 52, no. 4, pp. 3159–3170, 2016, doi: 10.1109/TIA.2016.2535117.
35. L. Bieber, J. Pfannschmidt, L. Wang, A. Nami, and W. Li, "A Hybrid Three-Level and Modular Multilevel Converter With DC Fault Blocking Capability

and Reduced Semiconductor Losses," *IEEE Trans. Power Deliv.*, vol. 35, no. 4, pp. 1895–1908, 2020, doi: 10.1109/TPWRD.2019.2956723.

36. R. Li, G. P. Adam, D. Holliday, J. E. Fletcher, and B. W. Williams, "Hybrid Cascaded Modular Multilevel Converter With DC Fault Ride-Through Capability for the HVDC Transmission System," *IEEE Trans. Power Deliv.*, vol. 30, no. 4, pp. 1853–1862, 2015, doi: 10.1109/TPWRD.2015.2389758.
37. M. M. C. Merlin *et al.*, "The Alternate Arm Converter: A New Hybrid Multilevel Converter With DC-Fault Blocking Capability," *IEEE Trans. Power Deliv.*, vol. 29, no. 1, pp. 310–317, 2014, doi: 10.1109/TPWRD.2013.2282171.

17

Application of CHB-MLI as a Three-Phase Star-Connected Nine-Level Shunt Active Power Filter

Jitendra Kumar Tandekar[1*]**, Amit Ojha**[2†] **and Shailendra Jain**[2‡]

[1]*Department of Electrical Engineering, Samrat Ashok Technological Institute, Vidisha, Madhya Pradesh, India*

[2]*Department of Electrical Engineering, Maulana Azad National Institute of Technology, Bhopal, Madhya Pradesh, India*

Abstract

The increasing use of power electronics devices in modern electrical power distribution systems is causing significant power quality issues. Non-linear characteristics of power electronic loads result in harmonic distortion, voltage dips and flicker, and current and voltage imbalance. These issues negatively affect system efficiency, power factor, communication lines, and other consumers. Active power filters (APFs) have been developed as a solution to these problems. Shunt active power filters, specifically voltage source inverter (VSI)-based filters, have proved to be effective in compensating for current harmonics and reactive power. Multi-level inverters, such as the cascaded H-bridge multi-level inverter (CHB-MLI), are a better alternative to conventional two-level inverters due to their reduced voltage stress and improved harmonic profile. The CHB-MLI-based VSI is applied for active filter applications, providing a dynamic performance under transient and steady-state operating conditions. The nine-level CHB-MLI-based SAPF system is simulated in MATLAB/Simulink, showing effectiveness in mitigating power quality issues caused by non-linear loads.

Keywords: Power quality, shunt active power filter, H-bridge multilevel inverter, voltage source inverter

**Corresponding author*: jitendra.tandekar@gmail.com

†*Corresponding author*: ojha.amit@gmail.com

‡*Corresponding author*: sjain68@gmail.com

Salman Ahmad, Farhad Ilahi Bakhsh, and P. Sanjeevikumar (eds.) Multilevel Converters, (339–360)

17.1 Introduction

Shunt active power filters have been widely used to improve power quality. However, the available literature related to SAPF shows that most of the work has been carried out with conventional SAPF for current compensation and power factor improvement [1–7]. Furthermore, SAPF has limitations in medium- and high-voltage applications [8, 9]. The work presented in this chapter aims to overcome these limitations by employing MLI-based SAPF.

The recent advancement is the use of MLI that offers several advantages compared to their conventional counterparts. With the help of MLI, staircase waveforms can be produced, which, in turn, approach the sinusoidal waveform with less harmonic distortion. However, the several sources on the DC side of the converter make MLI technology difficult to control because of the need to balance the several DC voltages [10–12].

High-power multilevel converter can be topologically classified as flying capacitor converter (FC), diode-clamped converter (NPC), and cascaded H-bridge converter (CHB) [13–15]. The CHB converter requires separated DC sources; still, this topology is an attractive option due to its several advantages, such as modularity, simplest composition, and reduced number of components as it does not require any extra clamping diodes or balancing capacitors. Additionally, the structure of multi-cells in CHB allows the nourishment of different charges in DC when it is used in an active rectifier application [16]. The use of CHB MLI has been successfully implemented commercially in available large drives and some static VAR compensators [17, 18].

However, the control of CHB-MLIs is a major challenge [19]. The controller must not only be capable of regulating and stabilizing the voltage levels of each DC-side capacitor [20, 21], but it must also make the current and voltages waveforms nearly sinusoidal and in-phase with each other.

The CHB-MLI should be capable of releasing or receiving a certain amount of energy during sudden fluctuations in the distribution system due to changing load. Hence, the energy storage elements (capacitive and inductive impedance) plays an important role in the operation of SAPF [11, 12, 20–23]. A traditional proportional–integral (PI) controller is used to control the DC link voltage. The PI controller for DC link voltage sets the amplitude of the CHB-SAPF's active current to regulate the DC link voltage based on its reference value [24]. The DC voltage of each H-bridge is balanced to ensure the normal state of the system. It also ensures that the losses of CHB MLI-SAPF are supplied by the source [25–29].

In this chapter, a simple modeling technique for CHB-MLI-based SAPF with instantaneous power theory is presented. Also, the procedure adopted for the development of CHB- MLIs is also presented. CHB-MLI such as five-level, seven-level, and nine-level-based SAPFs has been modeled.

17.2 Operating Principle of the CHB-MLI-Based SAPF

A three-phase star-connected CHB-MLI-based SAPF is illustrated in Figure 17.1. There are three major components of the SAPF system, namely:

- CHB-MLIs with separate DC capacitors
- Coupling inductors
- Controllers

At every stage, a coupling inductor operates both as a converter output-current filter and a reactive power coupler. The coupling inductors serve the purpose of filtering the current harmonics that is introduced because of the effect of modulation techniques employed in the converters.

The MLI is composed of several similar types of H-bridge converters, whose outputs are connected in series. Therefore, the output voltage is the addition of each individual H-bridge converters output and is given by Equation 17.1:

$$V_{gn} = V_{g1} + V_{g2} + \cdots + V_{g\,j_{HB}} \tag{17.1}$$

where g represents the phases (a, b and c) and j_{HB} is the number of per phase H-bridge converters.

As already known, a three-level of output voltages can be synthesized by a single H-bridge converter. Therefore, the number of output phase voltage levels is equal to $(2j_{HB} + 1)$. The highest number of line voltage level possible is $(4j_{HB} + 1)$. With every increase in the number of H-bridge converters, the output voltage waveform reaches near to sinusoidal.

Even though there is an enhancement in the power capacity of the system and improvement in the dynamic response, due to various practical concerns, the number of H-bridge converters employed cannot increase beyond a certain limit. One of the most critical concerns is the complexity of the control system. Figure 17.2 depicts the phase voltage waveform of the cascaded MLI.

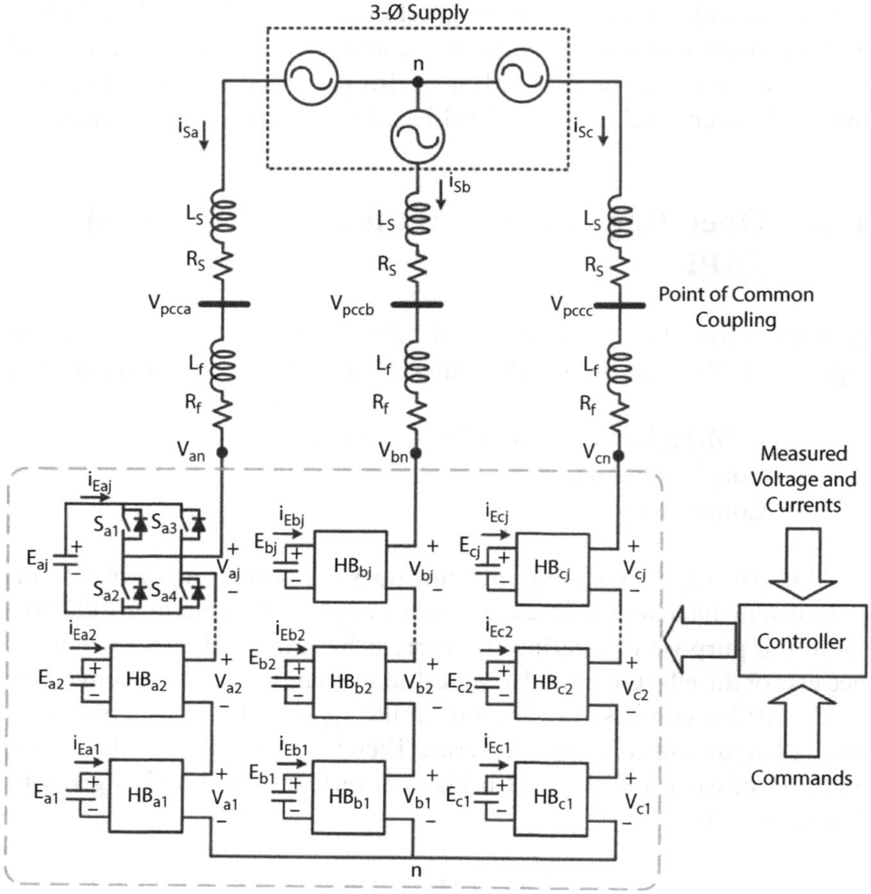

Figure 17.1 A general model of the proposed CHB-MLI-based SAPF system.

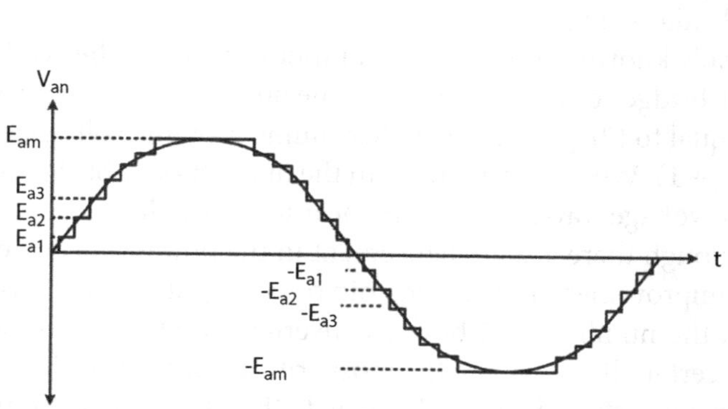

Figure 17.2 Output phase voltage waveform of CHB-MLI.

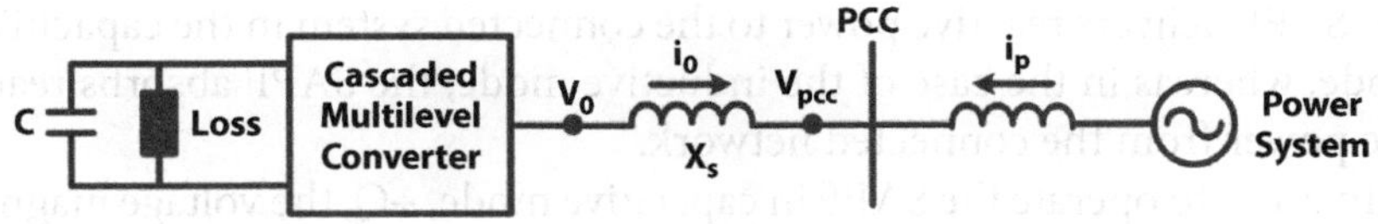

Figure 17.3 Single line diagram of CHB-MIL-based SAPF.

Figure 17.3 shows the single-line diagram of the SAPF system. Usually, the MLI is represented by ideal VSI and coupling impedance connecting it to the point of common coupling.

The transfer of active and reactive power between the CHB-MLI and the power system can be controlled by altering the magnitude and displacement angle of the converter output voltage. In case of a lossless converter, the controlled output voltage will have the same phase as that of the supply system.

The phasor diagrams filter output voltage and filter output current and voltage at the PCC in all quadrants of the PQ plane as shown in Figure 17.4.

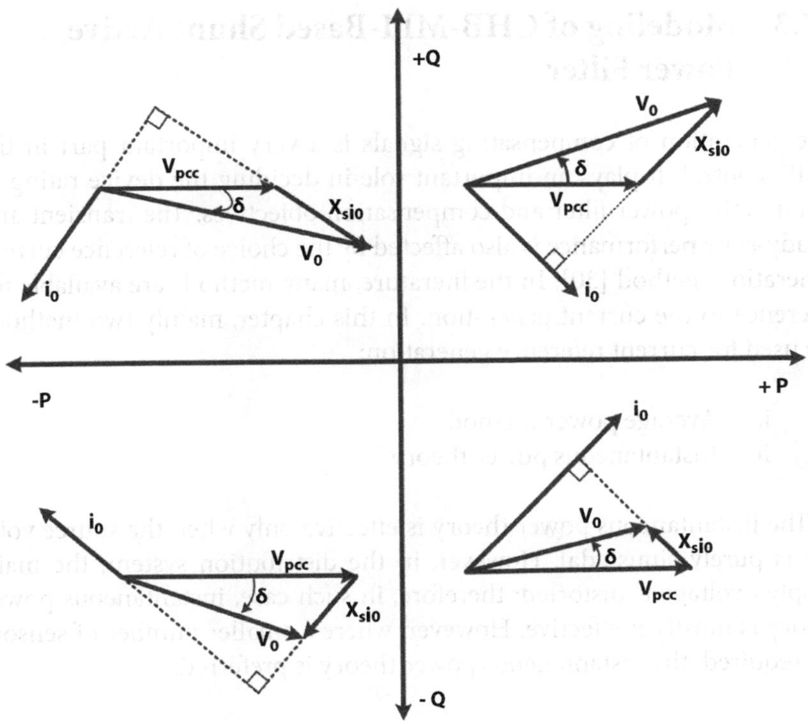

Figure 17.4 Phasor diagram for power exchanges in SAPF applications.

The SAPF delivers reactive power to the connected system in the capacitive mode, whereas in the case of the inductive mode, the SAPF absorbs reactive power from the connected network.

In order to operate the SAPF in capacitive mode, +Q, the voltage magnitude of the converter output should be greater than that of the voltage level of connected system, whereas in inductive mode, -Q, the voltage magnitude of the converter output should be less than that of the voltage level of the connected system.

The PQ plane describes two main control strategies of the CHB-MLI for SAPF applications:

- The transfer of reactive power between the CHB-MLI and the power system can be controlled by altering the magnitude of the converter output voltage.
- The transfer of active power in between the CHB-MLI and the power system can be controlled by altering the displacement angle of the converter output voltage.

17.3 Modeling of CHB-MLI-Based Shunt Active Power Filter

The generation of compensating signals is a very important part in the SAPF control. It plays an important role in deciding the device rating of shunt active power filter and compensation objectives. The transient and steady-state performance is also affected by the choice of reference current generation method [30]. In the literature, many methods are available for reference to the current generation. In this chapter, mainly two methods are used for current reference generation:

i. Average power method
ii. Instantaneous power theory

The instantaneous power theory is effective only when the source voltage is purely sinusoidal. However, in the distribution system, the main supply's voltage is distorted; therefore, in such case, instantaneous power theory is mostly ineffective. However, where a smaller number of sensors are required, the instantaneous power theory is preferred.

17.3.1 Generation of Reference Current Employing Average Power Method

The average power method provides precise results, even if the supply voltage waveform is distorted. The complete schematic of the reference current generation is presented in Figure 17.5.

v_{Sa}, v_{Sb}, and v_{Sc} are the sensed three-phase voltages and can be expressed as

$$\left.\begin{aligned} v_{Sa} &= V_m Sin\omega t \\ v_{Sb} &= V_m Sin(\omega t + 120^{\bullet}) \\ v_{Sc} &= V_m Sin(\omega t - 120^{\bullet}) \end{aligned}\right\} \quad (17.2)$$

V_m is the peak value of the source voltages, and it is calculated using Equation 17.3.

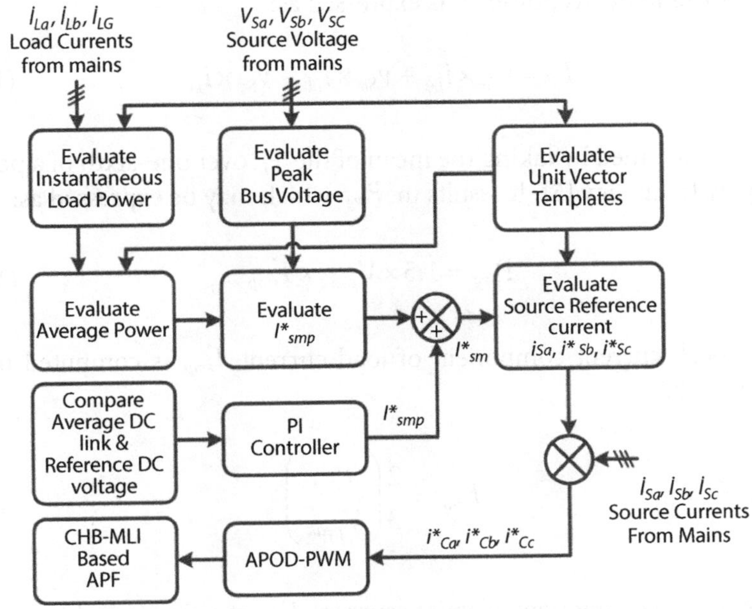

Figure 17.5 Complete schematic of reference current generation.

$$V_m = \sqrt{\frac{2}{3}\left(v_{Sa}^2 + v_{Sb}^2 + v_{Sc}^2\right)} \tag{17.3}$$

In order to obtain unit vector templates, the sensed voltages are divided by the peak value of voltage as expressed in Equation 17.4.

$$\left.\begin{aligned} U_{Sa} &= \frac{v_{Sa}}{V_m} \\ U_{Sb} &= \frac{v_{Sb}}{V_m} \\ U_{Sc} &= \frac{v_{Sc}}{V_m} \end{aligned}\right\} \tag{17.4}$$

The peak source current I_{sm}^* has two components, namely, I_{smp}^* and I_{smd}^*. The values of these two components are estimated as follows. The active source component consequent to the load I_{smp}^* is evaluated through the average power P_{avg}.

The instantaneous power P_L is expressed as:

$$P_L = v_{Sa} \times i_{La} + v_{Sb} \times i_{Lb} + v_{Sc} \times i_{Lc} \tag{17.5}$$

P_{avg} is obtained by taking the mean of the P_L over one-sixth of a period of supply frequency [5]. It results in P_{avg} which may be expressed as:

$$P_{avg} = 1.5 \times V_{peak} \times I_{smp}^* \tag{17.6}$$

The peak current component of load current I_{smp}^* is computed using Equation 17.7:

$$I_{smp}^* = \frac{2}{3}\left(\frac{P_{avg}}{V_{peak}}\right) \tag{17.7}$$

The second component of source current I_{smd}^* is estimated with the help of DC voltage error. For this, the average DC link voltage V_{dc} of the SAPF is sensed and compared with the desired reference DC voltage V_{dc_ref} to generate the DC voltage error V_{dc_error}, whose value can be represented as:

$$V_{dc\_error} = V_{dc\_ref} - V_{dc} \tag{17.8}$$

For maintaining the DC link voltage constant, the error is processed through the DC voltage PI controller. The control output is expressed as:

$$I^*_{smd} = \left\{ k_P + \frac{k_I}{s} \right\} V_{dc\_error} \tag{17.9}$$

where k_P is the proportional gain and k_I is the integral gain of the PI controller. I^*_{smd} is the output of the PI controller. The control signal of this PI controller decides to maintain the average voltage on the DC link at a constant as shown in Figure 17.6.

I^*_{smd} is the component responsible for proving the power losses occurring inside the SAPF. These two components, when added, give the I^*_{sm} component.

The reference supply currents i^*_{Sa}, i^*_{Sb}, and i^*_{Sc} are obtained through the product of I^*_{sm} with unit vector templates derived from sensed bus phase voltages, namely, v_{Sa}, v_{Sb}, and v_{Sc}. The complete block diagram of the scheme is shown in Figure 17.5. The obtained reference of the APF currents are i^*_{Ca}, i^*_{Cb}, and i^*_{Cc} calculated using the error between the instantaneous supply current and generated reference currents as:

$$\left.\begin{aligned} i^*_{Ca} &= i^*_{Sa} - i_{Sa} \\ i^*_{Cb} &= i^*_{Sb} - i_{Sb} \\ i^*_{Cc} &= i^*_{Sc} - i_{Sc} \end{aligned}\right\} \tag{17.10}$$

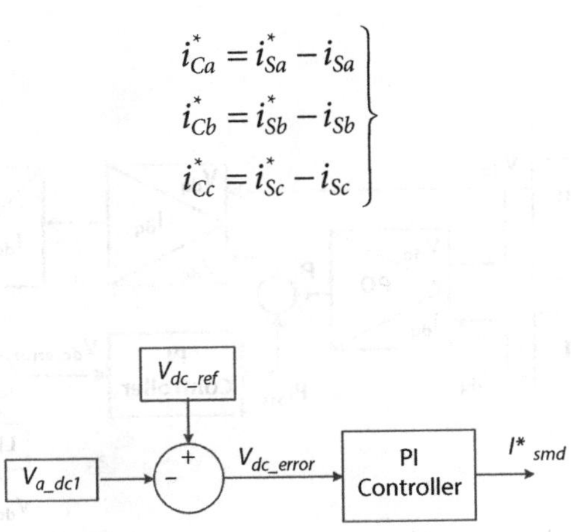

Figure 17.6 DC link capacitor voltage control scheme.

These reference signals, when compared with level-shifted triangular carriers f_C, generate the gate pluses for switches in CHB-MLI, which is discussed in the previous chapter.

17.3.2 Reference Current Generation Using Instantaneous Power Theory

To generate such reference current signals, it utilizes the p–q algorithm, also known as active and reactive power compensation method, based on the principle of α–β modulation [30]. Source voltages and the load currents are represented by Equations 17.11 and 17.12 which are acting as feedback. Figure 17.7 demonstrates the instantaneous power computational scheme for generating the suitable reference current in the system.

$$\left.\begin{aligned} v_{Sa} &= Sin(\omega t) \\ v_{Sb} &= Sin(\omega t - 120^{\bullet}) \\ v_{Sc} &= Sin(\omega t + 240^{\bullet}) \end{aligned}\right\} \tag{17.11}$$

Similarly for current,

$$\left.\begin{aligned} i_{Sa} &= Sin(\omega t) \\ i_{Sb} &= Sin(\omega t - 120^{\bullet}) \\ i_{Sc} &= Sin(\omega t + 240^{\bullet}) \end{aligned}\right\} \tag{17.12}$$

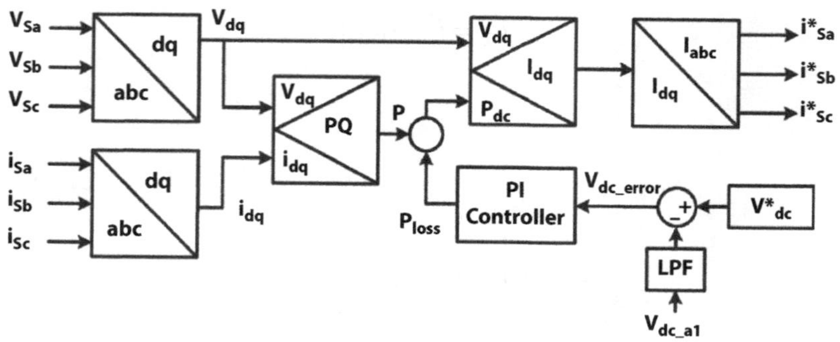

Figure 17.7 Reference current generation by instantaneous power theory.

From Equation 17.11:

$$v_d = \sqrt{\frac{2}{3}}\left(v_{Sa} - \frac{v_{Sb}}{2} - \frac{v_{Sc}}{2}\right) \tag{17.13}$$

$$v_q = \sqrt{\frac{2}{3}}\left[\sqrt{3}\left(\frac{v_{Sb}}{2} - \frac{v_{Sc}}{2}\right)\right] \tag{17.14}$$

Similarly from Equation 17.12:

$$i_d = \sqrt{\frac{2}{3}}\left(i_{Sa} - \frac{i_{Sb}}{2} - \frac{i_{Sc}}{2}\right) \tag{17.15}$$

$$i_q = \sqrt{\frac{2}{3}}\left[\sqrt{3}\left(\frac{i_{Sb}}{2} - \frac{i_{Sc}}{2}\right)\right] \tag{17.16}$$

$$P = v_d * i_d + v_q * i_q \tag{17.17}$$

$$Q = v_d * i_q - v_q * i_d \tag{17.18}$$

$$P_{dc} = P_{loss} + P \tag{17.19}$$

For P_{dc} estimation, the actual voltage V_{dc_a1} across CHB is sensed and compared with the desired reference DC voltage V_{dc}^{*} to generate the DC voltage error V_{dc_error} as:

$$V_{dc\_error} = V_{dc}^{*} - V_{dc\_a1} \tag{17.20}$$

For keeping a continuous DC bus voltage, the error is passed through a proportional–integral controller. Thus, the control output power is stated as:

$$P_{loss} = \left(Kp + \frac{K_I}{s}\right)V_{dc\_error} \tag{17.21}$$

where K_P is the constant denoting proportional gain and K_I is the symbolic denotation for an integral gain of the proportional–integral (PI) controller, and P_{dc} is controlled by the output power of the PI controller.

The reference direct and quadrature component of the current is shown in Equation 17.21.

$$i_d^* = \frac{v_d * P_{dc}}{v_d^2 + v_q^2}, i_q^* = \frac{v_q * P_{dc}}{v_d^2 + v_q^2}, \tag{17.22}$$

$$\left.\begin{aligned} i_a^* &= \sqrt{\frac{2}{3}} * i_d \\ i_b^* &= \sqrt{\frac{2}{3}}\left(-\frac{i_d}{2} + \frac{\sqrt{3}}{2} i_q\right) \\ i_c^* &= \sqrt{\frac{2}{3}}\left(-\frac{i_d}{2} - \frac{\sqrt{3}}{2} i_q\right) \end{aligned}\right\} \tag{17.23}$$

The instantaneous output power is fragmented into mean and variable terms, utilizing the second-order equation to form a low pass filter. For the specific purpose, Butterworth low- pass filter is chosen to segregate AC ripples out of the constant DC.

17.4 Nine-Level CHB-MLI-Based Shunt Active Power Filter

CHB has universally adopted topology in the field of multilevel inverter for high-power application due to its modular arrangement. It can produce high power output, which eliminates the requirement of a coupling transformer, required for interfacing the point of common coupling of the distribution network. The m-level-cascaded multilevel converters consist of (m-1)/2 single-phase full bridges where each bridge has its own separate DC link. The wave shape can be improved by increasing the number of output levels. Three-phase star-connected nine-level CHB-MLI-based APF is shown in Figure 17.8. In this, i_{Sa},i_{Sb}, and i_{Sc} are the three-phase AC grid currents, i_{La},i_{Lb}, and i_{Lc} are nonlinear load currents drawn by an

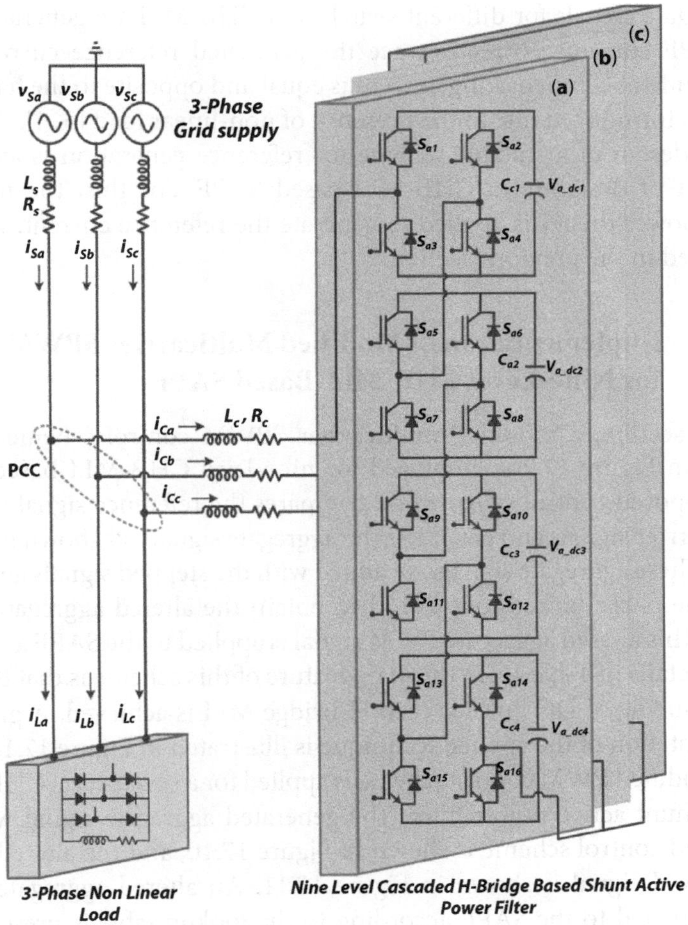

Figure 17.8 Schematic diagram of three-phase star-connected nine-level CHB-MLI-based SAPF.

uncontrolled rectifier feeding an inductive load, and i_{fa}, i_{fb}, and i_{fc} are the filter currents given by the APF.

$$\left.\begin{aligned} i_{fa} &= i_{Sa} - i_{La} \\ i_{fb} &= i_{Sb} - i_{Lb} \\ i_{fc} &= i_{Sc} - i_{Lc} \end{aligned}\right\} \tag{17.24}$$

The gate signals for different switches in CHB-MLI are generated such that APF currents correctly trace the generated reference current. The magnitude of compensating current is equal and opposite to the harmonic current, introduced due to the presence of non-linear load [3, 4].

The design of a suitable scheme for reference generation is an essential part of designing a CHB-MLI-based SAPF. For this, the instantaneous power theory is applied to generate the reference current, which is described in the previous section.

17.4.1 Implementation of Modified Multicarrier SPWM for Nine-Level CHB-MLI-Based SAPF

In this section, a modified multicarrier SPWM control scheme as presented in Figure 17.9 is employed for nine-level CHB-MLI-based SAPF. The proposed control scheme first compares the reference signal with the multicarrier signals and generates the aggregate signals, as shown in Figure 17.10. These aggregate signals are added with the stepped signals generated from the phase-locked loop (PLL) to obtain the altered aggregate PWM signal. This altered aggregate PWM signal is applied to the SAPF according to the details in Table 17.1. The main feature of this scheme is that the voltage balancing of DC link for each H-bridge MLI is achieved. A graphical representation of the applied technique is illustrated in Figure 17.11.

A modified PWM control scheme is applied for a seven-level CHB-MLI-based shunt active power filter. The generated aggregate signal with the proposed control scheme is shown in Figure 17.10, and the altered aggregate PWM signal is shown in Figure 17.11. An altered aggregate PWM signal applied to the SAPF according to the lookup table is presented in Table 17.1.

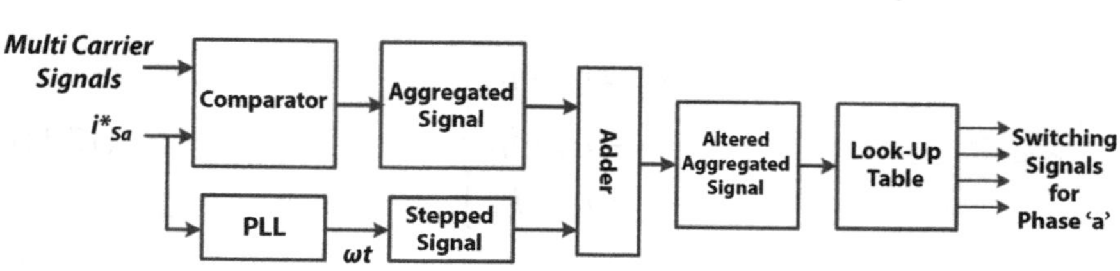

Figure 17.9 PWM generated by the altered aggregate signal.

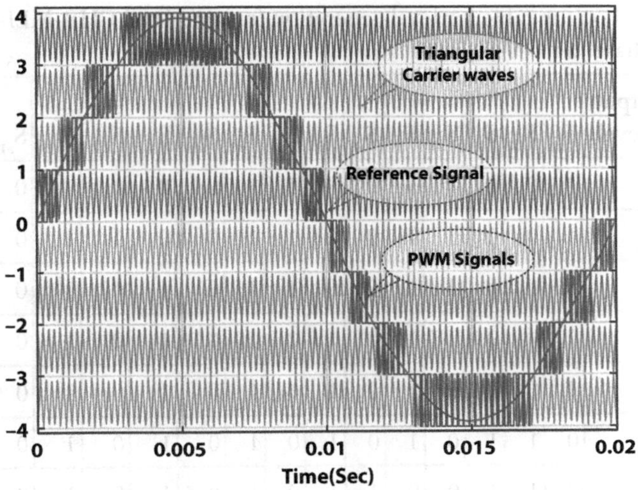

Figure 17.10 Aggregated signal APOD-PWM for nine-level MLI.

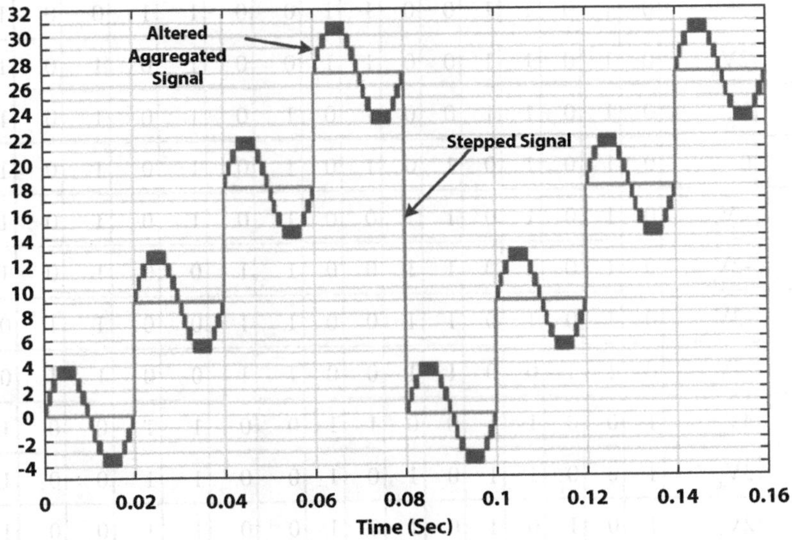

Figure 17.11 Altered aggregate PWM signal for nine-level MLI.

Table 17.1 Lookup table for level generation for nine-level CHB-MLI-based SAPF of phase "a".

State	Output level	S_{a1}	S_{a2}	S_{a3}	S_{a4}	S_{a5}	S_{a6}	S_{a7}	S_{a8}	S_{a9}	S_{a10}	S_{a11}	S_{a12}	S_{a13}	S_{a14}	S_{a15}	S_{a16}
4	$4V_{dc}$	1	0	0	1	1	0	0	1	1	0	0	1	1	0	0	1
3	$3V_{dc}$	1	0	0	1	1	0	0	1	1	0	0	1	1	0	1	0
2	$2V_{dc}$	1	0	0	1	1	0	0	1	1	0	1	0	1	0	1	0
1	V_{dc}	1	0	0	1	1	0	1	0	1	0	1	0	1	0	1	0
0	0	1	0	1	0	1	0	1	0	1	0	1	0	1	0	1	0
-1	$-V_{dc}$	0	1	1	0	1	0	1	0	1	0	1	0	1	0	1	0
-2	$-2V_{dc}$	0	1	1	0	0	1	1	0	1	0	1	0	1	0	1	0
-3	$-3V_{dc}$	0	1	1	0	0	1	1	0	0	1	1	0	1	0	1	0
-4	$-4V_{dc}$	0	1	1	0	0	1	1	0	0	1	1	0	0	1	1	0
13	$4V_{dc}$	1	0	0	1	1	0	0	1	1	0	0	1	1	0	0	1
12	$3V_{dc}$	0	1	0	1	1	0	0	1	1	0	0	1	1	0	0	1
11	$2V_{dc}$	0	1	0	1	1	0	0	1	1	0	0	1	0	1	0	1
10	V_{dc}	0	1	0	1	1	0	0	1	0	1	0	1	0	1	0	1
9	0	0	1	0	1	0	1	0	1	0	1	0	1	0	1	0	1
8	$-V_{dc}$	0	1	0	1	0	1	1	0	0	1	0	1	0	1	0	1
7	$-2V_{dc}$	0	1	0	1	0	1	1	0	0	1	1	0	0	1	0	1
6	$-3V_{dc}$	0	1	0	1	0	1	1	0	0	1	1	0	0	1	1	0
5	$-4V_{dc}$	0	1	1	0	0	1	1	0	0	1	1	0	0	1	1	0
22	$4V_{dc}$	1	0	0	1	1	0	0	1	1	0	0	1	1	0	0	1
21	$3V_{dc}$	1	0	0	1	1	0	1	0	1	0	0	1	1	0	0	1
20	$2V_{dc}$	1	0	1	0	1	0	1	0	1	0	0	1	1	0	0	1
19	V_{dc}	1	0	1	0	1	0	1	0	1	0	0	1	1	0	1	0
18	0	1	0	1	0	1	0	1	0	1	0	1	0	1	0	1	0
17	$-V_{dc}$	1	0	1	0	1	0	1	0	0	1	1	0	1	0	1	0

(*Continued*)

Table 17.1 Lookup table for level generation for nine-level CHB-MLI-based SAPF of phase "a". (*Continued*)

State	Output level	S_{a1}	S_{a2}	S_{a3}	S_{a4}	S_{a5}	S_{a6}	S_{a7}	S_{a8}	S_{a9}	S_{a10}	S_{a11}	S_{a12}	S_{a13}	S_{a14}	S_{a15}	S_{a16}
16	$-2V_{dc}$	1	0	1	0	1	0	1	0	0	1	1	0	0	1	1	0
15	$-3V_{dc}$	0	1	1	0	1	0	1	0	0	1	1	0	0	1	1	0
14	$-4V_{dc}$	0	1	1	0	0	1	1	0	0	1	1	0	0	1	1	0
31	$4V_{dc}$	1	0	0	1	0	0	0	1	1	0	0	1	1	0	0	1
30	$3V_{dc}$	1	0	0	1	1	0	0	1	0	1	0	1	1	0	0	1
29	$2V_{dc}$	1	0	0	1	1	1	0	1	0	1	0	1	1	0	0	1
28	V_{dc}	0	1	0	1	0	1	0	1	0	1	0	1	1	0	0	1
27	0	0	1	0	1	0	1	0	1	0	1	0	1	0	1	0	1
26	$-V_{dc}$	0	1	0	1	0	1	0	1	0	1	0	1	0	1	1	0
25	$-2V_{dc}$	0	1	1	0	0	1	0	1	0	1	0	1	0	1	1	0
24	$-3V_{dc}$	0	1	1	0	0	1	1	0	0	1	0	1	0	1	1	0
23	$-4V_{dc}$	0	1	1	0	0	1	1	0	0	1	1	0	0	1	1	0

17.4.2 Simulation of Nine-Level CHB-MLI-Based SAPF

In this section, different simulation results of nine-level CHB-MLI-based SAPF are presented. The various design parameters considered for realizing the CHB-MLI-based SAPF is also presented in Table 17.2.

Figure 17.12 depicts the source voltages (v_{Sa}, v_{Sb}, and v_{Sc}) five-level inverter voltages (v_{fa}, v_{fb}, and v_{fc}) source currents (i_{Sa}, i_{Sb}, and i_{Sc}), load currents (i_{La}, i_{Lb}, and i_{Lc}), SAPF currents (i_{fa}, i_{fb}, and i_{fc}), and all DC link voltages of each H-bridge (V_{dc_a1}, V_{dc_a2}, V_{dc_a3}, V_{dc_a4}, V_{dc_b1}, V_{dc_b2}, V_{dc_b3}, V_{dc_b4}, V_{dc_c1}, V_{dc_c2}, V_{dc_c3}, and V_{dc_c4}). In the simulation, the instant at which the filter is switched on is t_1 (0.2 s). It can be observed that when the filter is switched on, the source current becomes sinusoidal from non-sinusoidal wave shape, and the DC link voltage attains a steady value within a few number of cycles.

The steady-state waveform of the various parameters discussed above for the duration from 0.55 s to 0.65 s is presented in Figure 17.13, and Figure 17.14 depicts the THD profile of the load current i_{La} and source

Table 17.2 Set parameters of cascade H-bridge multilevel inverter-based shunt active power filter.

Parameter	Set value
Source voltage/frequency, v_{Sa} / f_S	11 kV/50 Hz
Source inductance, L_{Sa}	1 mH
Source resistance, R_{Sa}	0.51 Ω
AC side filter inductance, L_{fa}	3.3 mH
AC side filter resistance, R_{fa}	0.6 Ω
DC link capacitor, $C_{dc\_abc(1-6)}$	2,200 μF
Reference voltage, V_{dc_a1}	1,800 V
Carrier frequency, C_f	5 kHz

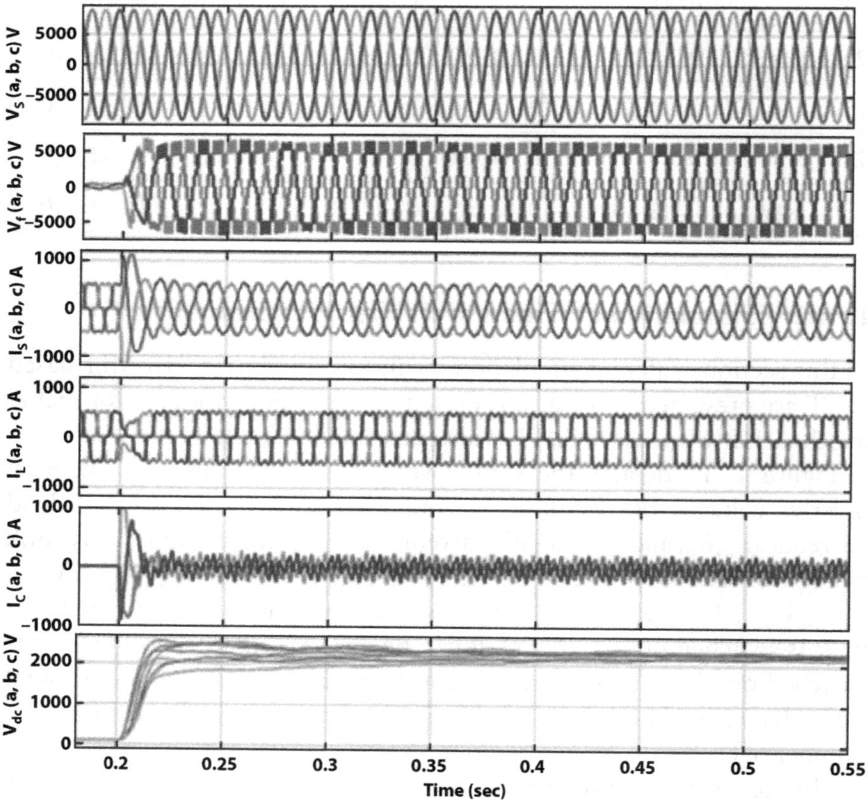

Figure 17.12 Source voltage, converter phase voltage, source current, load current filter current, and DC link voltages of each H-bridge.

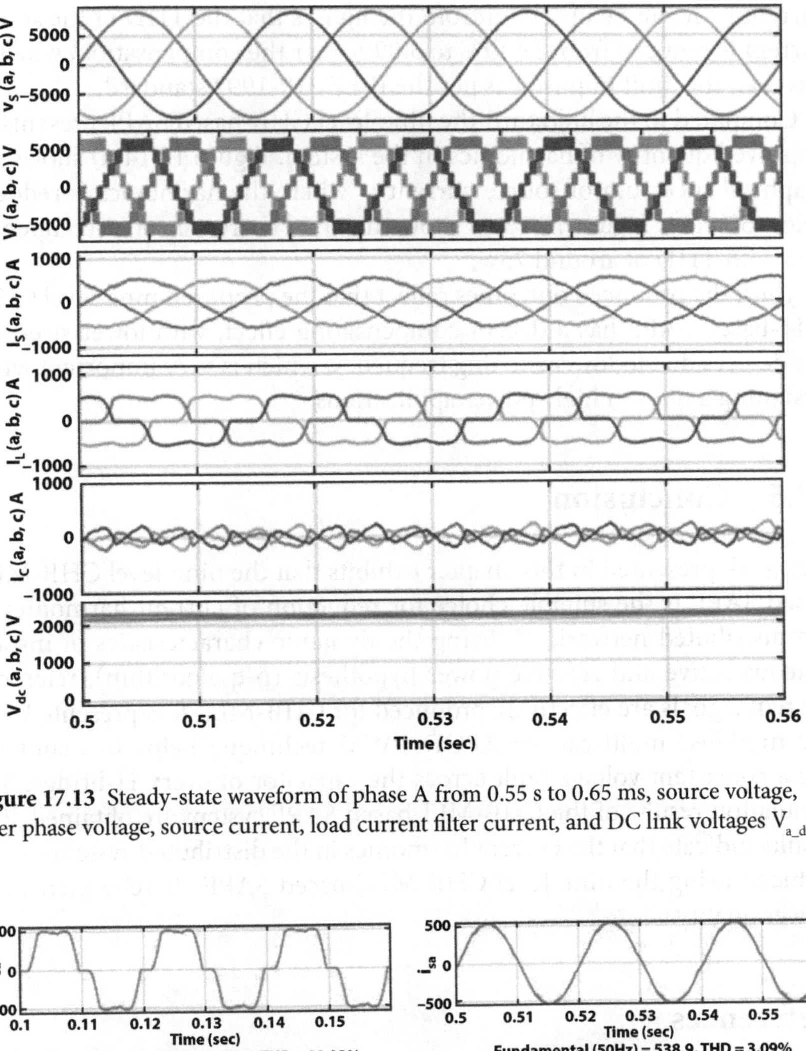

Figure 17.13 Steady-state waveform of phase A from 0.55 s to 0.65 ms, source voltage, filter phase voltage, source current, load current filter current, and DC link voltages V_{a_dc1}.

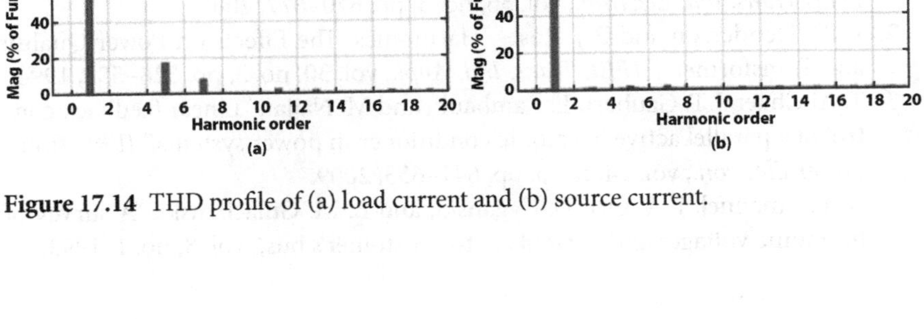

Figure 17.14 THD profile of (a) load current and (b) source current.

current i_{S}. It can be observed from the figures that the THD of the source current is reduced from 19.12% to 3.09% after the compensation, which is less than the limit imposed as per the IEEE 519-1992 standard.

Compared to the literature, the nine-level CHB-based SAPF presents an improved quantity of harmonics in the system. Figure 17.14(a) shows the graphical spectrum of source current i_{sa} where the harmonics is reduced below 5%, and Figure 17.14(b) shows the load current i_{la} after compensation with THD of around 25%.

Thus, the produced outcomes reflect that the proposed nine-level CHB-MLI-based SAPF has a decent compensating effect, with lower stress on the devices due to low switching frequency, which is very important while designing a filter in high-power applications.

17.5 Conclusion

The work presented in this chapter exhibits that the nine-level CHB-MLI-based SAPF is the suitable choice for reduction of current harmonics in the distributed network. Utilizing the dynamic characteristics of instantaneous active and reactive power hypothesis (p-q algorithm), reference current signals are effectively produced for CHB-MLI. It is presented that the modified multi-carrier APOD PWM technique helps in maintaining a consistent voltage built across the capacitor of every H-bridge. The simulation results of the CHB-MLI-based SAPF system are obtained. The results indicate that the current harmonics in the distributed system can be reduced using the nine-level CHB-MLC-based SAPF, thereby increasing power quality.

References

1. V. F. Corasaniti, M. B. Barbieri, P. L. Arnera, and M. I. Valla, "Hybrid active filter for reactive and harmonics compensation in a distribution network," *IEEE Trans. Ind. Electron.*, vol. 56, no. 3, pp. 670–677, 2009.
2. R. D. Henderson and P. J. Rose, "Harmonics: The Effects on Power Quality and Transformers," *IEEE Trans. Ind. Appl.*, vol. 30, no. 3, pp. 528–532, 1994.
3. T. Al Chaer, J. P. Gaubert, L. Rambault, and M. Najjar, "Linear feedback control of a parallel active harmonic conditioner in power systems," *IEEE Trans. Power Electron.*, vol. 24, no. 3, pp. 641–653, 2009.
4. A. E. Emanuel, J. A. Orr, D. Cyganski, and E. M. Gulachenski, "A survey of harmonic voltages and currents at the customer's bus," vol. 8, no. 1, 1993.

5. S. Rajasekar, A. Senthilkumar, Y. Shasi Kumar, and P. Ajay-D-VimalRaj, "Power quality enhancement using cascaded multilevel inverter based shunt hybrid active power filter," *Int. J. Eng. Sci. Technol.*, vol. 3, no. 9, pp. 37–46, 2012.
6. "A New Approach for Harmonic Compensation Using Single- phase Hybrid Active Power Filter.pdf.".
7. M. Azizur Rahman, "Analysis of current controllers for voltage-source inverter," *IEEE Trans. Ind. Electron.*, vol. 44, no. 4, pp. 477–485, 1997.
8. R. D. Henderson and P. J. Rose, "Harmonics : The Effects On Power Ouality And Transformers Robert."
9. S. K. Jain, P. Agarwal, and H. O. Gupta, "A Control Algorithm for Compensation of Customer-Generated Harmonics and Reactive Power," vol. 19, no. 1, pp. 357–366, 2004.
10. Kai Sun, Lipei Huang, Hongyan Ma, and Qing Wei, "Analysis of phase current reconstruction precision for PWM-VSI," no. 100, pp. 231–231, 2010.
11. K. Takeda, M. Ichinose, M. Taniguchi, and H. Miyata, "The alternative pulse reduction algorithm for three-phase voltage-source inverter," *2010 Int. Power Electron. Conf. - ECCE Asia -, IPEC 2010*, vol. 02, no. 2, pp. 3087–3092, 2010.
12. K. Selvajyothi and P. A. Janakiraman, "Reduction of voltage harmonics in single phase inverters using composite observers," *IEEE Trans. Power Deliv.*, vol. 25, no. 2, pp. 1045–1057, 2010.
13. T. Ahmed, K. Nishida, and M. Nakaoka, "Deadbeat current control of LCL-filter for grid connected three-phase voltage source inverter," *Proc. Int. Conf. Power Electron. Drive Syst.*, no. December, pp. 459–467, 2011.
14. K. Lee and G. Nojima, "Quantitative power quality and characteristic analysis of multilevel pulse width modulation methods in medium voltage high power industrial drives," *IEEE Energy Convers. Congr. Expo. Energy Convers. Innov. a Clean Energy Futur. ECCE 2011, Proc.*, pp. 3314–3321, 2011.
15. J. C. Vasquez, J. M. Guerrero, M. Savaghebi, J. Eloy-Garcia, and R. Teodorescu, "Modeling, analysis, and design of stationary-reference-frame droop-controlled parallel three-phase voltage source inverters," *IEEE Trans. Ind. Electron.*, vol. 60, no. 4, pp. 1271–1280, 2013.
16. Y. A. R. I. Mohamed, "Mitigation of dynamic, unbalanced, and harmonic voltage disturbances using grid-connected inverters with LCL filter," *IEEE Trans. Ind. Electron.*, vol. 58, no. 9, pp. 3914–3924, 2011.
17. S. Mikkili and A. K. Panda, "Electrical Power and Energy Systems Real-time implementation of PI and fuzzy logic controllers based shunt active filter control strategies for power quality improvement," *Int. J. Electr. Power Energy Syst.*, vol. 43, no. 1, pp. 1114–1126, 2012.
18. M. Angulo, D. A. Ruiz-Caballero, J. Lago, M. L. Heldwein, and S. A. Mussa, "Active power filter control strategy with implicit closed-loop current control and resonant controller," *IEEE Trans. Ind. Electron.*, vol. 60, no. 7, pp. 2721–2730, 2013.

19. F. S. Tidjani, "Control strategy for improving the power flow between Home Integrated Photovoltaic System, Plug- in Hybrid Electric Vehicle and Distribution Network," *IECON 2014 - 40th Annu. Conf. IEEE Ind. Electron. Soc.*, pp. 2003–2009.
20. S. Rahmani, A. Hamadi, K. Al-Haddad, and L. A. Dessaint, "A combination of shunt hybrid power filter and thyristor-controlled reactor for power quality," *IEEE Trans. Ind. Electron.*, vol. 61, no. 5, pp. 2152–2164, 2014.
21. R. Panigrahi, B. Subudhi, and P. C. Panda, "Model predictive-based shunt active power filter with a new reference current estimation strategy," no. June 2014, pp. 221–233, 2015.
22. A. Marcus, N. Lima, and S. Member, "A Robust DC-Link Voltage Control Strategy to Enhance the Performance of Shunt Active Power Filters Without Harmonic Detection Schemes," vol. 62, no. 2, pp. 803–813, 2015.
23. S. S. Patnaik and A. K. Panda, "Cascaded Three-level Inverter based Shunt Active Filter for Power Conditioning Application," 2013.
24. B. Singh, S. S. Murthy, L. Fellow, R. Sekhara, and R. Chilipi, "STATCOM-Based Controller for a Three-Phase SEIG Feeding Single-Phase Loads," vol. 29, no. 2, pp. 320–331, 2014.
25. L. Shridhar, B. Singh, and C. S. Jha, "Transient performance of the self regulated short nt self excited induction generator," *IEEE Trans. Energy Convers.*, vol. 10, no. 2, pp. 261–267, 1995.
26. "Voltage compensation of an induction generator with long-shunt connection," *IEEE Trans. Energy Convers.*, vol. 4, no. 3. pp. 526–530, 1989.
27. M. B. Brennen and A. Abbondanti, "Static Exciters for Induction Generators," *IEEE Trans. Ind. Appl.*, vol. IA-13, no. 5, pp. 422–428, 1977.
28. G. K. Kasal and B. Singh, "Voltage and frequency controllers for an asynchronous generator-based isolated wind energy conversion system," *IEEE Trans. Energy Convers.*, vol. 26, no. 2, pp. 402–416, 2011.
29. B. Singh, S. S. Murthy, and S. Gupta, "STATCOM-based voltage regulator for self-excited induction generator feeding nonlinear loads," *IEEE Trans. Ind. Electron.*, vol. 53, no. 5, pp. 1437–1452, 2006.
30. S. Meo and A. Perfetto, "Comparison of different control techniques for active filter applications," *ICCDCS 2002 - 4th IEEE Int. Caracas Conf. Devices, Circuits Syst.*, vol. 45, no. 5, pp. 722–729, 2002.

Index

Also of Interest

Check out these other related titles from Scrivener Publishing

PHOTOVOLTAIC SYSTEMS TECHNOLOGY, Edited by Mohammed Aslam Husain, MD Waseem Ahmad, Farhad Ilahi Bakhsh, Sanjeevikumar Padmanaban, and Hasmat Malik, ISBN: 9781394166428. Discover comprehensive insights into the latest advancements in solar PV technology, including power electronics, maximum power point tracking schemes, and forecasting techniques, with a focus on improving the performance of PV systems.

ELECTRIC VEHICLE DESIGN: Design, Simulation and Applications, Edited by Krishan Arora, Suman Lata Tripathi, and Himashu Sharma, ISBN: 9781394204373. This book will serve as a definitive guide to conceptual and practical knowledge about the design of hybrid electrical vehicles (HEV), battery electrical vehicles (BEV), fuel cell electrical vehicles (FCEV), plug-in hybrid electrical vehicles (PHEV), and efficient EV charging techniques with advanced tools and methodologies for students, engineers, and academics alike.

INDUSTRIAL CONTROL SYSTEMS, Edited by Vipin Chandra Pal, Suman Lata Tripathi, and Souvik Ganguli, ISBN: 9781119829256. This volume serves as a comprehensive guide in the journey of industrial control systems with a multidisciplinary approach to the key engineering problems in the 21st century.

MODERN AUTOMOTIVE ELECTRICAL SYSTEMS: Theory and Applications, Edited by Pedram Asef, Sanjeevikumar Padmanaban, and Andrew Lapthorn, ISBN: 1119801047. Presenting the concepts and advances of modern automotive electrical systems, this volume, written and edited by a global team of experts, also goes into the practical applications for the engineer, student, and other industry professionals.

CLEAN AND RENEWABLE ENERGY PRODUCTION, Edited by Surajit Mondal, Adesh Kumar, Rupendra Kumar Pachauri, Amit Kumar Mondal, Vishal Kumar Singh, and Amit Kumar Sharma, ISBN: 9781394177723. This exciting new volume explores the recent advances in renewable energy production, the policies surrounding renewable energy, and practical applications.

URBAN ENERGY SYSTEMS: Modelling and Simulation for Sustainable Smart Cities, Edited by Deepak Kumar, ISBN: 9781119847441. Presenting the latest technologies and practices in this ever-changing field, this groundbreaking new volume covers the gambit for providing solutions and practical applications of smart and efficient energy systems.

ENERGY STORAGE TECHNOLOGIES IN GRID MODERNIZATION, Edited by Sandeep Dhundhara, Yajvender Pal Verma, and Ashwani Kumar, ISBN: 9781119872115. Written and edited by a team of experts, this exciting new volume discusses the various types of energy storage technologies, the applications of energy storage systems, the performance improvement of modern power systems, their role in the real-time operation of power markets, and the operational issues of modern power systems, including renewable-based generating sources.

SMART GRIDS FOR SMART CITIES VOLUME 1, Edited by O.V. Gnana Swathika, K. Karthikeyan, and Sanjeevikumar Padmanaban, ISBN: 9781119872078. Written and edited by a team of experts in the field, this first volume in a two-volume set focuses on an interdisciplinary perspective on the financial, environmental, and other benefits of smart grid technologies and solutions for smart cities.

SMART GRIDS FOR SMART CITIES VOLUME 2: Real-Time Applications in Smart Cities, Edited by O.V. Gnana Swathika, K. Karthikeyan, and Sanjeevikumar Padmanaban, ISBN: 9781394215874. Written and edited by a team of experts in the field, this second volume in a two-volume set focuses on an interdisciplinary perspective on the financial, environmental, and other benefits of smart grid technologies and solutions for smart cities.

SMART GRIDS AND INTERNET OF THINGS, Edited by Sanjeevikumar Padmanaban, Jens Bo Holm-Nielsen, Rajesh Kumar Dhanaraj, Malathy Sathyamoorthy, and Balamurugan Balusamy, ISBN: 9781119812449. Written and edited by a team of international professionals, this groundbreaking new volume covers the latest technologies in automation, tracking, energy distribution and consumption of Internet of Things (IoT) devices with smart grids.

SMART GRIDS AND GREEN ENERGY SYSTEMS, Edited by A. Chitra, V. Indragandhi and W. Razia Sultana, ISBN: 9781119872030. Presenting the concepts and advances of smart grids within the context of "green" energy systems, this volume, written and edited by a global team of experts, goes into the practical applications that can be utilized across multiple disciplines and industries, for both the engineer and the student.

SMART GRIDS AND MICROGRIDS: Concepts and Applications, Edited by P. Prajof, S. Mohan Krishna, J. L. Febin Daya, Umashankar Subramaniam, and P. V. Brijesh, ISBN: 9781119760559. Written and edited by a team of experts in the field, this is the most comprehensive and up to date study of smart grids and microgrids for engineers, scientists, students, and other professionals.

MICROGRID TECHNOLOGIES, Edited by C. Sharmeela, P. Sivaraman, P. Sanjeevikumar, and Jens Bo Holm-Nielsen, ISBN: 9781119710790. Covering the concepts and fundamentals of microgrid technologies, this volume, written and edited by a global team of experts, also goes into the practical applications that can be utilized across multiple industries, for both the engineer and the student.

INTEGRATION OF RENEWABLE ENERGY SOURCES WITH SMART GRIDS, Edited by A. Mahaboob Subahani, M. Kathiresh and G. R. Kanagachidambaresan, ISBN: 9781119750420.
Provides comprehensive coverage of renewable energy and its integration with smart grid technologies.

Encyclopedia of Renewable Energy, by James G. Speight, ISBN: 9781119363675. Written by a highly respected engineer and prolific author in the energy sector, this is the single most comprehensive, thorough, and up to date reference work on renewable energy.

Green Energy: Solar Energy, Photovoltaics, and Smart Cities, edited by Suman Lata Tripathi and Sanjeevikumar Padmanaban, ISBN: 9781119760764. Covering the concepts and fundamentals of green energy, this volume, written and edited by a global team of experts, also goes into the practical applications that can be utilized across multiple industries, for both the engineer and the student.

Energy Storage, edited by Umakanta Sahoo, ISBN: 9781119555513. Written and edited by a team of well-known and respected experts in the field, this new volume on energy storage presents the state-of-the-art developments and challenges in the field of renewable energy systems for sustainability and scalability for engineers, researchers, academicians, industry

Energy Storage 2nd Edition, by Ralph Zito and Haleh Ardebili, ISBN: 9781119083597. A revision of the groundbreaking study of methods for storing energy on a massive scale to be used in wind, solar, and other renewable energy systems.

Hybrid Renewable Energy Systems, edited by Umakanta Sahoo, ISBN: 9781119555575. Edited and written by some of the world's top experts in renewable energy, this is the most comprehensive and in-depth volume on hybrid renewable energy systems available, a must-have for any engineer, scientist, or student.

Progress in Solar Energy Technology and Applications, edited by Umakanta Sahoo, ISBN: 9781119555605. This first volume in the new groundbreaking series, Advances in Renewable Energy, covers the latest concepts, trends, techniques, processes, and materials in solar energy, focusing on the state-of-the-art for the field and written by a group of world-renowned experts.

A Polygeneration Process Concept for Hybrid Solar and Biomass Power Plants: Simulation, Modeling, and Optimization, by Umakanta Sahoo, ISBN: 9781119536093. This is the most comprehensive and in-depth study of the theory and practical applications of a new and groundbreaking method for the energy industry to "go green" with renewable and alternative energy sources.

DESIGN AND DEVELOPMENT OF EFFICIENT ENERGY SYSTEMS, edited by Suman Lata Tripathi, Dushyant Kumar Singh, Sanjeevikumar Padmanaban, and P. Raja, ISBN: 9781119761631. Covering the concepts and fundamentals of efficient energy systems, this volume, written and edited by a global team of experts, also goes into the practical applications that can be utilized across multiple industries, for both the engineer and the student.

INTELLIGENT RENEWABLE ENERGY SYSTEMS: Integrating Artificial Intelligence Techniques and Optimization Algorithms, edited by Neeraj Priyadarshi, Akash Kumar Bhoi, Sanjeevikumar Padmanaban, S. Balamurugan, and Jens Bo Holm-Nielsen, ISBN: 9781119786276. This collection of papers on artificial intelligence and other methods for improving renewable energy systems, written by industry experts, is a reflection of the state of the art, a must-have for engineers, maintenance personnel, students, and anyone else wanting to stay abreast with current energy systems concepts and technology.

SMART CHARGING SOLUTIONS FOR HYBRID AND ELECTRIC VEHICLES, edited by Sulabh Sachan, Sanjeevikumar Padmanaban, and Sanchari Deb, ISBN: 9781119768951. Written and edited by a team of experts in the field, this is the most comprehensive and up to date study of smart charging solutions for hybrid and electric vehicles for engineers, scientists, students, and other professionals.

Printed in the USA/Agawam, MA
September 18, 2026

910661.120